DESIGN APPROACHES AND TOOLS IN EDUCATION AND TRAINING

Design Approaches and Tools in Education and Training

Edited by

Jan van den Akker
University of Twente,
The Netherlands

Robert Maribe Branch
The University of Georgia,
U.S.A.

Kent Gustafson
The University of Georgia,
U.S.A.

Nienke Nieveen
University of Twente,
The Netherlands

and

Tjeerd Plomp
University of Twente,
The Netherlands

In collaboration with Interuniversitair Centrum voor Onderwijskundig Onderzoek (ICO)

KLUWER ACADEMIC PUBLISHERS
DORDRECHT / BOSTON / LONDON

A C.I.P. Catalogue record for this book is available from the Library of Congress.

ISBN 0-7923-6139-3

Published by Kluwer Academic Publishers,
P.O. Box 17, 3300 AA Dordrecht, The Netherlands.

Sold and distributed in North, Central and South America
by Kluwer Academic Publishers,
101 Philip Drive, Norwell, MA 02061, U.S.A.

In all other countries, sold and distributed
by Kluwer Academic Publishers,
P.O. Box 322, 3300 AH Dordrecht, The Netherlands.

Printed on acid-free paper

Printed in the Netherlands.

TABLE OF CONTENTS

Preface

The broad field of education and training is in rapid evolution. In our 'knowledge-intensive and learning society', the expectations about the contribution of education and training are still rising. Moreover, the potential of information and communication technology (ICT) creates many challenges. These trends do not only affect the aims, content and processes of learning, they also appear to have a strong impact on design and development approaches in research and professional practices.

The major purpose of this book is to present and discuss current thinking and promising examples of design and development approaches and tools in education and training. The editors of this book hope that it will serve as a resource and reference work that stimulates the advancement of our field. It is intended to be useful in academic settings as well as for professionals in design and development practices.

The book has two parts. The first part offers an overview and analysis of design and development approaches, both in (research) literature as well as in professional practices. Moreover, it includes a series of more specific design approaches in various domains. The second part provides an overview and analysis of computer-based design and development tools, including a large selection of specific examples.

The two parts have some initial chapters that can be seen as foundational to the remaining contributions. Part One has two of such introductory chapters. Chapter 1 deals with principles and methods of development research. It analyzes how research activities can be connected to design and development activities in order to strengthen the knowledge base of educational design and development. Chapter 2 provides a comprehensive overview and discussion of a number of different perspectives on educational design and development. It offers a conceptual framework that helps to clarify and articulate the perpectives in many chapters later on in the book. The nature of these chapters is rather varied in terms of conceptual orientation, inclusion of practical examples or providing empirical research evidence.

The introductory text of Part Two, chapter 13, provides structure to the subsequent chapters in the second half of the book, as it classifies and analyzes all presented (computer-based) tools for a wide variety of design and development contexts.

The readers can find further help for orientation in the abstracts and key words at the beginning of each chapter. As a simple aid to facilitate communication between authors and readers, a list of authors' names, affiliations and e-mail address is added (at the end of the book).

The organizational context of this book stems from the Dutch Interuniversity Centre for Educational Research (Dutch acronym: ICO), in which nine Dutch universities participate. The research activities within that Centre are clustered in four divisions. This book relates to research within the division on 'design-oriented research'. During the last five years faculty members and PhD students from various universities have been engaged in research projects and workshops within this broad area.

This volume marks the transition to a next stage of ICO, in which the original four broad divisions have been replaced by a more focused research planning around ten specific research themes. Most of the researchers that participated in the design-oriented division (and that have written chapters for this volume) will continue their collaboration in one of those themes: "ICT Tools for Designing". The overall research question within that theme is: How can the role of ICT tools for various design activities in education be optimized? These tools may support different groups of designers: (a) professional designers of products in a wide range of domains (such as curriculum, instruction, media, assessment); (b) teachers who design their own instruction and who are involved in professional development; and (c) students, especially in learning processes that emphasize constructivist principles. The research approach within this theme will often be characterized as (prototypical) development research.

The current book, with its broad first part and more specific second part, already illustrates this trend towards a more specific focus on ICT tools. We hope that the highlights of the first ICO-years will offer inspiration for the forthcoming years. Moreover, we definitely intend to keep the interaction and cooperation with research colleagues from abroad (clearly illustrated by many contributions to this book) as active and productive as before.

Obviously, editing and producing a book like this, with its many authors from different institutions, is no small task. The editors want to express their gratitude to all authors for their professional cooperation. Also, we feel indebted to the two reviewers who have provided us with valuable feedback on both the original book proposal as well as the draft manuscript.
A special word of thanks is for Petra Zuithof for her very competent secretarial assistance in the preparation of the book.

Jan van den Akker
Robert Branch
Kent Gustafson
Nienke Nieveen
Tjeerd Plomp

Chapter 1
Principles and Methods of Development Research

Jan van den Akker
University of Twente

Keywords: Development research, Formative research, Research methods

Abstract

This chapter discusses the role of research in relation to educational design and development activities. The first part of the chapter focuses on the rationale and basic principles of development research by outlining motives for conducting formative research, analyzing definitions and aims of various types of development research, and discussing several of its key characteristics. The second part of the chapter deals with methods of development research, exploring some of its typical problems and dilemmas, and discussing several challenges for further action and reflection.

1. INTRODUCTION

This book offers a broad overview and many examples of design and development approaches and tools for education and training purposes. These texts share a common characteristic: they interpret education primarily as a 'design science', emphasizing its problem-oriented and interdisciplinary orientation. That perspective is also illustrated through the research approaches and methods chosen in many of those studies. Although 'Design & Development Research' would perhaps serve as a more adequate term to represent all intended research types, we have chosen the term 'Development Research' as an umbrella term, because it is shorter and also because the words 'design' and 'development' are often used interchangeably (see part I of this book).

Although research in relation to design and development issues is strongly emerging in various sub-domains of education, few efforts have yet been made to articulate its major principles and methods (the best exception being the well documented analysis by Richey and Nelson, 1996). It is the goal of this chapter to contribute to that discussion. The first part of the chapter (sections 2, 3, and 4) will focus on the rationale and basic principles of development research by: outlining motives for conducting development research, analyzing definitions and aims of various types of development research, and discussing several of its key characteristics. The second part of the chapter (sections 5 and 6) will focus on methods of development research, exploring its typical problems and dilemmas, and discussing several challenges for further action and reflection.

2. MOTIVES FOR DEVELOPMENT RESEARCH

Various motives for initiating and conducting development research can be mentioned. A basic motive stems from the experience that 'traditional' research approaches (e.g. experiments, surveys, correlational analyses), with their focus on descriptive knowledge, hardly provide prescriptions with useful solutions for a variety of design and development problems in education. Probably the greatest challenge for professional designers is how to cope with the manifold uncertainties in their complex tasks in very dynamic contexts. If they do seek support from research to reduce those uncertainties, several frustrations often arise: answers are too narrow to be meaningful, too superficial to be instrumental, too artificial to be relevant, and, on top of that, they usually come too late to be of any use. Designers do appreciate more adequate information to create a solid ground for their choices and more timely feedback to improve their products. Moreover, the profesional community of developers as a whole would be helped by a growing body of knowledge of theoretically underpinned and empirically tested design principles and methods.

Another reason for development research stems from the highly ambitious and complex nature of many reform policies in education worldwide. These reform endeavors usually affect many system components, are often multi-layered, including both large-scale policies and small-scale realization, and are very comprehensive in terms of factors included and people involved. Those radical 'revolutions', if promising at all, cannot be realized on the drawing table. The scope of diverse needs is often very wide, the problems to be addressed are usually ill-specified, the effectiveness of proposed interventions is mostly unknown beforehand, and the eventual success is highly dependent on implementation processes in a broad variety of contexts. Therefore, such reform efforts would profit from more evolutionary (interactive, cyclic, spiral) approaches, with integrated research activities to feed the process (both forward and backward). Such an approach would provide more opportunities for 'successive approximation' of the ideals and for more strategic learning in general. Especially those endeavors that aim at exploring and exploiting the potential of information and communication technologies in education (see the many examples in this book), are very much in need of research to overcome their inherent complexities. Policy makers rarely commission such research, at least compared to investments for development itself. Such negligence may not only be caused by naïveté (underestimating the need for clarification of the task and for empirical feedback) or frugality, but also by the still hazy image of development research. Its characteristics and added value demand better theoretical articulation, more empirical evidence, wider application in development practices, and a more prominent place in professional and scientific publications.

A motive also worth mentioning here, refers to the rather dubious reputation of educational research in general. Not only does one hear complaints about the lack of relevance of educational research from policy makers and practitioners, also many researchers themselves share that opinion (although usually with somewhat other nuances in their wording of the problem). They strive after more direct contributions to educational improvement processes with a noticeable impact. Moreover, more direct interaction with practice might offer them valuable incentives for sharpening their theoretical insights and transforming their descriptive knowledge in design principles. Thus, last but not least, there is also a distinct scientific interest at stake.

3. DEFINITIONS AND AIMS OF DEVELOPMENT RESEARCH

3.1 Conceptual confusion

Thusfar in this chapter the Development Research label has been used to refer to various kinds of research approaches that are related to design and development work. In the literature one comes across many different, more specific labels for such research. A (not exhaustive) list:

- *Design* studies; *Design* experiments; *Design* research;
- *Development/Developmental* research;
- *Formative* research; *Formative* inquiry; *Formative* experiments; *Formative* evaluation;
- *Action* research;
- *Engineering* research.

The situation is rather confusing. Some of these terms, such as 'action research' or 'formative evaluation', have longer roots in literature and professional practice, most others are of more recent date. Some labels seem related to specific sub-domains of education (e.g. action research was predominantly used in the area of Teacher Education, and formative evaluation in the areas of Curriculum and Instructional Development) but they become also fashionable in other areas (e.g. 'action research' in the area of learning and instruction). Also, 'new' terms (such as 'design experiments') come to the surface. Clearly, we are dealing with an emerging trend, characterized by a proliferation of terminology and a lack of consensus on definitions.

Of course, we do not aspire to solve these conceptual puzzles here for once and for all. However, in order to create a basis for later discussion on methodological issues, we need to establish a basic structure. As a starting point, we will explore the use of these concepts in different sub-domains of education.

3.2 Development research in different sub-domains

Ideas about the role of development research appear to differ among various sub-domains. Let us take a closer look at a selection of areas where these ideas are practiced and discussed.

- *Curriculum*

In discussing methodological issues in curriculum research, Walker has used various related terms, such as 'formative research' (Walker, 1992) and 'development research' (Walker & Bresler, 1993). The major goal of such research is to inform the decision making process during the development of a product/program in order to improve the product/program being developed and the developers' capabilities to create things of this kind in future situations. The emphasis in his writings is on the supportive role of research for development.

Comparable proposals have been formulated by van den Akker and Plomp (1993) who defined 'development research' by its twofold purpose: (i) supporting the development of prototypical products (including providing empirical evidence for their effectiveness), and (ii) generating methodological directions for the design and evaluation of such products. In this approach, the scientific contribution (knowledge growth) is seen as equally important as the practical contribution (product improvement). In recent years these ideas have been exemplified in a number of doctoral dissertations at the University of Twente, in a variety of curricular contexts, such as: Voogt (1993) and Keursten (1994) in the area of courseware development for various school subjects; Kessels (1993), focusing on design standards in the context of corporate education; van den Berg (1996) and Roes (1997) addressing scenarios for teacher in-service education; Nieveen (1997), exploring the potential of computer support for curriculum developers; Visser (1998) on communication support tools in distance education; and Thijs (1999) on teacher development in developing countries.

- *Media & Technology*

In the rapidly growing area of educational media and technology, development research has a prominent place. For example, Flagg (1990) underlined the role of formative evaluation for (media) program improvement and offered a number of exemplary studies. Richey and Nelson (1996) have provided the most comprehensive overview and analysis of 'developmental research' in the sub-domain of educational technology so far. They mention as its ultimate aim: "improving the processes of instructional design, development, and evaluation ... based on either situation-specific problem solving or generalized inquiry procedures" (o.c., p. 1213). They make a clear distinction between performing a process and studying that process. In the latter instance, these studies often have a reconstructive nature.

- *Learning & Instruction*

Also in the broad sub-domain of learning and instruction, pleas for development research are increasingly made. Referring to earlier proposals of Brown (1992) and Collins (1992) to invest more in 'design experiments', Greeno, Collins and Resnick (1996) underline a significant shift in the relationship between theoretical and practical work in educational psychology. They highlight the "kind of research that includes developmental work in designing learning environments, formulating curricula, and assessing achievements of cognition and learning and, simultaneously, on efforts to contribute to fundamental scientific understanding" (o.c., p. 41). Researchers should not only concentrate on the question of whether a theory yields coherent and accurate predictions, but also ask whether it works: do the theoretical concepts and principles inform practices in productive ways. In the same vein, Reigeluth and Frick (1999) argue for 'formative research', referred to as "a kind of developmental or action research that is intended to improve design theory for designing instructional practices or processes" (o.c., p. 633).

- *Teacher Education & Didactics*

Two related sub-domains where types of development research are already relatively well-established are teacher education (encompassing both pre- and in-service education as well as the broader phenomenon of professional development) and didactics (especially linked to several school subjects). In the teacher education area the concept of 'action research' is rather popular. It refers to practical inquiries where teachers (often in collaboration with others) investigate and reflect on their own teaching and students' learning. The primary goal is usually to contribute to the teachers' professional learning and/or bringing about change in a specific educational setting (Elliott, 1991; Hollingsworth, 1997). In the area of didactics the emphasis tends to be on 'developmental research' as an interactive, cyclic process of development and research in which theoretical ideas of the designer feed the development of products that are tested in classroom settings, eventually leading to theoretically and empirically founded products, learning processes of the developers, and (local) instructional theories (see e.g. Gravemeijer, 1994, for the mathematics domain, and Lijnse, 1995, for the science domain).

The four sub-domains mentioned above (all of them often related to product/program development, in one way or another) show the most visible and increasing role of development research in education (and also illustrate the conceptual confusion). Of course, there may also be examples in other sub-domains, but they often do not (yet) represent a strong trend in research approaches in those areas. Also, there are exemplary studies that very well reflect development research approaches but that can not easily be classified under a certain sub-domain. For example, the work in the Jasper Project of the Cognition and Technology Group at Vanderbilt (1997) is most remarkable as it reflects a broad range of sophisticated research procedures in various stages and on various aspects of that project, thus contributing to nearly all kinds of (previously mentioned) aims.

3.3 A typology of development research activities

The previous section revealed that one can distinguish a broad variety of activities, with different emphases in their primary aims, under the main umbrella of development research. Also, on a rather abstract level, one can distill a very general aim of all approaches: reducing uncertainty of decision making in designing and developing (educational) interventions. The term 'intervention' then serves as common denominator for products, programs, materials, procedures, scenarios, processes, and the like. That general aim can be specified in two more specific goals that apply to those approaches in various degrees: (a) providing ideas (suggestions, directions) for optimizing the quality of the intervention to be developed; (b) generating, articulating and testing design principles. These principles can be of a 'substantive' nature, referring to characteristics of the intervention (what it should look like), or of a 'procedural' nature (how it should be developed).

The goals differ in their relative contribution to 'practice' and 'science'. While the first one (optimization of intervention) is especially oriented towards practical ends in a given situation, the last aim (design principles) more strongly reflects scientific or scholarly aspirations, since it is more explicitly oriented towards production of knowledge of a generalizable nature.

An additional objective is also visible in various approaches: stimulating professional development of participants. This motive even appears to be in the forefront of many 'action research' activities.

However, in an effort to structure various types of development research, we find it difficult to include action research in our framework. First, because it is a rather volatile term that has known changing connotations over the last decades in various contexts (see Hollingsworth, 1997). Second, because in practice it often refers to activities that almost exclusively convey the 'action' component and lack an explicit scholarly orientation on contributions to knowledge that is accessible to others. In our view, an approach which is to be subsumed under the heading of development research must possess more balance between development and research.

Another way to differentiate between various types of development research is to focus on the temporal relation between design and development activities on one hand, and the research activities on the other: Is the research concentrated in the stages before, during, or after the bulk of design and development activities? Most of the examples mentioned thus far focused on activities that occur either before actual design (explorative studies to analyze problems-in-context and to screen related examples of interventions in order to generate design ideas) or throughout the entire, cyclic development work (in order to offer suggestions for improvement and to test design principles). In these processes the personal linkage between those who conduct the design and development activities and those who perform the research is usually strong. In their overview of 'developmental research' Richey and Nelson (1996) distinguish between 'type I' and 'type II' approaches. Type I refers to an approach in which the roles of designer and researcher (partly) coincide within a specific development context. Such research usually occurs throughout the complete development cycle. In type II, however, that relationship is more loose: the researchers are not involved in the design and development process themselves, but they study those processes (including tools and models applied) as practiced by others, in order to come to conclusions concerning design principles of generalizable nature. The emphasis in timing of those reconstructive activities is during, but especially after those design and development practices.

We prefer a somewhat more meaningful labeling of the various approaches. A first approach worth mentioning is: *Explorative Design Studies*. These activities, preceding the actual development work, aim at clarifying the design problem-in-context and at generating tentative design ideas. Such explorations can be very valuable in directing the development work. However, we tend not to include these studies under the heading of ‘scientific’ research, because they do not aim at statements of a more or less generalizable nature. That intention is very apparent in the two main types of development research:

- *Formative Research*. Research activities performed during the entire development process of a specific intervention, from exploratory studies through (formative and summative) evaluation studies; aimed at optimization of the quality of the intervention as well as testing design principles.
- *Reconstructive Studies*. Research activities conducted sometimes during, but oftentimes after the development process of several interventions; focused on the articulation and specification of design principles

Obviously, this typology does not pay justice to many possible differences and linkages between various sorts of development research. The boundaries and labels are not very strict (for example, exploratory design studies often convert to formative research) but they do make a distinction between approaches with relatively different emphases. The typology offers a useful starting point for more detailed and nuanced discussions on characteristics and methods of development research in the remaining part of this chapter.

4. KEY CHARACTERISTICS OF DEVELOPMENT RESEARCH

We will try to delineate the key characteristics of development research by exploring two questions:

(i) What are differences between development research - as a research endeavor - and typical design and development approaches in professional practices?

(ii) In what respect does development research differ from other research approaches?

The next two sections will address these questions.

4.1 Differences between development research and professional design and development

Development research is often initiated for complex, innovative tasks for which only very few validated principles are available to structure and support the design and development activities. Since in those situations the image and impact of the intervention to be developed is often still unclear, the research focuses on realizing limited but promising examples of those interventions. The aim is not to elaborate and implement complete interventions, but to come to (successive) prototypes that increasingly meet the innovative aspirations and requirements. The process is often cyclic or spiral: analysis, design, evaluation and revision activities are iterated until a satisfying balance between ideals and realization has been achieved.

To what extent do these development research activities differ from what is typical for design and development approaches in professional practices? What are the implications of the accountability of researchers to the 'scientific forum'? At the risk of exaggerating the differences, let us outline some of them, based on what is known about routinized standard-patterns in professional contexts. Of course, a lot of activities are more or less common for both approaches, so the focus will be on those additional elements that are more prominent in development research than in common design and development practices.

(1) Preliminary investigation

A more intensive and systematic preliminary investigation of tasks, problems, and context is made, including searching for more accurate and explicit connections of that analysis with state-of-the-art knowledge from literature. Some typical activities include: literature review; consultation of experts; analysis of available promising examples for related purposes; case studies of current practices to specify and better understand needs and problems in intended user contexts.

(2) Theoretical embedding

More systematic efforts are made to apply state-of-the-art knowledge in articulating the theoretical rationale for design choices. Moreover, explicit feedback to assertions in the design rationale about essential characteristics of the intervention (substantive design principles) are made after empirical testing of its quality. This theoretical articulation can increase the 'transparency' and 'plausibility' of the rationale. Because of their specific focus, these theoretical notions are usually referred to as 'mini'- or 'local' theories, although sometimes connections can also be made to 'middle-range' theories with a somewhat broader scope.

(3) Empirical testing

Clear empirical evidence is delivered about the practicality and effectiveness of the intervention for the intended target group in real user settings. In view of the wide variation of possible interventions and contexts, a broad range of (direct/indirect; intermediate/ultimate) indicators for 'success' should be considered.

(4) Documentation, analysis and reflection on process and outcomes

Much attention is paid to systematic documentation, analysis and reflection on the entire design, development, evaluation and implementation process and on its outcomes in order to contribute to the expansion and specification of the methodology of design and development.

Relating this outline to the differentiation in various types of development research in section 3.3, it seems that all of them have been touched upon. In reality, the emphasis will differ. Formative research is the most comprehensive type of development research, including all of the components mentioned above, but concentrating on theoretical embedding and empirical testing. The label of exploratory design studies is in place when the emphasis is on preliminary investigation, while reconstructive studies focus on documentation, analysis and reflection.

4.2 Development research compared to other research approaches

Differences between development research and other research approaches are best perceived from their different aims and context. We will explore some of these features below.

Successive approximation of interventions in interaction with practitioners

More than most other research approaches, development research aims at making both practical and scientific contributions. In the search for innovative 'solutions' for educational problems, interaction with practitioners (in various professional roles: teachers, policy makers, developers, and the like) is essential. The ultimate aim is not to test whether theory, when applied to practice, is a good predictor of events. The interrelation between theory and practice is more complex and dynamic: is it possible to create a practical and effective intervention for an existing problem or intended change in the real world? The innovative challenge is usually quite substantial, otherwise the research would not be initiated at all. Interaction with practitioners is needed to gradually clarify both the problem at stake and the characteristics of its potential solution. An iterative process of 'successive approximation' or 'evolutionary prototyping' of the 'ideal' intervention is desirable. Direct application of theory is not sufficient to solve those complicated problems. One might state that a more 'constructivist' development approach is preferable: researchers and practitioners cooperatively construct workable interventions and articulate principles that underpin the effects of those interventions.

Another reason for cooperation is that without involvement of practitioners it is impossible to gain clear insight in potential implementation problems and to generate measures to reduce those problems. New interventions, however imaginative their design, require continuous anticipation at implementation issues. Not only for 'social' reasons (to build commitment and ownership of users) but also for 'technical' benefits: to improve their fitness for survival in real life contexts. Therefore, rigorous testing of practicality is a *conditio sine qua non* in development research.

Nature of knowledge from development research

A couple of remarks are to be made about the knowledge claims based on development research.

The major knowledge to be gained from development research is in the form of (both substantive and methodological) 'design principles' to support designers in their task. Those principles are usually heuristic statements of a format such as: "If you want to design intervention X [for the purpose/function Y in context Z], then you are best advised to give that intervention the characteristics A, B, and C [substantive emphasis], and to do that via procedures K, L, and M [procedural emphasis], because of arguments P, Q, and R."

Obviously those principles cannot guarantee success, but they are intended to select and apply the most appropriate (substantive and procedural) knowledge for specific design and development tasks.

It is not uncommon in formative research that such knowledge, especially the substantive knowledge about essential characteristices of an intervention, can partly be extracted from a resulting prototype itself. That is one of the reasons that makes it so profitable to search for and carefully analyze already available interventions to generate ideas for new design tasks (something easier to be done, of course, when the intervention is a concrete product than a more process-oriented scenario). However, the value of that knowledge will strongly increase when justified by theoretical arguments, well-articulated in providing directions, and convincingly backed-up with empirical evidence about the impact of those principles. Moreover, those heuristic principles will be additionally powerful if they have been validated in successful design of more interventions in more contexts. Chances for such knowledge growth will increase when development research is conducted in the framework of research programs, because projects can then build upon one another.

5. METHODS OF DEVELOPMENT RESEARCH

Methods of development research are not necessarily different from those in other research approaches. However, there are some specific features that are worth discussing here to further clarify the image of development research. The first one has to do with the central role of formative evaluation procedures in formative research.

The second aspect refers to several typical methodological problems and dilemmas for development researchers.

5.1 Formative evaluation as key activity

As has become clear in the previous sections, formative evaluation holds a prominent place in development research, especially in formative research. The main reason for this central role is that formative evaluation provides the information that feeds the cyclic learning process of developers during the subsequent loops of a design and development trajectory. It is most useful when fully integrated in a cycle of analysis, design, evaluation, revision, et cetera, and when contributing to improvement of the intervention.

It is hardly useful to summarize here the major methods and techniques of formative evaluation in relation to development work. Many adequate sources are already available, e.g. Brinkerhoff et al. (1983), Flagg (1990), Herman (1987), Tessmer (1993), Thiagarajan (1991); see also Nieveen (1997) for a synopsis. However, a few typical characteristics of formative evaluation within the context of development research approaches deserve some elaboration.

Priority on information richness and efficiency

Formative evaluation within development research should not only concentrate on locating shortcomings of the intervention in its current (draft) version, but especially generate suggestions in how to improve those weak points. Richness of information, notably salience and meaningfulness of suggestions in how to make an intervention stronger, is therefore more productive than standardization of methods to collect and analyze data. Also, efficiency of procedures is crucial. The lower the costs in time and energy for data collection, processing, analysis and communication, the bigger the chances on actual use and impact on the development process. For example, samples of respondents and situations for data collection will usually be relatively small and purposive compared to sampling procedures for other research purposes. The added value of getting 'productive' information from more sources tends to decrease, because the opportunities for 'rich' data collection methods (such as interviews and observations) are limited with big numbers . To avoid an overdose of uncertainty in data interpretation, often triangulation (of methods, instruments, sources, and sites) is applied. These arguments especially hold true for early stages of formative evaluation, when the intervention is still poorly crystallized.

Shifting emphasis in quality criteria

The basic contribution of formative evaluation is to quality improvement of the intervention under development. Quality, however, is an abstract concept that requires specification. During development processes, the emphasis in criteria for quality usually shifts from validity, to practicality, to effectiveness (cf. Nieveen's chapter 10 in this book). Validity refers to the extent that the design of the intervention is based on state-of-the-art knowledge ('content validity') and that the various components of the intervention are consistently linked to each other ('construct validity'). Practicality refers to the extent that users (and other experts) consider the intervention as appealing and usable in 'normal' conditions. Effectiveness refers to the extent that the experiences and outcomes with the intervention are consistent with the intended aims.

The methods and techniques for evaluation will usually be attuned to that shift in criteria. For example, validity can adequately be evaluated through expert appraisal, practicality via micro-evaluations and try-outs, and effectiveness in field tests. In later stages of formative evaluation, methods of data collection will usually be less intensive but with an increasing number of respondents (e.g. achievement test for many students compared to in-depth interview with a few experts).

5.2 Problems and dilemmas in development research

What are some typical problems and dilemmas faced by researchers engaging in development research? Let us briefly explore some of them, notably as often experienced in formative research, where the combination of practical and scientific aims is most prominent.

Tension in role division between development and research

An often expressed problem in formative research is the tension in the division of roles between developers and researchers. Although that problem can be minimized when working in bigger teams in which some role variation is possible, a tension can easily arise between designers who are eager to pursue their ideals in creating innovative interventions on the one hand, and researchers who tend to critically seek for correctness of decisions and empirical proof of outcomes, on the other hand. Such controversy between 'subjective and imaginative involvement' and 'objective and critical distance' is not necessarily negative. It can also be treated as a productive force that contributes to balanced solutions. As a rule of thumb, one might stipulate that in early stages of formative research, progress is helped by a dominance of the creative designers' perspective, while at later stages a shift to a stronger voice for the more critical researchers' position is preferable.

Isolating 'critical' variables versus comprehensive and complex design

A typical difference between formative research and many other sorts of research is that one can hardly isolate, manipulate and measure separate variables in the same study. On the contrary, it is the very nature of formative research to investigate comprehensive interventions that deal with many interrelated elements at the same time. That makes it very hard to apply, for example, experimental approaches. Although that limitation is disappointing to some researchers, it should not be of great concern to formative researchers. The aim of formative research is different, as already extensively argued, and that requires other, more comprehensive and flexible approaches that offer less strict methodological precision, but more adequate answers to design problems. Otherwise, it should be noted that experimental approaches are not entirely impossible in the context of development research. Especially if the design aims toward instructional interventions, summative evaluation via experimental methods may be appropriate and feasible at the end of the trajectory, when it makes sense to measure the effectiveness of 'mature' interventions with larger numbers of students. For that matter, it is not surprising that educational psychologists in particular like to speak about 'design experiments'.

The apparent lack of rigor and control in methodology of development research is sometimes aggravated by unforeseen events or forces in the environment of the development task and context. Especially in long development trajectories of interventions that are easily susceptible to changes in external policies or organizational conditions, one may be forced to adapt research foci or procedures in the methodology approach initially chosen in order to address those changes (cf. Smith, 1993; Thijs, 1999).

Generalization of findings

Since data collection in formative research is usually limited to small (and purposive) samples, efforts to generalize findings cannot be based on statistical techniques, focusing on generalizations from sample to population. Instead one has to invest in 'analytical' forms of generalization: readers need to be supported to make their own attempts to explore the potential transfer of the research findings to theoretical propositions in relation to their own context. Reports on formative research can facilitate that task of analogy reasoning by a clear theoretical articulation of the design principles applied and by a careful description of both the evaluation procedures as well as the implementation context (Miles and Huberman, 1994; Yin, 1994). Especially a 'thick' description of the process-in-context may increase the 'ecological' validity of the findings, so that others can estimate in what respects and to what extent transfer from the reported situation to their own is possible. Another option that may stimulate exploration of possibilities for (virtual) generalization is to organize interactive meetings with experts from related contexts to discuss the plausibility of the research findings and recommendations for related tasks and contexts.

6. MAJOR CHALLENGES FOR DEVELOPMENT RESEARCH

As previously indicated, development research, as a relatively new and upcoming research approach, has its potentials and limitations. At the end of this chapter, it seems worthwhile to mention a few challenges for those who are interested in further exploration and improvement of its methodology.

For example, a challenging trend for designers is the increasing prominence of prototyping approaches (see the contributions of Moonen, and Nieveen in chapters 8 and 10 of this book). Various questions arise: What does (rapid/evolutionary) prototyping imply for efficiency of the development process? Will it affect the balance between creative and systematic features of the approach? Does it reduce the relevance of preliminary investigations? To what extent does it influence the relationship between methodology (as prescribed in literature) and actual design activities in professional practices (can 'theory' keep up with 'practice', or will the gap even widen)?

In relation to this trend: will the many emerging technological tools and environments for learning, communication, designing and performance support (see Part II of this book) strongly change development approaches and outcomes? Will they reduce the distance between design, delivery, and utilization of educational interventions?

Many challenges are also apparent with respect to evaluation methodology. What are appropriate tactics for increasing the information richness and efficiency of data collection procedures and instruments? How may the linkages between data collection, processing, and analysis be optimized? How can the communication about evaluation findings and the subsequent utilization for improvement of interventions be furthered? What are the most relevant indicators of quality, success and impact of interventions? What are promising approaches to further the generalizibility of research findings? How can the utilization of evaluation findings to design tasks in other settings be facilitated? Many useful suggestions and examples of such tactics are already offered in various publications (e.g. Miles and Huberman, 1994; Walker, 1992), but additional support is quite welcome.

An overall reflection is that research-based progress to expand and sharpen our knowledge on design and development is greatly enhanced through interdisciplinary approaches with purposive cross-fertilization between the many specialized sub-domains in educational science and technology. Moreover, it is our own experience that joint development research efforts of professionals in various roles offer fine opportunities for professional learning and capacity building. Such activities have the potential to sometimes produce outcomes with 'interocular' significance: results that hit you between the eyes (Scriven, 1996).

REFERENCES

Brinkerhoff, R.O., Brethouwer, D.M., Hluchyi. T. and Nowakoski, J.R., 1983, *Program evaluation: A practitioner's guide for trainers and educators*, Kluwer-Nijhoff, Boston.

Brown, A.L., 1992, Design experiments: Theoretical and methodological challenges in creating complex interventions in classroom settings, *Journal of the Learning Sciences, 2*, 141-178

Cognition and Technology Group at Vanderbilt, 1997, *The Jasper project: Lessons in curriculum, instruction, assessment, and professional development*, Lawrence Erlbaum Associates, Mahwah, NJ.

Collins, A., 1992, Towards a design science of education, In Scanlon, E. and O'Shea, T. (Eds.), *New directions in educational technology* (pp.15-22). Springer, Berlin.

Elliott, J., 1991, *Action research for educational change*, Open University Press, Milton Keynes.

Flagg, B.N., 1990, *Formative evaluation for educational technologies*. Lawrence Erlbaum Associates, Hillsdale, NJ.

Gravemeijer, K., 1994, *Developing realistic mathematics education*, Doctoral dissertation, Freudenthal Institute, Utrecht.

Greeno, J.G., Collins, A. and Resnick, L., 1996, Cognition and learning, In Berliner, D.C. and Calfee, R.C. (Eds.), *Handbook of educational psychology* (pp. 15-46), Macmillan, New York.

Herman, J.L. (Ed.), 1987, *Program evaluation kit*, Sage, Newbury Park, CA.

Hollingsworth, S., 1997, *International action research: A casebook for educational reform*, Falmer Press, London.

Kessels, J.W.M., 1993, *Towards design standards for curriculum consitency in corporate education*, Doctoral dissertation, University of Twente, Enschede.

Keursten, P., 1994, *Courseware-ontwikkeling met het oog op implementatie: De docent centraal* [Courseware development from an implementation perspective: A central role for the teacher], Doctoral dissertation, University of Twente, Enschede.

Lijnse, P., 1995, "Developmental research" as a way to empirically based "didactical structure" of science, *Science Education, 79*, 189-199.

Miles, M.B. and Huberman, A.M., 1994, *Qualtative data analysis: An expanded sourcebook*, Sage, London.

Moonen, J., 1996, Prototyping as a design method, In Plomp, T, and Ely, D.P. (Eds.), *International encyclopedia of educational technology* (pp. 186-190), Pergamon, Oxford.

Nieveen, N.M., 1997, *Computer support for curriculum developers: A study on the potential of computer support in the domain of formative curriculum evaluation,* Doctoral dissertation, University of Twente, Enschede.

Reigeluth, C.M. and Frick, Th.W., 1999, Formative research: A methodology for creating and improving design theories, In C. Reigeluth (Ed.), *Instructional-design theories and models. A new paradigm of instructional theory,* Volume II (pp. 633-651), Lawrence Erlbaum Associates, Mahwah, NJ.

Richey, R.C. and Nelson, W.A., 1996, Developmental research, In D. Jonassen (Ed.), *Handbook of research for educational communications and technology* (pp. 1213-1245), Macmillan, London.

Roes, M.A., 1997, Nascholing op basis van lesvoorbeelden in de context van curriculumvernieuwing [Inservice education based on exemplary lesson materials within the context of curriculum renewal], Doctoral dissertation, University of Twente, Enschede.

Scriven, M., 1996, In Cooley, W., Gage, N. and Scriven, M., "The vision thing", Educational research and the AERA in the 21st century, Competing visions of what educational researchers should do, *Educational Researcher, 26*(4), 18-21.

Smith, N.L., 1993, Flexibility in the evaluation of emergent programs, *Studies in Educational Evaluation, 16*, 209-229.

Tessmer, M., 1993, *Planning and conducting formative evaluations: Improving the quality of education and training,* Kogan Page, London.

Thijs, A., 1999, Supporting science curriculum reform in Botswana: The potential of peer coaching, Doctoral dissertation, University of Twente, Enschede.

Thiagarajan, S., 1991, Formative evaluation in performance technology, *Performance Improvement Quarterly, 4*(2), 22-34.

van den Akker, J. and Plomp, T., 1993, *Development research in curriculum: Propositions and experiences*, Paper presented at AERA meeting, April 12-16, Atlanta.

van den Berg, E., 1996, *Effects of inservice education on implementation of elementary science*, Doctoral dissertation, University of Twente, Enschede.

Visser, L., 1998, The development of motivational communication in distance education support, Doctoral dissertation, University of Twente, Enschede.

Voogt, J.M., 1993, Courseware for an inquiry-based science curriculum: An implementation perspective, Doctoral dissertation, University of Twente, Enschede.

Walker, D.F., 1992, Methodological issues in curriculum research, In Jackson, P. (Ed.), *Handbook of research on curriculum* (pp. 98-118), Macmillan, New York.

Walker, D. and Bresler, L., 1993, *Development research: Definitions, methods, and criteria*, Paper presented at AERA meeting, April 12-16, Atlanta.

Yin, R.K., 1994, Case study research: Design and methods (2nd edition), Sage, Beverley Hills, CA.

Chapter 2

Educational Design and Development: An Overview of Paradigms

Irene Visscher-Voerman
University of Twente

Kent Gustafson
The University of Georgia

Tjeerd Plomp
University of Twente

Keywords: Curriculum development, Instructional development, Design paradigms, Design rationality, Instrumental approach, Deliberative approach, Prototyping approach, Artistic approach

Abstract

This chapter will provide four different perspectives on educational design and development: instrumental, communicative, pragmatic, and artistic. The perspectives will be compared, and upon technical as well as social aspects of design will be their different views described. We will argue that designers should be aware of the usefulness of these perspectives and use the one that is most adequate for a certain design situation.

1. INTRODUCTION

Over the years, many authors have tried to conceptualize and visualize educational design and development processes, either starting from theoretical notions and ideas or from observations of practice. Although technical aspects of the development processes, such as task analysis and selection of instructional strategies, are central to many of these conceptualizations, social aspects and other perspectives are increasingly being stressed, such as the relationship between developer and client. Below are described four perspectives on how the educational design and development processes might be conducted. Although they are discussed separately, designers may draw on more than one perspective.

In many cases, educational development is depicted as an *instrumental* process that starts with the formulation of specific goals and outcomes for the product to be developed (e.g., Dick & Carey, 1996; Gagné, Briggs & Wager, 1992; Gustafson & Branch, 1997). These goals and outcomes function as guidelines during the development process. In these models, mostly stemming from the field of instructional design, the focus is on the technical aspects of the development approach. Vignette 1 provides an example of this instrumental, planning-by-objectives approach.

Other authors stress the social aspects of the development process. They depict development as a *communicative* process in which the product is developed in close interaction and communication between the development team and others involved (e.g. Kessels, 1993; Kessels & Plomp, 1997; Walker, 1990.) Team members and other stakeholders formulate their ideas and wishes for the product to be developed and discuss these ideas to conceptualize the problem and to generate promising versions of materials. The communication between developers, clients and users functions as a legitimizing factor for decisions. They discuss and assess the merits of these early versions, and revise them across repeated trials. This focus on deliberative and communicative aspects of development processes is illustrated in vignette 2.

In several publications, (e.g. Moonen, 1996; Tessmer, 1994; Tripp & Bichelmeyer, 1990) development is depicted as a *pragmatic* process in which developers create their product by quickly building, testing, and revising several prototypes or early versions of the product. An important characteristic of this approach is that prototypes often are used to get a better understanding of the design problem and to create final product specifications. Prototyping is seen as a practical endeavor in which intended users are directly involved in test and analysis activities. Vignette 3 illustrates this approach.

Yet, others (e.g. Eisner, 1979, 1996) argue that developers act like *artists*, who, based on their unique expertise and experience, create their products in response to the specific situation in which they work. Educational development, in this view, is both a practical and artistic undertaking that cannot be reduced to a specific method. From this artistic perspective, many of the important decisions about the product are made during implementation, in response to context demands. Vignette 4 illustrates this artistic approach.

These four general descriptions and their accompanying vignettes illustrate four different development processes. Some may perceive these approaches as different only in detail, though essentially the same (Visscher & Schulten, 1997). According to these 'generalists', differences between development approaches are only differences in focus, originated by contextual factors and by the distinctive purpose for which the approaches are employed. We, however, argue in this chapter, that these approaches represent different paradigms of educational design and development, and that the vignettes demonstrate that development processes typically can be conducted in fundamentally different ways. In this chapter we first describe in more detail each of the four paradigms, after which we discuss the fundamental differences among them. These differences are related to the social as well as the technical aspects of the development process such as for example, the roles of team members and other stakeholders in the process, the role and function of analysis and evaluation activities, and the role and function of prototypes.

2. DEVELOPMENT PARADIGMS

In this section, we describe and illustrate the four development paradigms. If we turn to contemporary philosophy, we see that these development paradigms are embedded in four main movements (Visscher & Schulten, 1997; see also Coyne, 1995; Self, 1997). These main philosophical movements (modernism, critical theory, pragmatism, and post-modernism) reflect basically different stances towards 'truth' and 'truthful acting' in sciences, or as to what constitutes 'good design' and 'good designing' in the field of design and development. In earlier papers these different stances have been called 'rationalities' (e.g. Visscher, 1997; Visscher & Schulten, 1997). Each of the development paradigms is named according to its underlying rationality and characterized by an activity typical for the approach it represents:

- instrumental paradigm: planning-by-objectives;
- communicative paradigm: communication to reach consensus;
- pragmatic paradigm: interactive and repeated tryout and revision;
- artistic paradigm: creation of products based on connoisseurship.

2.1 Instrumental paradigm: Planning-by-objectives approach

The majority of curriculum and instructional development models in the literature can be characterized as a form of planning-by-objectives. As the term indicates, the formulation of objectives is central to these models. The approach they represent devotes (often extensive) time to exploring and formulating the intended outcomes of the project, resulting in concrete project goals and learning objectives. Thus, problem and needs analysis is done at an early stage of the development process. Next, the means are determined that are necessary to reach these concrete goals and objectives. The focus on ends and on the means to reach them shows the instrumental rationality that underlies this approach. Although great emphasis is placed on the analysis of goals and means, the specific activities that are employed can differ in scope and time. Depending on the specific project they can vary, for example, from extensive interviews and observations with potential learners and subject matter experts, to a short interview with the client. Having specified the goals, the developer specifies the means that will be used to reach the goals. This design phase results in a blueprint of the product to be made. The objectives that have been formulated at the beginning of the project serve as a reference throughout the entire process of development. Formative evaluation is then conducted to check whether the objectives are being met and to prevent deviations from the objectives. Vignette 1 presents an example of a planning-by-objectives approach within the instrumental paradigm.

In the instrumental paradigm, logical reasoning and working systematically are considered essential elements to guarantee that the best solution is reached. That the emphasis is on instrumental or technical reasoning in this paradigm needs little debate since both proponents and opponents of planning-by-objectives models proclaim this. Early advocates of this approach include Bobbitt (1924) and Tyler (1949). Current examples of models within this paradigm are Dick and Carey, (1996); Gagné, Briggs and Wager, 1992; Kemp, Morrison and Ross (1994).

According to the instrumental paradigm, the formulation and specification of ends is the first and most essential step in the development process. Therefore, in a development approach within this paradigm, great emphasis is placed on preliminary data collection and analysis. This is not to say that most of the time is devoted to this phase, but rather that considerable weight is attached to the results of the analysis in guiding the remainder of the project. Instrumental developers may strive for getting the answers to their analysis questions as quickly as possible, but they at least want to have them before moving on.

Karen has been developing several kinds of educational media for seven year. She works at a national center that specializes in educational research and development. One of her colleagues, who also works for a language institution, is organizing a conference on language instruction. He asks Karen to develop a video program that can be used as part of the introduction to this conference.
Over several years, Karen has developed a checklist with questions that she wants to get answered before she decides whether or not to accept projects such as this. In a conversation she explores her client's wishes and needs. She asks him questions like: "Do you really need a video? What is the target group? What is the purpose of the video? How should topics be visualized? On what locations should the video be recorded? What is the budget? How will tasks and responsibilities be divided between her and the client? What is the spoken introduction about? How will the video be related to this introduction?" Only when she has clear answers to these questions, is she willing to start the project. Knowing exactly where to go or what to make and what to do to get there, is a first and essential step in her process. Karen puts the answers to these questions on paper and checks them with her client so that they both agree upon where they are going. They will use this document in a later stage for the formative evaluation.
The client gives Karen full responsibility for the video to be made. Based on her client's wishes, Karen estimates how many days the video production will take and hires an external agency that will be responsible for the technical production. To manage the project, Karen makes a detailed plan, indicating what needs to be done, who will do it, how much time it will take, and when it should be ready. In this plan she deliberately reserves three opportunities for reflection and control in order to keep track of the process and to assure herself that she will end the project in time, with a product that will meet the needs and pre-specified objectives. She checks this plan with her client and the others who have a stake in the project to make sure that they will be able to work within that framework.
Because Karen has overall responsibility for the video and its content, she starts with a short but intense exploration of the content by reading books suggested by the client and by interviewing some of the persons who are likely to be interviewed in the video. She asks these persons information about the topic in general, but also wants to hear what specific ideas they have about the content of the video to be made. Based on this information, and relying on her development expertise, Karen makes a global script for the video. She evaluates this script with her client. They decide that a few changes are needed to meet the pre-specified goals. Having changed the script according to the client's wishes, Karen plans visits to the different locations that will be filmed and prepares the persons who will be interviewed in the video. After these visits, she turns the global script into a final script and, having checked it with her client, sends that to the involved preparation for the filming. During the production, she accompanies the video crew to assure the final script is being followed.
The video production agency first makes a draft version of the video based on Karen's specific directions. The client does not evaluate this concept version because of time constraints. Karen, however, has obtained his full authority to proceed. She checks the draft version against the information from the conversation, and formulates some areas of improvement for the video. Then the final version of the video is produced.
Karen views development as an ongoing process wherein successive products are more detailed than their previous products, but do not show radical differences in form or functionality. Working in small steps, with formative evaluation, helps her to show her client she is working on his problem and to reach the specified goal. She values clients that do not change their demands during the course of the project. In her interaction with clients and others Karen tries to solve problems as soon as possible and complete the project in a timely and efficient manner within predetermined specifications.

Vignette 1. Educational development as an instrumental process

2.2 Communicative paradigm: consensus approach

In several publications, the social aspect of design and development receives major emphasis (e.g. Kessels, 1993; Walker, 1990; Zwanenburg, 1993). These authors argue that whenever people are involved, there will be different perceptions and opinions about the problem to be solved and about the product to be made.

Walker's deliberative approach for generic curriculum development (1990) is a clear example of the communicative rationality. Walker characterizes the development process as follows:

"In the broad outline, the process is one of formulating a platform of ideas, using these ideas to conceptualize the problem and to generate promising versions of materials, assessing the merits of promising early versions, and revising them until they cannot be improved further" (p. 472).

First, developers formulate a platform of ideas. This platform consists of all relevant topics known to them, such as a definition of the problem, material constraints, fundamental didactic or psychological premises, and a tentative concept of the product and how it might be developed. They verify whether or not these first ideas are "correct" by analyzing existing materials, by designing aspects of the new materials, or by observing real learning situations. The findings are then used by the design team to further elaborate project goals and outcomes and to refine ideas; for example the adjustment of an opinion or the elaboration of a certain perspective. Prototypes are made and discussed to feed the discussion and to shape ideas as a basis for revision decisions. The repertoire of the developers, based upon their state-of-the-art knowledge and practical knowledge and experience, functions as a starting point for the development.

Kessels' model for the design of corporate training (1993) also fits the communicative paradigm well. He argues that developers should obtain the perceptions of all the stakeholders involved and integrate them in the product to be made, because the people involved will have to implement the solution. Regular communication among team members and other stakeholders is essential to reach consensus as to what the problem is and how it should be solved. In both Walker's and Kessels' model, extensive communication functions as a legitimizing factor for decision making. Vignette 2 presents an example of the communicative approach to educational development.

In the communicative paradigm, the social part of the development process receives considerable attention. Establishing a shared frame of reference and reaching consensus among all those involved are important elements. The development process is seen as an inter-subject social activity among all those involved in the process. Communication functions as a legitimizing factor. The communicative paradigm is rooted in the naturalistic approach of Schwab (1970). Schwab was reacting to a perceived uniformity of development procedures as proposed by those we have called instrumentalists. According to Schwab, curricular decisions are too complex to be solved by simple algorithms. Schwab views the development of a curriculum as a process of "trial and error." Ideas are collected, partly elaborated, and then, depending upon their coherence, used as an example or non-example during further development of the product. This is a continuous process of deliberation of means and ends, of possibilities, and alternatives (Schwab, 1970). Others have portrayed the communicative development process in a more systematic way than Schwab. For example, Kessels (1993) indicates in his 'relational approach' that developers should try to reach what he calls "external consistency." That is, assuring the involvement of all who are affected so that ultimately there will be congruence among their opinions. Walker (1990) indicates that certain rules must be followed to guarantee that the process of deliberation is fair. He comments:

"Those who would master the arts of deliberation must develop sensitivity to the moral and legal rights and obligations of all affected by a decision and learn to arrange deliberation in ways that consider the particular human and institutional context without undue threat to those involved" (p. 201).

Marc works as an external training advisor for a bureau that specializes in employee training and education. The national center for professional health care asks him to develop an educational program. This program is needed, because health care workers differ from each other in how quickly and accurately they diagnose diseases and how they treat their patients. They want patients to receive the same procedures and services in different centers. Although Marc has developed many major programs and has an idea of what could be done in this project, he does not say so to his client. He first forms a project team with the client and several individuals from the different health care centers who are familiar with this problem. In the first team meetings, Marc deliberately takes time to discuss the question and to explore what kind(s) of solution(s) the team thinks useful. He brings some materials and solutions from his former projects and discusses the pros and cons of these solutions for the current situation. He also asks the team to develop and discuss different solution scenarios. He works this way, because in his experience, the concerns that are identified in the beginning might not turn out to be the real problem after all, and he does not want to loose time by focusing on the 'wrong' problem and solution.

Once the team has agreed that they should develop a curricular program for all current and near-term future health care workers, Marc encourages them to discuss what the function of the health care centers should be and how the daily activities of the health care workers could impact this function. He wants to use these as criteria to decide what the content of the program will be. Because there are representatives of different health care centers, there is considerable initial disagreement about the function of the centers. Marc stimulates them to formulate and explain their own opinions, and tries to steer the process in such a way that they end up with a solution with which all are satisfied. Between meetings, he asks team members to write reports of the meeting or to formulate plans for the continuation of the project. He also regularly asks them to discuss and evaluate the decisions made in the meeting with their colleagues back at work. He thinks that these follow-up discussions will impact their feeling of ownership and increase the likelihood they will continue to use the solution once he has left the scene. Marc wants the client and others to have the feeling that it is their product, and that they invented it themselves.

At some point in the process, the team members arrive at consensus about the content and method and make decisions about the outline and framework of the training program. The questions that are addressed are: "What are necessary changes in the workplace? What skills are needed to meet these changes? How can workers be measured to determine whether they have reached these skills? And what learning activities should be used that help workers reach those skills?" When the team reaches consensus about what to do, he sees that as the best alternative, even if there is another solution that might seem theoretically better. He thinks that if the local people will not support the program, and if they do not encourage their workers to apply the learned skills, then the effort will have been a waste.

Marc only wants to develop what the client sees as the best alternative. He argues that there may be a 'best theoretical alternative' but if the client thinks that that alternative will not work (for whatever reason), it is better to decide to go for the less optimal solution. This implies that the final solution may be quite different from his original ideas, but as long as the client is satisfied and works with it, this is the preferred solution.

Vignette 2. Educational development as a process of communication

2.3 Pragmatic paradigm: Interactive and repetitive try-out and revision

In the pragmatic paradigm, both the practical environment in which the product will be implemented and its users are central to the design process. Prototyping approaches fall into this category. According to advocates of these approaches, developers can only be satisfied with what they make if it works and if it is useful. To judge this, the multiple perspectives of the users are needed to judge an educational product as 'true' and useful. Therefore, early in the development process, several prototypes or interim versions of the product are evaluated with users. Thus, developers quickly start to develop a preliminary version of the eventual product. This preliminary version -a prototype- is evaluated formatively with the users either by discussing it (usually in the beginning of the project) or by trying it out (often later in the project). During these evaluations, the potential of the product becomes more concrete, helping the developers to make better decisions about the specifications of the product. Because of this highly iterative nature, some authors refer to these processes as rapid prototyping (Rieber, 1994; Tripp & Bichelmeyer, 1990).

In the pragmatic paradigm, the use of concrete prototypes is seen as an essential means for identifying the requirements of the product in interaction with client, experts, and, especially, members of the user population. Depending on user satisfaction and available time and finances, the process of development, evaluation, and revision is repeated several times. One of the main reasons for using a prototyping approach is the belief that the best way to test the value of a product and its functionality is with users. Especially in projects where goals are unclear, for example, in cases of a new product, or a new context, or when clients are unable to formulate exact wishes, this approach is seen as very useful. By extensively testing products during development, implementation failures and systems problems will become clear (Moonen, 1996; Nieveen, 1997; Tripp & Bichelmeyer, 1990).

The prototyping approach originated in the field of application software development. More recently it has also been used in the field of educational software, particularly for interface design (Hix & Hartson, 1993) and to a limited extent in the field of electronic learning materials development (Keursten, 1994; Moonen, 1996). Vignette 3 presents an example of the pragmatic paradigm.

2.4 Artistic Paradigm: Connoisseurship approach

Eisner's work (1979) clearly finds its place within the artistic paradigm. Although his work originates in the field of curriculum development, it can be applied to other fields of educational development as well. Eisner portrays social reality as negotiated, subjective, constructed, and having multiple perspectives. He maintains that there are many ways in which individuals construct meaning. He considers each person who makes decisions about educational products as very much like an artist choosing among an almost limitless variety of ways of representing his or her view of reality (Marsh & Willis, 1995). Eisner suggests that "the process [of curriculum or educational planning] is far more convoluted, circuitous, and adventitious, than one might be led to believe by reading the formal literature" (p. 116).

Eisner (1979) believes that evaluation is fundamentally the same natural process as the way in which people constantly engage in attempting to make sense of the world around them and of their own lives. This kind of public portrayal of valuing and reasoning is part of formal evaluation, but even in its absence, evaluation is always going on informally. This way of thinking is well reflected in the work of Schön (1983) who provided a framework for describing design processes. Applied to education, these skills are called educational connoisseurship, which Eisner defines as:

Jim is a multimedia specialist who has worked in the field for six years. His agency, which he founded five years ago, specializes in developing educational simulations and web-based environments. One day he is approached by a large national gas company that conducts research on gas and provides training programs for employees of gas companies. This company wants to know to what extent multimedia programs could be integrated in their training programs. They ask Jim to develop three simulations that can be used flexibly in two of their regular training programs, and that can be adapted by the trainer using them.

In a first meeting, Jim talks with the clients about what they want him to make. They describe the training context in which the simulations will be used. Because they have no experience with the use of multimedia programs they are not well able to specify the requirements for the simulations. They agree with Jim's proposal to start an iterative design process in which Jim will develop just one simulation and will ask regular feedback from them and from the trainer, as well as from prospective trainees. Based on this first simulation, they will then be able to specify the requirements for the second and third simulations. The clients ask the trainer to select a specific topic for the first simulation and then discuss the mathematical model that underlies the topic with Jim.

Jim returns home and implements the mathematical model in a first rough prototype. He makes only two parameters functional to give the clients and trainer an impression of what the program will be able to do. For the interface, he uses buttons that he designed earlier for another project. After a week, he takes his portable computer to the clients and invites them and the trainer to evaluate the program. The trainer soon comes up with some possible improvements: it turns out that Jim has selected parameters for the simulation that the trainer never uses in his training and that he also uses other parameters that Jim has left out.

Again, Jim goes home and revises the prototype based on the trainer's suggestions. Two weeks later, they meet again to evaluate the second prototype. Since the trainer is beginning to see the possibilities of the program, he asks Jim to add and integrate special conditions for the simulation as well. Having integrated this in the third prototype, Jim asks the trainer and three prospective trainees to work with it. It turns out that the trainees have difficulties with the buttons: Jim has used scrollbars to change parameters, while the trainees are used to sliders in their regular work. Jim shows them examples on his computer of sliders that he used in simulations in other projects. They then discuss how these could be changed to make them fit their specific situation.

Having integrated the findings in a fourth prototype, it is tested again by the trainer and trainees. Although they agree about the functionality of the product, they still see some minor improvements to the interface of the program. Jim includes these in the final product.

Jim thinks that an iterative design process with several prototypes is essential in these kinds of situations. He argues that it is often not possible to specify the functional and/or technical specifications in the beginning of the project. Clients are often only able to tell him what they want, once they have seen at least one, and perhaps several, prototypes.

Vignette 3. Educational development as a pragmatic process

"The art of appreciating what is educationally significant. It is a means through which the shape of the context and the configurations within it can be recognized so that intelligent decisions about that context can be made" (p. x).

While applying the connoisseurship approach, developers first render the qualities perceived in the situation, either in a formal or informal evaluation. Then they try to interpret and understand what they have rendered, by using, among other things, ideas, concepts, models and theories from the social sciences and from history. Also, they assess the educational importance or significance of the events or objects described or interpreted. This results in a new action. Schön (1983) refers to this phenomenon as conversation-with-the-situation and reflection-in-action. Eisner (1979) argues that the skills of developers are very much artistic in the sense that, although they can be learned and heightened through practice and experience, they cannot be reduced to a specific method. According to Eisner, the development process is: "both a practical and an artistic undertaking; it requires prudence, wisdom, and practical insight into the realities of elementary and secondary school classrooms. It also requires a sense of taste, design, and fitness. The parts must all hang together. They need to make sense, they have to be aesthetic" (p. xi).

On the one hand, depicting developers as artists does justice to the fact that they bring their own values, ideas, and judgment to the development process and that they use their creativity while shaping the solution. On the other hand, it tends to place these developers on a pedestal, which makes it hard to reveal the underlying structure of the development process. Vignette 4 thus depicts a development process in which artistic and creative elements can be found.

> Josephine (Jo) is a licensed health care provider who also has an interest in medical education. This interest in education includes both continuing education for medical professionals and patient education for preventative and therapeutic purposes. Over the past few years Jo has attended numerous education conferences and has come to believe that interactive multimedia programs are particularly powerful for shaping attitudes and presenting complex information. She has produced a number of multimedia programs and is recognized in the medical profession as being highly competent in this area.
>
> The rising incidence of AIDS and its profound impact on patients and their families (as well as on medical care providers and society at large) convinces Jo that a multimedia program should be developed for use by all these different audiences. Jo talks to a number of different people about her idea and gradually conceives the general framework for an extensive computer-based multimedia program. Her idea is to follow the cases of real individuals with AIDS over the progression of the disease and tie the cases to extensive databases of information. The databases would relate to each stage of the disease from contraction, through early and late stages of treatment, to impending death and its impact on the patient, family, friends, and medical staff who provide their care. The idea is that different audiences would access some common information, but that specialized information would also be available. Thus, physicians might look at some of the underlying pathology of AIDS while patients might access similar information written in a less technical language. Extensive information about the psychological impact of the disease on the patient, family, friends, and medical staff would also be included.
>
> This general framework evolved over time, partially as a result of discussions with different people, and partially from what Jo describes as "ah-ha's" as she continues to think about the idea. When she finally feels comfortable with the framework and has a sense of what would be required to produce the program Jo approaches a philanthropic foundation to seek funding. Impressed with the idea and Jo's reputation for doing creative work, the foundation provides substantial funding.
>
> The program takes almost two years to develop and basically follows Jo's original framework. However she continues to experience new insights about the design of the program and make modifications right up until each section is finalized. Jo makes all the design decisions, based on her own "intuition and experience," although she "tests" many of her ideas by talking about or showing them to others and watching their reaction. The final program is reviewed by content specialists to assure accuracy and subjected to extensive "debugging" before being released. Meanwhile Jo is already busy thinking about another multimedia program aimed at physicians and is beginning to work out its general framework.

Vignette 4. Educational development as an artistic process

3. THE PARADIGMS COMPARED

In previous sections of this chapter, we have described four educational design and development paradigms, based on their underlying rationalities. And as we have described, a developer's rationality reflects his or her specific stance toward what constitutes 'a good design' and 'good designing."

What then, is 'a good design' and what is 'good designing' according to the different paradigms and their underlying rationalities? Actually, this question asks for the specification of quality criteria within the different approaches; quality criteria relating to the product ('what is a good design') and to the process ('what is good designing'). From the descriptions and vignettes in the previous sections, the criteria for the different paradigms can be derived. These criteria are summarized in Table 1.

Paradigm	A design product is good...	The design process is good...
Instrum.	If it meets a pre-specified, absolute standard. If there is a clear relationship between goals, learning situations and processes, and outcomes.	If it starts with a thorough needs and problem analysis, resulting in a specification of the solution. If it occurs rationally, logically, and systematically.
Commun.	If it meets the standard as discussed and agreed upon by the team and other stakeholders. If the standard is shared by all those involved.	If it contains activities aimed at, and resulting in, reaching clarity and consensus among team members about what to make and how to make it. If decisions are based upon deliberation and consensus.
Pragmatic	If it as been proven useful and effective with and for users.	If design and evaluation activities are intertwined. If prototypes are regularly tested with users for their usefulness and effectiveness.
Artistic	If it meets the developer's own subjective quality criteria.	If the developer relied on his/her own judgment and has used that judgment well.

Table 1. Quality criteria for product and process in the different paradigms

A developer's rationality carries over into specific actions and activities during the development process. Thus, developers with different rationalities: possess different conceptions of what they call a good design and a good design process; differ in how they conduct development processes; and have different reasons for why they are doing it. This implies that specific technical activities such as analysis and evaluation, or the design of interim products or prototypes, are common aspects of each development process, but may have different functions and receive differing amounts of emphasis within the different paradigms. Take, for example, the activity of formative evaluation which many regard as an important activity in the design process (e.g. Nieveen, 1997; Tessmer, 1994; Maslowski & Visscher, 1997). Each description and paradigm shows several evaluation activities, but the function of the specific activities turns out to be different. In instrumental processes, formative evaluation is primarily aimed at checking to what extent the product is reaching the previously formulated criteria or standard. Based on the evaluation data, weaknesses in the product can be changed and the product can be improved. In pragmatic approaches, evaluation is primarily used to test whether or not the prototype is working for its users. Not only the quality of the product, but also the design specifications themselves are subjected to evaluation. In communicative approaches, evaluation of a prototype is done to stimulate the discussion and sense making within the group and to reach agreement about the product to be made or the process to be followed. And in artistic approaches, the function of (ongoing) evaluation is to feed the developer's reflection while working.

These differences in functions of formative evaluation, as well as salient differences in analysis and the design of prototypes are summarized in Table 2. The differences can also be derived from the descriptions and vignettes in the previous paragraphs.

Paradigm	Function of analysis	Function of evaluation	Function of prototype
Instrum.	To specify needs and goals at the beginning of the project. Great weight is attached to this phase.	To check whether needs and goals as specified in the analysis phase are being met.	Prototype resembles final product as much as possible. It is evaluated to confirm that it works and to find any bugs. Needed changes are result of inadequate analysis or design.
Commun.	To have team and stakeholders specify what they want to make/want to be made, and how they want to make it/want it to be made. Analysis occurs during 'platform'-phase, but also along the way.	To assess existing and newly made materials, and to support deliberation and discussion among team members and other stakeholders.	Prototype can be existing material or a new idea or perspective. It is evaluated to support deliberation among stakeholders and thus to get ideas about what to make and how to make it more clear.
Pragmatic	Most analysis activities are conducted by means of evaluating prototypes.	To test part of a product, or an interim product in practice to get insights about its quality and to specify further design requirements.	Global outlines of (the specific parts of) a product can function as a prototype. They are evaluated to identify requirements for design and to determine usefulness and effectiveness of product. Prototype might even be discarded or rejected.
Artistic	Analysis is an integrated part of connoisseurship. It occurs continuously.	Evaluation is an integrated part of connoisseurship that is concurrent with analysis.	The product gradually takes form in the manner of a painting. Developers may test their ideas on users just to see whether they are reaching the client with their ideas.

Table 2. The function of three development activities in the four paradigms.

We started this chapter by stating that social aspects of design are increasingly being stressed in different publications. Developers are aware that they make products for people and with people and, by their products, they may change people. Zwanenburg (1993) calls this the intangible part of the product. Thus, clients and other stakeholders are increasingly committed not only to the product, but also to the process. Instrumental developers, for example, may use their project management techniques to communicate their plans to the client and to get their approval. The attention for the social aspect is obvious in communicative and pragmatic processes, where stakeholders and/or users are given a voice in the decision making process and are asked to think through or work through the product. In the artistic approach, a shift is made away from objectivistic and technical reasoning toward the unique experience and ability of the developer. In many cases, the developer may create a product alone. In other cases, developers may ask clients or other stakeholders to test their ideas, not for the purpose of seeking their input, but rather, to see whether their product is achieving desired emotional or learning effects.

Although the social aspect of development can be divided into a large number of different themes, we focus on two aspects here; the role of the developer and his/her relation with the client or stakeholders. These differences are summarized in Table 3.

Paradigm	Role of the developer	Relation with client
Instrum.	Developer is an expert, using his/her objective, expert knowledge to develop a product. He/she is held mainly responsible for what is made.	Client is mainly 'used' to get information and to obtain approval.
Commun.	Developer is a facilitator, helping clients specify their own needs and wishes and facilitating the development of the product. Developers and stakeholders have a shared responsibility for what is being made.	Client and stakeholders are given an equal say in the process.
Pragmatic	Developers work out prototype, with or without the client, and test them in interaction with users.	Users have substantial influence on the shape of the product.
Artistic	Developer is an artist who uses his/her subjective, unique knowledge and reflection to shape the product.	Client and stakeholders may or may not be consulted.

Table 3. Function of several social aspects in the four development paradigms

4. CONCLUDING REMARKS

The framework presented here examines some of the fundamentally different ways in which developers view the world and how they engage in the development of educational products. The four paradigms are not completely mutually exclusive. That is, to some extent, several activities associated with each paradigm are present, at least to some degree, in each of the other paradigms. Although these activities may be present in several paradigms, their function and use can be very different. For example, any developer who works for clients will communicate with them about the product to be made, but that does not make him/her a communicative developer in the sense described here. Similarly, although a developer might work with several prototypes, this does not make him/her a prototyper per-se. And the same can be said for creativity; hopefully it is present in any project and is not the exclusive domain of the artistic developer.

We have seen that developers with *different* rationalities differ in what they perceive as good products and good development processes and differ in the way they conduct the process. This, however, does not imply that developers with the *same* rationalities are all alike in how they shape the process. We think that the remark that all development processes differ from each other and that each process is unique (e.g., Wedman & Tessmer, 1993) still holds. We suspect that even developers with the same rationality may vary widely in how they conduct the process. The differences within a paradigm are, however, more on a tactics-level than on a general design level and can likely be attributed to available resources, type of problem, complexity of situation, etc. In this respect, we would also caution that the vignettes presented in this chapter illustrate only one of many possible variations on how developers might operate even when staying largely within one of the paradigms. For example, instrumental developers could rely on: formal interviews; informal meetings with clients or experts or users; observations; document analysis; literature study; or a variety of other sources of data to make the problem and project goals clear. In some cases, they will become subject matter experts themselves, while in other cases external subject matter experts will exercise complete control over the content based on predetermined specifications. Pragmatic developers can differ regarding: how many prototypes they test; the number of users they test; the kinds of prototypes they make (paper or computer based); and whether they prepare a global outline of the whole product first (horizontal prototype) or a specific part in detail (vertical prototype) (Hix & Hartson, 1993). The same point can be made about the artistic paradigm in that teams can exercise great creativity rather than a single individual having sole responsibility. Of course, the team will have to develop some means (perhaps extended deliberation) to arrive at a consensus when competing ideas must be reconciled.

Although we suspect that developers' rationalities largely depend on their personal preferences, we would argue that developers may not use the same rationality for different projects. In fact we would argue that they should not be tied to a single rationality and paradigm. Different circumstances will likely call for different development approaches and the competent professional should be able to select the one that is most appropriate. Or better still, developers should be able to combine elements from each of the paradigms into a unique model for a specific project. An example of this type of synthesis might be when rapid prototyping is necessary for designing a computer interface for a set of computer-based lessons for a college course on economics to be delivered over the World Wide Web. Since the content is likely to be determined by a team of subject matter experts, adoption of a communicative process by the developer seems appropriate and a creative effort will be necessary to make the course stimulating and effective.

Ultimately, we believe *all* of the paradigms and their accompanying perspectives, tools, and techniques can and do play useful roles in designing effective, efficient, relevant, and engaging educational experiences. We believe that all practicing professionals should be aware of the value of each paradigm and use the one that is most appropriate for the specific situation. To do less is to be less than a complete and competent practitioner. After all, would you want to have your medical care being provided by a physician who could only use a limited set of diagnostic tests and procedures and then prescribe from only a limited number of treatments? Speaking for ourselves, we want to be cared for by physicians who can be instrumental, communicative, pragmatic, and yes, even artistic in their practice of medicine. Shouldn't clients expect as much from educational developers as well?

REFERENCES

Bobbitt, F., 1924, *How to make a curriculum,* Mifflin Company, Boston, MA, Houghton.

Coyne, R., 1995, *Designing information technology in the Postmodern Age,* MIT Press, Cambridge, MA.

Dick, W. and Carey, L., 1996, *The systematic design of instruction* (4th ed.), Harper Collins College Publishers, New York.

Eisner, E.W., 1979, 1996, *The educational imagination: On the design and evaluation of school programs,* Macmillan, New York.

Gagné, R., Briggs, L. and Wager, W., 1992, *Principles of instructional design* (4th ed.), Holt, Rinehart, and Winston, New York.

Gustafson, K. and Branch, R., 1997, *Survey of instructional development models* (3rd ed.). ERIC Clearinghouse on Information Resources, Syracuse University, Syracuse, NY.

Hix, D. and Hartson, H., 1993, *Developing user interfaces: Ensuring usability through product & Process,* John Wiley & Sons, New York.

Kemp, J., Morrison, S. and Ross, G., 1994, *Designing effective instruction,* Merrill, New York.

Kessels, J.W.M., 1993, *Towards design standards for curriculum consistency in corporate education.* Doctoral dissertation, University of Twente, Enschede.

Kessels, J.W.M. and Plomp, Tj., 1997, The importance of relational aspects in the systems approach, In Romiszowski, A.J and Dills, C.R. (Eds.), *Instructional development paradigms* (pp. 93-126), Educational Technology Publications, Englewood Cliffs, NJ.

Keursten, P., 1994, *Courseware-ontwikkeling met het oog op implementatie: de docent centraal* [Courseware development from an implementation perspective: focus on the teacher]. Doctoral dissertation, University of Twente, Enschede.

Marsh, C. and Willis, G., 1995, *Curriculum: alternative approaches, ongoing issues,* Merrill, Prentice Hall, Englewood Cliffs, NJ.

Maslowski, R. and Visscher, A.J., 1997, *Methoden en technieken voor formatieve evaluatie in sociaal-wetenschappelijke ontwerpsituaties* [Methods and techniques for formative evaluation in social scientific design situations], University of Twente, Enschede.

Moonen, J.C.M.M., 1996, Prototyping as a design method, In Plomp, Tj. and Ely, D.P. (Eds.), *International encyclopedia of educational technology* (2nd ed., pp. 186-190), Pergamon, Cambridge.

Nieveen, N.M., 1997, *Computer support for curriculum developers: A study on the potential of computer support in the domain of formative curriculum evaluation,* Doctoral dissertation, University of Twente, Enschede.

Rieber, L.P., 1994, *Computers, graphics and learning,* Brown and Benchmark, Dubuque, IA.

Self, J., 1997, From Constructionism to Deconstructionism: Anticipating Trends in Educational Styles, *European Journal of Engineering Education*, *22*(3), 295-307.

Schön, D.A., 1983, *The reflective practitioner: How professionals think in action,* Basic Books, New York.

Schwab, J.J., 1970, *The practical: A language for curriculum,* National education association, Washington, DC.

Tessmer, M., 1994, Formative evaluation alternatives, *Performance Improvement Quarterly, 7*(1), 3-18.

Tripp, S.D. and Bichelmeyer, B., 1990, Rapid prototyping: An alternative instructional design strategy, *Educational Technology and Development, 38*(1), 31-44.

Tyler, R.W., 1949, *Basic principles of curriculum and instruction,* The University of Chicago Press, Chicago.

Visscher, J.I.A., 1997, *Hoe werken professionele ontwerpers in de onderwijs- en opleidingspraktijk?* [How do professional developers work in training and education practice?] Paper presented at the ORD, Leuven, Belgium, 21-23 May.

Visscher, J.I.A. and Schulten, E., 1997, Design approaches in training and education: Insights from practice, Paper presented at the AERA, Chicago, IL, March.

Walker, D., 1990, *Fundamentals of curriculum,* Harcourt Brace College, Fort Worth.

Wedman, J. and Tessmer, M., 1993, Instructional designers' decisions and priorities: A survey of design practice, *Performance Improvement Quarterly, 6*(2), 43-57.

Zwanenburg, M.A., 1993, *Methodologie van sociaal ontwerp: Voorbereiding op epistemologische aspecten* [Methodology of social design: Preparation to epistemological aspects] (OSF report 4), University of Twente, Enschede.

Chapter 3

Systems Thinking in Instructional Design

Gordon Rowland
Ithaca College

Ann-Marie Adams
Cornell Cooperative Extension

Key words: Instructional design, Systems thinking

Abstract

The word *system* in Instructional Systems Development tends to connote systematic process. Yet systems scientists suggest this is a limiting interpretation. What would it mean to incorporate more powerful notions of system, for example wholeness and interdependence, in designing instruction? Perhaps experienced practitioners do so and can offer us direction. The study reported here asked designers to identify what they considered to be key relations among components of an instructional system—those things that made the relations among parts important to consider. The results reveal that components are seen as highly interdependent and suggest that simple ISD models and procedures fail to capture much of the systemic nature of the process. Possibilities for supporting practice and developing expertise are implied.

1. BACKGROUND

The process of creating instructional events, activities, lessons, and courses is often given the formal label Instructional Systems Design or Instructional Systems Development (ISD). In this phrase, *systems* may connote a number of things, but tends to most often be interpreted as systematic process, that is, an orderly pattern in which one set of considerations precedes another (e.g., Hug & King, 1984). Considerations are typically clustered in stages or steps that define major entities relevant to instruction such as who the learners are, what they need to learn, and what instructional methods will be used, and these stages are arranged in a logical sequence.

If as systems scientists suggest, the *system* emerges more from the relationships among parts than from the parts themselves, then the system of ISD is most evident not in the separate steps or activities, but in the logic or orderly pattern among them. For example, the early definition of goals and the use of these goals in subsequent decisions would appear to be a key feature, as would empirical justification in a final evaluative stage (Dick & Carey, 1996).

This logic of ISD is often illustrated in models, in which component stages are identified and arranged, say in a series of boxes with lines or arrows between. An essentially linear pattern is frequently represented, although authors' descriptions suggest a more complex process involving iterations and looping. The nature of such iterations, for example, what causes them to be initiated and how they occur, is not very clear, a fact that is not surprising given our relatively poor understanding of design practice. One could argue that the basic linear process is maintained while adjustments are a matter of judgment and depend on experience, or one could describe the entire activity as being less predictable and composed more of *ad hoc* decision making. The accuracy and utility of such positions may be more a matter of context and/or moment-to-moment situational details. Regardless, the fact is we have little explicit knowledge of what instructional designers think about in their practice. Therefore, arguing for or against a particular position regarding the logic of ISD, that is, the nature of the system, tends to be a matter of speculation with more theoretical than empirical basis. Pieters and Bergman (1995) is a notable exception. They surveyed graduates of their educational technology program and found support for teaching a basic systematic process.

To shed light on such issues, relatively direct methods such as protocol analysis could be employed (e.g., Rowland, 1992; Perez & Emery, 1995). A specific focus on systemic relations in such analyses could be quite revealing. However, these methods are relatively costly in terms of time and effort on the part of both researcher and subject. Another, perhaps less powerful, option is to rely on retrospective reports, for example, to use surveys and interviews in which practitioners are asked what they think about when they design. The problem here is that retrospective reports tend to be unreliable (Ericsson & Simon, 1984).

Attempting to work around these difficulties and obtain a fresh perspective, we decided to ask about the product of ISD rather than the process, that is, about the instruction or instructional system created by designing rather than about designing itself. This seems justified since evidence does exist that what is designed influences how designing occurs, as well as that designers are at least as product-oriented as process-oriented in their thinking (Hubka & Eder, 1987). Furthermore, this seems reasonable because each element of process tends to be fairly clearly associated with a particular product: needs analysis defines a need; learner analysis defines the learners; task analysis defines the task and content; media selection specifies media; etc.

We asked, therefore, how does the concept of system, in terms of relations among parts, manifest itself in instructional designers' thinking. To approach this in a way that went beyond stepwise activity, we asked about the relations among components of the design product. For example, we asked how designers think about relations between the content and activities that might be involved in a lesson, rather than about the processes that went into their selection and creation.

2. METHOD

We chose to employ a pair of surveys sent and returned via electronic mail. We recruited participants by contacting colleagues at various institutions, asking if they might be able to refer us to practitioners with at least five years experience and who were currently active as instructional designers.

In the first survey, we asked respondents a few demographic questions: years of ID experience; current level of activity; and nature of preparation/education for doing ID-related work. We then asked them to identify what they felt were the key components of an instructional system. We provided a short list of possibilities and asked respondents to edit this list and provide a brief definition for each component they named. The operational definition of *instructional system component* reflected in the directions and questions was (1) something seen as one of about 5-7 important elements of an instructional system, (2) something that needs to be decided (e.g., a variable), (3) something frequently cited by authors, theorists, and practitioners as a primary category (e.g., an element identified in a model, a decision as opposed to a process), and (4) an entity, not an activity.

In the second survey, we asked how these components were related to one another. We paired each component that the respondent had listed and defined with all other components he or she had listed. Again, because pilot testing resulted in only "do X before Y" types of responses, we shifted attention away from process, in this case by asking "What key question or issue arises from the relationships between these components?" Following this, we asked respondents to indicate which relationships they felt to be most important, then asked them to reflect on the complexity of relationships, for example, if they felt there were particular interrelationships among sets of three or more components.

We analyzed responses to Survey 1 and identified a common set of components, then mapped relationships within this set of components in two-dimensional space and explored them for implications.

3. RESULTS

We sent survey 1 to twenty-two designers and nineteen (86%) were returned. Respondents had between 5 and 30 years of experience as instructional designers, with a mean of 14.6 years, and they reported spending between 5 and 50 hours per week engaged in instructional-design activities, with a mean of 23.5 hours. Clearly, this was a group actively engaged in ID practice and with much experience.

In terms of preparation, 4 respondents had a Doctorate, 7 Masters degrees, 1 Bachelors degree, 4 had "some courses," while 3 learned ID entirely on the job. Quite a wide range of backgrounds. We did not ask directly about jobs, but from position titles, addresses, and comments, we could infer that the group worked in a wide range of contexts, also. These included both education (e.g., university) and training (e.g., aerospace industry), and internal (i.e., preparing training for their own people) and external (independent ID company) organizations.

Respondents offered a fairly consistent set of instructional system components, and these components were readily clustered into 7 categories: Problem Situation; Goal; Learners; Content; Instructional Setting; Instructional Method; Assessment. These particular terms were not always chosen, but the definitions provided by respondents made clustering fairly easy. For example, “target audience,” “learners,” “participants,” and “target population” all clearly referred to those who would learn through their being part of the system.

We should note two things about these components. First, a few respondents expressed minor difficulty in identifying them, but this difficulty was not in generating them, but rather in limiting their number to approximately 5 to 7, as requested in the survey item. That is, they felt it necessary to maintain a larger set of issues in the forefront as opposed to clustering them under broader headings. Second, these components could be seen to extend the basic journalistic questions of who, what, when, where, and why:

Problem Situation	Why? (in terms of larger performance system)
Goal	Why? (in terms of instructional system itself)
Learners	Who?
Content	What?
Instructional Setting	When and where?
Instructional Method	How?
Assessment	How well?

In an attempt to check for patterns followed by “groups,” we clustered responses according to demographics—level of current activity; years of experience; preparation; context of work. No patterns appeared, so we interpreted responses to represent those of experienced instructional designers as a whole.

We sent Survey 2 to the nineteen Survey 1 respondents and asked them to relate each component that they identified in Survey 1 to every other component. (To be clear, note that Survey 2 did not address the seven categories above. In order to stay grounded in the respondents’ experiences it included the components that they listed in their responses to Survey 1.) This was a far more extensive task, but it too had a good response rate. Of 19 surveys sent, 14 (74%) were returned. Without exception, respondents were able to provide clear statements relating each component to every other, and, in many cases, backed up these statements with either general comments or descriptions of specific cases where relationships had been important. These statements were then grouped according to component pairs, which we constructed from the seven categories above. For example, a statement about *target population and strategy* was grouped with others having to do with *learner and instructional method*, the latter combining the categories we created based on Survey 1 responses. These relational statements are summarized below. For each, we offer the one or two issues or key questions that seem to best represent responses, then offer a brief explanation.

Problem Situation & Goal: Will meeting the goal satisfy a larger system need (i.e., reduce gaps between current and desired performance)?

This relation is one of connection to the larger system. Does the instructional system (at least in terms of its stated goal) serve the needs of the organization? For example, can learners successfully apply what they have learned to on-the-job performance? It also connects the instructional system to business requirements. For example, do the goals (goal statements) truly benefit the situation and respond to the issues in it?

Problem Situation & Learners: How are learners involved in and affected by the problem?

This relation has a number of facets, perhaps because of the range of definitions of problem situation offered by respondents, that is, as anything from training need to business problem. It includes learners' involvement in defining the problem, who the appropriate learners are, and their characteristics and competencies in relation to a need. Notice how the relation is two-way, asking both "Who needs to learn?" (Problem situation→Learners) and "What do the learners perceive to be the problem?" (Learners→Problem situation). The arrows here and below can be read as "affects" or perhaps in the extreme, "determines."

Problem Situation & Content: What knowledge and skills are required to meet the need, that is, to reduce a gap in performance?

This relation highlights the need to carefully select what knowledge and skills are included in the instructional system, making sure that they directly relate to the gap in performance.

Problem Situation & Instructional Setting: To what extent does the instructional setting promote transfer to the performance arena and thus respond better to the need?

This relation implies selection of the instructional setting by a comparison to the performance environment, that is, a search for correspondence, and a cost-benefit analysis that asks such questions as how close can we get for a reasonable cost, and how close do we need to get. The issue of transfer is raised. Note that this question cannot be answered without a careful analysis of the performance environment.

Problem Situation & Instructional Method: To what extent are instructional methods relevant to the situation and the issues that triggered creation of the instructional system?

Assuming, at least for the moment, that designing constitutes a form of inquiry, this relation concerns the validity of instruments, that is, how well methods address the research question, or perhaps, the issues driving the study. It focuses attention outside of the instructional system, as did transfer above in regard to setting, asking how the means that are used can be seen to connect with the performance issues in the situation. What seems important here is to go beyond face validity, suggesting assessment not only against the goal but against the problem situation.

Problem Situation & Assessment: How, how well, and at what levels is the extent to which the instructional system removes the need (solves the problem or improves performance) determined?

This relation asks whether the instructional system works. It may include a variety of assessments, from the quality of ways in which the problem was identified and the success of the instructional system was determined, to the effectiveness of the instructional system itself and the quality of feedback and revision mechanisms. Essentially it implies evaluation of all processes and products.

Goal & Learners: (1) Are these the right learners, that is, do they have sufficient entry knowledge/skill to succeed, and are they the ones who really need to reach the goal? (2) Are goals/objectives written in a way that matches the learners in terms of their abilities, knowledge and skills?

Responses suggested that goals and learners need to be considered in tandem. For example, writing a goal prior to identification and analysis of learners was described as a theoretical ideal not well matched to practical constraints. In real situations, learners can rarely be selected or interchanged based on a single criterion. Adding complexity to this is the fact that the designer must consider multiple goals and relations among them, for example, learner versus sponsor, organizational versus individual.

Goal & Content: What content is necessary and sufficient to meet the goal?

This relation addresses relevance and completeness—the "right stuff" in both focus and amount—and suggests that content/task analysis is a focused activity, guided by a clear understanding of the learners and desired performance.

Goal & Instructional Setting: To what extent do the instructional setting and associated technologies, tools, and resources support reaching the goal?

This relation addresses selection or adaptation of settings. Comparison of setting options and consideration of their adaptation are suggested, as is a critical examination of the goal. That is, the goal appears to be given prominence here, but must itself be changeable when the constraints of the setting are prohibitive.

Goal & Instructional Method: To what extent do the behaviors elicited by instructional methods match those implied by the goal?

In most responses an order was implied: the goal is established then methods are selected to match. Other responses did not suggest an order, but rather a sort of congruence between the two components. Both if-then and what-if questions were evident: If goal X is desired, then what method should we use? What goal would be achieved if we used method Z?

Goal & Assessment: Does assessment truly measure how well the instructional system accomplishes the goal?

This relation makes it clear that instructional systems are goal-driven. Assessment is seen as directly related to goal statements, and the quality of instruments is judged based on their correspondence to them.

Learners & Content: Is the content appropriate for the learners in terms of such things as quantity, level, emphases, frame of reference, and learning skills?

This relation focuses on appropriateness and implies a close relation between learner and task/content analysis. That is, responses suggested that it is necessary to perform task/content analysis with the learners in mind. Otherwise you cannot know where to start or stop, which areas to focus on, and how content can be best arranged/represented.

Learners and Instructional setting: Is this the best setting for this audience?

A fairly clear directional relationship is evident here (Learners→Setting) and deals with such characteristics of the setting as visibility, acoustics, heat, realism, convenience, familiarity, accessibility/proximity, and tools, and their appropriateness for learners.

Learners & Instructional Method: How good is this method for these learners?

This relation matches methods (strategies, tactics, activities, interfaces, etc.) to audience characteristics (prior knowledge/skills, reading level, interests, experience and comfort level with various strategy and media options, learning style, views and attitudes, maturity, etc.).

Learners & Assessment: How well does the assessment match the learners' needs, experiences, and abilities?

This question focuses on the features of tests, for example, on the way they are worded, and relates those features to the learners in terms of general characteristics, experiences, especially experience with testing options, and the use of knowledge/skills on the job.

Content & Instructional Setting: To what extent does the setting support learning the content?

When setting is broadly defined to include such things as space, time, resources, and tools, this relation concerns the match with what is to be learned, that is, a type of setting is more or less appropriate for learning a particular type of knowledge/skill. However, responses suggested that this is not a major concern, especially for adults.

Content & Instructional Method: How well will the instructional method lead to learning the content?

This relation means that particular methods or types of methods (strategies, activities, events, media, materials and sequence) are felt to be better or worse for different content or types of content. A few responses suggested that selection can be based on known principles, that is, method type X works best for content type Y. Others gave no sense of how the selection is made or designing is done.

Content & Assessment: To what degree is there congruence between what is taught and what is tested?

A two-way relation is evident here: (1) Do tests cover what is taught? and (2) Are those things that are deemed important enough to be tested included as content?

Instructional Setting & Instructional Method: What methods can and should be used in the setting?

Responses tended to suggest that the setting is fixed and methods are chosen based on constraints of that setting, although a couple implied the opposite, that is, choosing the best method and then finding an appropriate setting.

Instructional Setting & Assessment: (1) What options for assessment are available, feasible, appropriate, and desirable in the instructional setting? (2) How good can and will testing be if done in this setting?

A majority of responses indicated a relation in one direction or the other, that is, either Setting➔Assessment, or Assessment➔Setting. Other respondents cautioned that assessment done in the instructional setting is less important than assessment done in the performance setting, that is, in terms of application on the job.

Instructional Method & Assessment: How consistent are strategies for teaching and testing, for example do they involve the same types of behaviors on the part of the learners?

A fairly clear relation of consistency was evident here, although some responses focused on the evaluation of methods as opposed to evaluation of learners. Both appear important.

Trying to capture relationships and search for patterns, we constructed several visual maps. The most elegant of these combines the systems components and their interrelationships but is a bit rich in information for clear understanding at a glance. Consequently, we have included three versions here that build in richness. *Figure 1* shows the system components. *Figure 2* adds lines that represent the potential pair-wise relationships. Then *Figure 3* adds the key question/issue for each relationship along these lines. Notice that the space outside the large oval in these figures represents the larger performance situation to which the instructional system responds.

Before commenting further on the map in *Figure 3*, we will summarize the responses to two additional questions asked in Survey 2. One had to do with priority, that is, with which relationships were seen as most important. The other asked about complexity. In terms of priority, we chose to use respondents' components from Survey 1 in their own versions of Survey 2, rather than the seven categories as a standard set, and this made analysis difficult. In fact, the manipulations required here made suspect any sort of conclusions we might have drawn. The only thing we can say with any degree of confidence is that a pattern occurred in the citing of particular components. Respondents tended to give higher priority to relations involving the *problem situation* and the *learners*, and lower priority to relations involving *assessment* and *instructional setting*. Note, however, that the survey question asked respondents which relations would "make or break the effort," and many relations were cited. It seems reasonable to conclude, therefore, that these relations were not just generated for the study. They were seen as important considerations in designing.

In terms of complexity, a conclusion can be drawn from many comments by respondents: the pair-wise relations they were asked to consider in this study greatly underestimated their perception of the complexity of instructional systems.

4. DISCUSSION

Because the study was conceived to explore systems thinking, we will interpret results using a set of systems science concepts: dimensionality; complexity; and directionality. Dimensionality means essentially the number of types of things that are said to be part of a system, or more specifically, those things that make possible a more accurate system description and that, when varied, modify the system's behavior (Warfield & Christakis, 1987). Given the respondents' reported difficulty not in generating but in limiting components to about six, it seems safe to conclude that designers perceive instructional systems to be at least six-dimensional, and they agree, in general, on what those dimensions are.

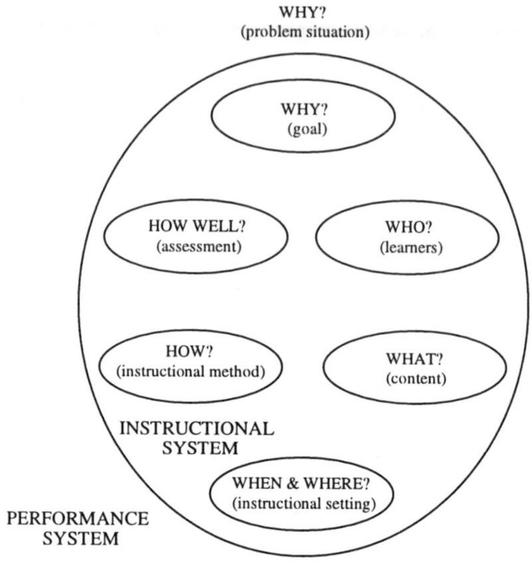

Figure 1. Instructional system components

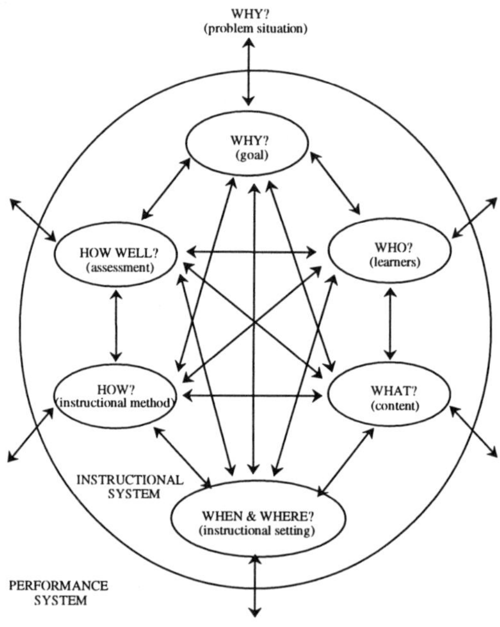

Figure 2. Pair-wise interrelationships among instructional system components

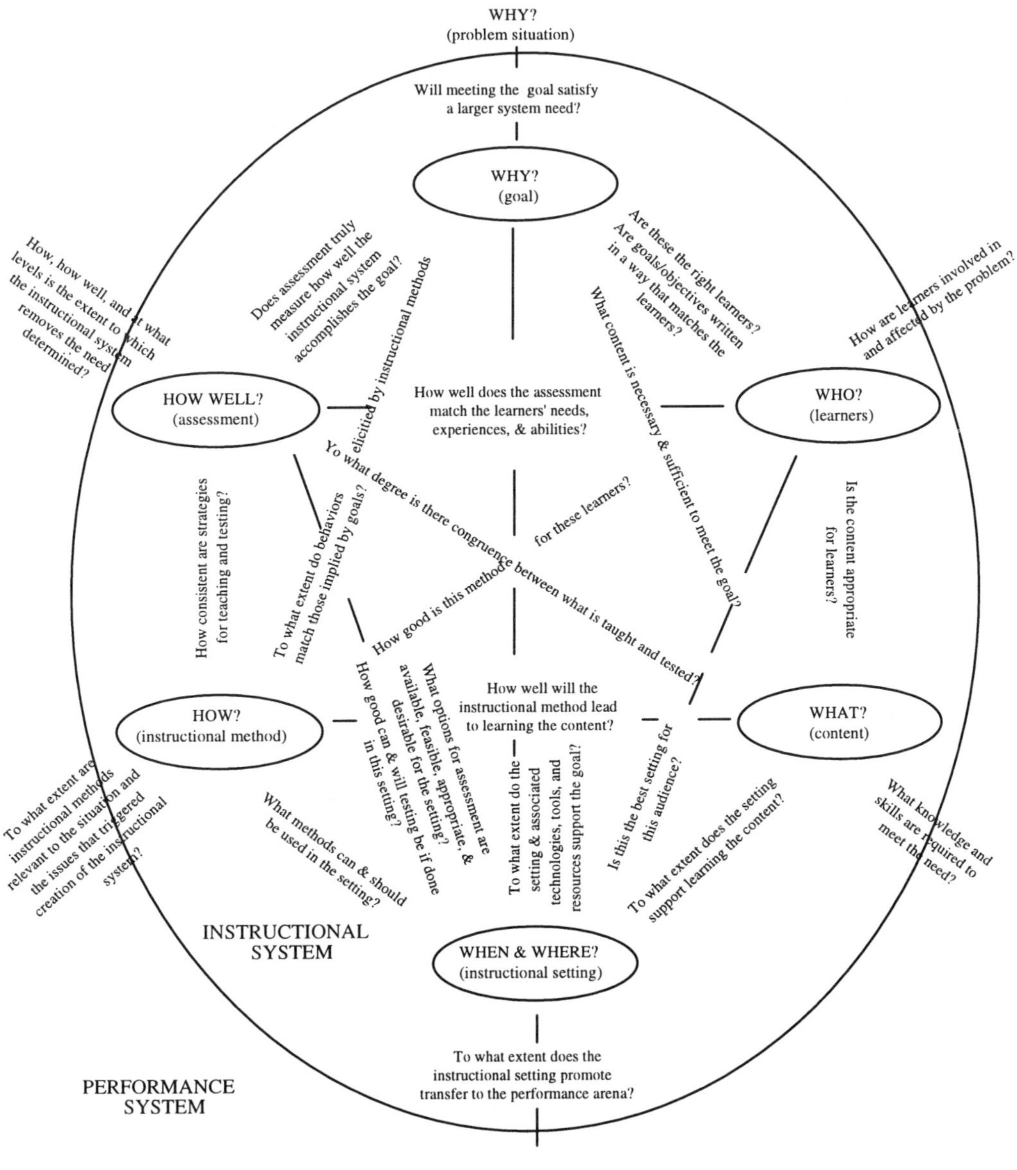

Figure 3. Key issues/questions in relations among instructional system components

A six-dimensional system is quite complex. In systems terms, where by definition interdependency is assumed, all combinations of parts must be considered. So a two-dimensional system with dimensions A and B requires consideration of three things (A; B;

and AB), a three-dimensional system with dimensions A, B, and C requires consideration of seven things (A; B; C; AB; AC; BC; and ABC), and a six-dimensional system requires consideration of 62 things. Note also that these figures assume that interdependencies are not directional, that is, that AB is equivalent to BA, something contradicted by several of the relations described above. As Christakis (1996) points out, Miller's (1956) often cited mental capacity of "seven plus or minus two" is exceeded when one moves past three dimensions. Even if this limit could be greatly exceeded, a six-dimensional system would seem to lie beyond human capacity for simultaneous consideration of all parts.

How instructional designers are able to manage this sort of complexity strikes us as a very important question, and we know of no studies that provide solid answers. Perhaps the issue lies at the root of why simple linear approaches are appealing, especially in education and in work situations where designers with less skill and experience are employed. And maybe this has some explanatory power for the emphasis given such things as performance objectives. Objectives are, in essence, three-dimensional statements (conditions-actions-standards) that serve to reduce complexity and guide decision making on further dimensions. Similarly, Reigeluth's (1983) Conditions-Methods-Outcomes framework may serve to reduce the complexity of ID as a whole. It does not seem accidental that we choose three-part frameworks or that we seek to fix some elements in order to prescribe others. The reasons for and the nature of these "phrases" in our "pattern language" (see Alexander, Ishikawa & Silverstein, 1977) are not very clear.

Perhaps the most surprising result of the study, at least surprising to us given our efforts to get beyond a simple "systematic process" interpretation of *system*, comes from a close examination of *Figure 3*. If you read the relationships as they pertain to individual components, a clear pattern is evident. Start with the problem situation, and read all statements along the lines leading to it (the spokes out from the large oval). Notice how problem situation (or need) is the constant in all cases, that is, other things are said to vary with respect to the problem situation. Now read all statements linked to goal. Nearly everything varies with respect to the goal except the problem situation. Now try it with learners, content, and so on (in *Figure 3*). There is a clear pattern of directionality in relations, and this moves first inside the oval from outside and then clockwise around the figure.

Interpreting this directional pattern, we first considered the possibility that it was an artifact of the research method, that is, we asked if the apparent directionality was caused by the way in which questions were asked rather than by how respondents truly perceive instructional system components to be related to one another. We examined questions and responses and found that in a clear majority of cases, the order of components in the responses was reversed from that given in the questions. This argues against the result being an artifact.

However, we did find that in constructing the individual versions of Survey 2, we had consistently fixed the first component and varied the second (e.g., goal and content, goal and learner, goal and instructional method, etc.) This may have caused respondents to cue on the first component as one that could remain fixed in their mind while other things varied, thus causing the directional patterns that give emphasis to the first component, and thus the overall clockwise pattern in *Figure 3*. In order to examine this possibility, we conducted a follow-up study in which the component pairs were randomized, both in terms of which component was stated first and the sequence of pairs. The response to this study was disappointing: nine practitioners were recruited; six returned Survey 1; three returned Survey 2. Nonetheless, we found no evidence of an artifactual source of the directional patterns. In the three Survey 2 responses, relations either clearly supported the summary statements we had made or were neutral with respect to them. Of course, drawing a conclusion from three responses is inappropriate.

If the patterns of *Figure 3* were to remain in further studies, this could be taken as support for a linear approach and for the use of essentially deterministic models. However, this would contradict respondents' comments in other areas, particularly their answers to the final questions of Survey 2, and it would go against studies of design processes where a more dynamic and iterative process has been observed (e.g., Rowland, 1992).

Perhaps underlying this apparent contradiction is the inference that logical dependency (i.e., directional relation) requires particularly ordered activity. That may not be the case. Designers may think in terms of directional relations but not act in a purely linear fashion. For example, what logical dependencies may require is not necessarily a linear pattern of engagement, with one component defined, then work begun on the next, but a linear pattern of component "fixing," that is, one definition fixed before the next is fixed. Work on defining a number of components could be performed simultaneously or in any expedient manner (as noted by Tessmer & Wedman, 1995), while the final decision-making process was specifically ordered. In this case we might interpret major sections of a linear model to represent a sequence of completions rather than a sequence of activities, all of which might occur in the final stages of a more extended dynamic process. The various components would begin as place holders and gain detail as a model of the case was constructed. This interpretation seems to eliminate the apparent contradiction, and it is congruent with recent methodological developments such as rapid prototyping and constraint-driven design (de Hoog, de Jong & de Vries, 1994).

The description above may constitute a more accurate representation of instructional designing, but such a conclusion requires greater empirical evidence. If it does prove more accurate, what might serve as a better descriptor than "systematic" or "deterministic process" is "emphasis-shifted and mutual-shaped defining of system components." The implication is a dynamic process of defining components wherein emphasis shifts from one component to another while all are continually seen as interdependent and thus as mutual shaping. The particular order of emphasis shifting may match linear ID models.

5. VALIDITY

There is cause to trust as well as to challenge the results of this study. First, the sample was quite small, although it represented a broad range of work contexts, and respondents were clearly active and experienced. Of course, experience should not be assumed equivalent to expertise. The lack of any apparent clusters of responses in either survey with respect to "groups," that is, with certain levels of preparation, activity, or experience, makes generalization more reasonable, also.

There appeared to be agreement on the components of an instructional system, but sample components were given in Survey 1, and these might have been simply accepted. Most respondents did not choose to just repeat the given components, but their responses certainly may have been shaped by them. These sample components were based on a careful review of ID literature, but may not represent the specific components that practitioners normally consider. Also, respondents' ability to state relations doesn't show that they think of such in designing. On the other hand, it does indicate that the components are reasonable. Respondents did not appear to have difficulty specifying components, considering relations, or indicating which were important. Indeed, as one respondent stated, the failure to consider any one of the relations he specified could break the effort.

In an effort to maintain a closer link to practice, the components listed by each respondent in Survey 1 were used in Survey 2. This was a tradeoff and made the question of priority of relations difficult to answer for the group as a whole. That is, comparisons of priority would have been easier had we used a standard set of component labels in Survey 2, but this would have taken the responses out of the designers' own languages.

Care was taken to pilot test drafts of instruments with practitioners, and two revisions preceded implementation. All analyses were independently repeated with nearly identical results by the second author, also. Finally, drafts of this report were forwarded to respondents who were asked the extent to which the description of the study matched their recollection of involvement, if results adequately represented their responses, and if inferences and conclusions were reasonable. Response to this request was small but positive.

A view of the study from systems science points out other limitations. First, systemic relations are often dynamic, so a static and singular description may gain clarity at the expense of richness and accuracy. Similarly, a system is poorly described by a single model from a single viewpoint. As implied above, *Figure 3* shows logical dependence–in a sense, structure–but it does not directly address process. Our comments on process are therefore inferential. Perhaps, the most serious concern would be the fact that the study violates general systems theory. As von Bertalanffy (1968) pointed out, general systems theory deals with the whole, with the "entire systems in which many variables interact and in which their organization produces strong interactions. It does not deal with isolated processes, with relations between two or a few variables or with linear causal relations" (quoted in Hug & King, 1984, p. 19). So the study may represent a step *toward* a systems view of ISD but remains quite a ways from it.

6. IMPLICATIONS AND NEXT QUESTIONS

The study explored perceptions of systemic relations in the products of instructional design, and provided some indirect evidence of systems thinking in process. Both offer opportunities for speculation and suggest questions for further research. We will offer some general implications first, then list a few of the questions raised.

Results may be useful in performance support and design education. The key issues/questions might offer performance support to practitioners, particularly those with less experience. They might serve as prompts to consider issues at particular times in designing. These key issues/questions might be useful in design education, also. They might serve to foster development from novice to expert status, for example. Pair-wise linking of components might represent a stage between the *part thinking* of novices and the *system thinking* of experts. Theory and model development might follow from work in this area, for example, the map in *Figure 3* might be adapted to serve as a general view or specific guide to designing. Caution should be exercised here, of course. The figure is product-oriented and purely descriptive. For example, specifically how the sort of emphasis-shifting and mutual shaping processes are or should be carried out was not addressed by the study. A next question, therefore is *How might mutual shaping and emphasis shifting occur and intertwine in designing instruction?*

A second more general question is *How do instructional designers manage complexity?* The study suggests a six or more dimensional system meaning that complexity is high, perhaps pushing component relations beyond simultaneous consideration. The sort of mechanisms or patterns designers use to manage this are unclear.

A related question is *How are design quality and complexity related?* Davies (1995) points out that quality must be designed in, rather than flaws inspected out. For example, integration is not something we can just test for during a formative evaluation process after most if not all substantive decisions are made because the quality of a design's overall conception cannot be transcended through inspection and refinement. If that is so, then how do designers create high-quality products, ones that become powerful wholes, when complexity is very high*?* Davies suggests that awareness and consideration of systemic relations may be reduced by a serial process and increased by concurrent processes, but how does a designer engage in concurrent processes, that is, think about and do many things at once, when the number of things to think about and do is very high?

Perhaps, as suggested above, there are particular processes and particular sets of components that are emphasized and balanced at different times, that is, emphasis shifted and mutually shaped, and this affords both manageable complexity and system integrity. Contrary to what might be assumed, this along with the notion of "fixing" components may enhance creativity.

The issue of complexity leads to questions of representation. We construct models to simplify, to highlight particular aspects of phenomena, so *What sort of models might be constructed to encourage systems thinking in designing?* Perhaps illustrating logical dependence would be as easy as reversing the direction of arrows in a linear model. The relation only to adjacent components would have to be mitigated, but the simple flip in perspective may have power.

Of course, single models do not have equal value in varying contexts. Tessmer and Wedman's (1990) work on "layers of necessity" suggests how adaptations might be made. A question here would be what systemic relations should be maintained across contexts and how could designers do so, in a sense asking *How might systems thinking vary across contexts?* With regard to layers of necessity, it also raises the question of whether it should be activities/processes or particular sets of systemic relations in the product that need to differentially considered across contexts. (Tessmer & Wedman's concept of layers of necessity can be seen as an application of dimensionality to process as opposed to product.)

The final question we will pose here is *What other aspects of systems thinking are or might be powerful aspects of designing?* The discussion above addressed only dimensionality, complexity, and directionality. It may prove fruitful to explore other system concepts such as interaction and coupling (Perrow, 1984), hierarchy, communication and control (Checkland, 1981). We believe that an expanded notion of *system*, one rooted in contemporary system science, would be useful in the study of instructional systems design.

REFERENCES

Alexander, C., Ishikawa, S. and Silverstein, M., 1977, *A pattern language: Towns, buildings, construction,* Oxford University Press, New York.

Bertalanffy, L. von, 1968, *General systems theory,* Brazilier, New York.

Checkland, P., 1981, *Systems thinking, systems practice,* John Wiley & Sons, New York.

Christakis, A.N., 1996, A People Science: The CogniScope™ System Approach, *Systems, 1*(1), 16-19.

Davies, I.K., 1995, Re-inventing ISD, In Seels, B.B. (Ed.), *Instructional design fundamentals: A reconsideration* (pp. 31-44), Educational Technology Publications, Englewood Cliffs, NJ.

Dick, W. and Carey, L., 1996, *The systematic design of instruction* (4th ed.), Harper Collins College Publishers, New York.

Ericsson, K.A. and Simon, H.A., 1984, *Protocol analysis: Verbal reports as data,* MIT Press, Cambridge, MA.

Hoog, R. de, Jong, T. de and Vries, F. de, 1994, Constraint-driven software design: An escape from the waterfall model, *Performance Improvement Quarterly, 7*(3), 48-63.

Hubka, V. and Eder, W.E., 1987, A scientific approach to engineering design, *Design Studies, 8*(3), 123-137.

Hug, W.E. and King, J.E., 1984, Educational interpretations of general systems theory, In Bass, R.K. & Dills, C.R. (Eds.), *Instructional development: The state of the art, II* (pp. 18-28), Kendall/Hunt, Dubuque, IA.

Miller, G.A., 1956, The magical number seven, plus or minus two, *Psychological Review, 63*, 81-97.

Perez, R.S. and Emery, C.D., 1995, Designer thinking: How novices and experts think about instructional design, *Performance Improvement Quarterly, 8*(3), 80-95.

Perrow, C., 1984, *Normal accidents: Living with high-risk technologies,* Basic Books, New York.

Pieters, J.M. and Bergman, R., 1995, The empirical basis of designing instruction, *Performance Improvement Quarterly, 8*(3), 118-129.

Reigeluth, C.M. (Ed.), 1983, *Instructional-design theories and models,* Lawrence Erlbaum Associates, Hillsdale, NJ.

Rowland, G., 1992, What do instructional designers actually do? An initial investigation of expert practice, *Performance Improvement Quarterly, 5*(2), 65-86.

Tessmer, M. and Wedman, J., 1995, Context-sensitive instructional design models: A response to design research, studies, and criticism, *Performance Improvement Quarterly, 8*(3), 38-54.

Tessmer, M. and Wedman, J., 1990, A layers-of-necessity instructional development model, *Educational Technology Research & Development, 38*(2), 77-85.

Warfield, J.N. and Christakis, A.N., 1987, Dimensionality, *Systems Research, 4*(2), 127-137.

Chapter 4

Educational Design and Development: A Study of Dutch Design Practices

Irene Visscher-Voerman
University of Twente

Keywords: Curriculum design, Curriculum development, Instructional design, Instructional development, Design approaches, Design practice

Abstract

This chapter will report some preliminary findings of a study on educational design and development practices in the Netherlands. The purpose of the study is to describe the strategies that professional education and training developers use in practice. Twenty-four developers were interviewed intensively and documents of their projects were analyzed. The data show that there are many differences between developers, but that there are also several common themes. This paper discusses the most important results in both the technical-professional and socio-professional domain, as well as the overall structuring of the process.

1. BACKGROUND AND CONCEPTUAL FOCUS OF THE STUDY

1.1 Background

In the literature, there can be found many procedural models for educational design and development. Many of these models were based on theoretical insights, and scarcely grounded in practice or based on empirical research, except a few (e.g. Kessels, 1993; Walker, 1990). In some cases, authors formulated their models starting from their own design experience. But also in these cases, empirical validation of the model is low (Andrews & Goodson, 1991). There is, however, an increasing attention for practice itself, as a way to deepen the insights into the development process and as a basis for further refinement of development models. Goel and Pirolli, (1992), Kessels (1993), Richey and Nelson (1996), Rowland (1992), Saroyan (1993), Schön (1983, 1987), Semb, Ellis, Fitch, Parchman and Irick (1995), Van den Akker, Boersma and Nies (1990), Wedman and Tessmer (1993), and Zemke (1985) have studied and reconstructed development practices, to gain insights in how development processes occur, or to see what factors influence the flow of the development. Holcomb,

Wedman, and Tessmer (1996) and Klimczak (1997) have identified factors that contribute to the success of instructional design products.

A general conclusion is that existing development models do not reflect development practice adequately and, therefore, cannot serve as guides to successful practice (see Holcomb, Wedman & Tessmer, 1996). One reason is that models have often been developed for one specific context or sector and, therefore, are less likely to work in other contexts. Another reason is that development models are often so abstract, that they do not provide developers with guidelines for how to deal with concrete, daily situations in which all kinds of constraints such as time, money, and unexpected circumstances can force developers to change their strategy quickly and adequately (Wedman & Tessmer, 1993).

Assuming that there exist many interesting, successful and relevant professional design practices, the University of Twente originated a research program in 1993 (*OSF-program*), aimed at describing and exploring differences and similarities between development approaches of professional developers in different social fields. The study described in this chapter is conducted within the framework of this OSF-program and focuses on 24 design and development practices, in six different sectors of training and education.

A first analysis of the data showed that developers possess different conceptions of what they call a good design and a good design process; differ in how they conduct development activities; and have different reasons for why they are doing it. Due to a continuous reflection on the data and on theory, a framework evolved, grounded in both theory and practice (Strauss & Corbin, 1994). This framework assumes that the ways in which developers view the world and how they engage in the development of educational products can be fundamentally different. The framework consists of four different paradigms (instrumental, communicative, pragmatic, and artistic) in which developers may be engaged. The framework helps describe and interpret developers' individual actions and rationalizations and thus contributes to a better understanding of the variety of development approaches applied by professionals. Chapter 2 of this book describes the four paradigms and highlights differences between developers from different paradigms.

1.2 Conceptual focus of the study

In 'traditional' design models (see for an overview Andrews & Goodson (1991), Edmonds, Branch & Mukherjee (1994), or Gustafson & Branch, 1997) development processes are usually seen as technical, rational, systematic processes in which analysis, design, evaluation, and implementation activities are conducted successively. The emphasis in these models is on 'technical-professional' activities (Goodlad, 1994), such as conducting a needs assessment; formulating learning goals; selecting instructional content, instructional strategies, and media; formulating test items. By applying problem solving and decomposition techniques, developers are supposed to reach products that are 'internally consistent' (Kessels, 1993). Several critiques have been formulated on these models and a change in thinking about educational design and development has evolved. This has been stimulated by the development of new technologies such as the WWW and computer-supported design tools, by new epistemological insights and insights in learning psychology such as constructivism, as well as by an increasing number of reconstruction studies of practice.

Reconstruction studies have shown that, although developers operate in the *technical-professional* domain, they do it differently as supposed by the models.

Firstly, they do not conduct each activity in each project, and secondly, they do not conduct the activities in the order prescribed (Holcomb, Wedman & Tessmer, 1996). The development process cannot be reduced to a rational and technical-professional process. The development process is situated in context and developers work with actors and stakeholders involved. Thus, developers should be skilled in the *socio-professional* domain of development as well (OSF-program, 1993), and accordingly, models should also promote this domain.

Accordingly, the concept of quality and what developers think important to do has changed over time. A shift can be noticed from a focus on a good product with internal consistency to product implementation and results. Several researchers plead that developers should pay consistent attention to the people who have a stake in the implementation process during the development process. By involving them in the process, developers increase the chance that they will get to feel responsible for the product, thereby increasing the chance of a successful implementation (Kessels, 1993; Tessmer & Harris, 1990). In the education sector, Keursten (1994) and Van den Akker (1988) investigated what developers should do to get teachers committed and to increase the chance of a good implementation by presenting them with procedural specifications. Roes (1997) and Van den Berg (1996) investigated the impact of having teachers practice the new teaching skills before implementation.

Such insights and empirical evidences result in other conceptualizations of the development process. Firstly, linear models are being replaced by more iterative, spiral, cyclic models, in which there is no formal distinction between phases of *analysis, design, evaluation*, and *implementation*. Examples of these kinds of models are prototyping models (e.g. Tripp & Bichelmeyer, 1990; Moonen, 1996) and the 'layers of necessity model' of Tessmer and Wedman (1995). In these models, there is less attention for the analysis phase to set project and development goals. Formative evaluation activities get a central role in improving the product during the process. Secondly, there is a lot of attention for future users and stakeholders in the process, and the context in which the product will be implemented. Implementation is not regarded as a separate phase at the end of the process, but as something that needs to be regarded and taken care of from the start of the project (Van den Akker, 1996). An example of a model with this implementation perspective is provided by Plomp (1982, 1992).

To deepen insights in the development process, this study focused on both the technical-professional and socio-professional domain and their interrelationship. Questions were asked such as: 1) What kind of technical-professional activities do developers conduct in the different phases of the development process and why do they do it that way?; 2) What activities do developers conduct in the socio-professional domain? What actors are involved in the development process or design team? What is their role? How do developers build a relationship with them? Who is responsible for what?; 3) How can the overall process be structured?

Because of the explorative nature of this study, there were no hypothesis formulated at the beginning of the project. Rather, these questions were chosen as a guideline to find out how developers work. The data show many differences between developers, which can partly be explained by the different paradigms that they employ. The data also provide evidence to assume that developers from different paradigms share common ideas, notions, tactics, or themes. This chapter explores several of these commonalties.

2. RESEARCH METHOD

2.1 Design of the study

A case study approach was used to collect data from 24 developers with professional, high-reputation developers within different training and education contexts. Interviews were the main source of information, supplemented with document analyses.

Case studies and interviews

In this study, the developers were seen as 'a case' and they were interviewed about one or more of their projects. Documents of these projects were studied. Data were gathered in several phases. In the first phase, 12 developers from 6 different sectors were interviewed. The first two interviews served as pilots, but their data could be used as well, because it was agreed upon with the developers that missing and additional information could be gathered from them. After the first phase, the interview data were analyzed globally. Based on the analysis, the framework with paradigms, as described in chapter 2, evolved. It was decided to change the interview instrument slightly and to add interview citations to the interview reports. In the second phase, 12 other developers were interviewed.

Professional, high-reputation developers

Although Van den Akker, Boersma and Nies (1990) show that some existing professional practices are far from optimal, there is reason to believe that there exist many relevant and successful practices. There are many developers who are competent professionals, and highly capable of dealing with unique and specific design situations efficiently and effectively. Descriptions of such developers could serve as an example or as a frame of reference for other developers and help them reflect on, or improve their own professional work. Therefore, this study focuses on high-reputation developers, persons who could serve as positive examples for others. The underlying assumption is that a developer's reputation is directly linked to the quality of his/her work. It is unlikely that developers gain a high reputation if they make ineffective products, or if they work very inefficiently.

Development processes are context-specific and partly dependent on the context in which they are conducted. To assure a great variation of development contexts in the study, developers from 6 different contexts were asked, namely school book development (SB); curriculum development (CR); development of educational multimedia (MM); Human Resource Development in internal training departments (IT); Human Resource Development by external training bureaus (ET); and development of distance education (DE).

2.2 Selection of participants

A 'purposive sampling' strategy (Warwick & Lininger in Johnson, 1990) was used to select developers. The term 'developer' was conceptualized as follows:

- works on concrete development products (micro or meso level);
- is responsible for the educational, instructional, and/or pedagogical aspects of the product;
- has more than three years of experience;
- spends more than 50% of his/her time on design and development work;

- is able to reflect on his/her work.

For each of the development sectors, three or four experts -either researcher or developer-, familiar with the sector, were asked to nominate high-reputation developers. The experts nominated several developers. For each developer they gave possible reasons for their high reputation:

- quality of the products ("has developed a leading method");
- satisfaction of clients ("clients worship him");
- development approach ("is very creative", "always comes up with new things", "does not only do, does also think");
- years of experience ("developer has a broad orientation, knows how to survive in this field", "he is an old hand in this field").

Based upon the nomination of these experts, 12 developers were selected for each phase and invited to participate in the research. All of them wanted to participate in the research. Only one developer did not meet the criteria of being involved in development work more than 50% of her daily task and had to be replaced. Three developers asked a colleague to participate in the interviews.

2.3 Data gathering

Data were gathered by means of interviews (primary data) and documents (secondary data). The data were summarized in a case report and sent to the developers for comments and validation (tertiary data).

Interviews

Each developer was interviewed twice, about one specific project she or he had recently completed. By focusing on a concrete project, developers were invited to illustrate and embed their general statements in a concrete project example. By specifying why they worked that way in the particular project, relevant factors could be detected that influence the way they work. Developers were also systematically asked to specify to what degree the strategies and activities conducted in the specific project reflect their general approach. Thus, the interviews yielded insights into the general development approach, abstracted from a specific context of the project. When appropriate, developers were also asked to specify why they deviated from their general approach in the project and what factors 'forced' them to conduct alternative activities. Four developers used examples from different projects to illustrate their approach.

A bottom-up approach was used in the interviews: the topics and their corresponding interview questions were addressed, but the flow of the interview depended largely upon what the developer had to report. If the emphasized specific topics that were not on the list, there was room to explore the topic. In fact, the interviewer was a "traveler" (Kvale, 1996, p. 4), who deliberately sought specific sites [topics], conversing with the local inhabitant(s) [developers] who also could point out interesting sites as well.

Each interview took at least one hour and a half. Demographic data were collected at the start of the first interview (e.g. years of development experience, professional background, job title). Having given a short overview of the project, the developer was invited to start describing the process from the early beginning. The second interview was used to clarify obscurities from the first interviews and addressed aspects that were not yet discussed in the first interview. It also focused more generally on the topics from the framework and addressed questions derived from the project documents.

Document analysis

Twenty developers had project documents available, such as materials, process reports, etc. During the interviews, they used these documents to illustrate their statements or to refresh their memory. Eighteen developers gave permission to take the documents for further analysis. The purpose of the document analysis was to get a clear(er) picture of the flow of the development process and the decisions taken. They were used as a 'check' of the interview data. No formal analysis scheme was used, but the relevant interview questions were applied. Only in a few cases did interview descriptions and documents differ from each other. This was asked in the second interview.

Report

All interviews were taped and typed into protocols. Based on the protocols, the following reports were made: demographic overview of the respondent (age, gender, organization, professional background, relevant experience); project profile (short summary of the project; including name of developer, name of client, project description, actors involved, characterization of the phases of development, the currency of the project and the developer's reflection upon project success); summary of the interviews: reports (adding up to 15 pages, including several çitations from the interviews). Each interview report had the same outline.

The project profile and interview report were sent to the developers, requesting their comments and/or specifications. All developers, except two, sent the interview report back. Most of them had looked at the reports very critically. Three of them added new insights, reflections, and even models. All these comments were again integrated in the interview report. These improved reports, accorded by the developer, were used as the basis for analysis.

Analysis

Each topic from each interview was put together in a different document. Because of the structure of the reports, the reports needed not be coded at this point and therefore, no analysis program was used. Per topic, the data was reduced in matrixes, as supposed by Miles and Huberman (1994). Conclusions were written and illustrated with interview excerpts.

3. RESULTS

Some preliminary results are presented in this section. First, an overview is given by the developers and their basic paradigms, followed by common themes in the technical-professional activities and social-professional domain.

3.1 Overview of developers and paradigms

In each of the six sectors, four developers were interviewed, summing up to 24, 17 male, 7 female. Their experience ranged from three till more than 20 years.

The developers were characterized according to one or another paradigm based upon their design activities and their motivations. Developers with much attention for the formulation of specific goals and outcomes for the product were labeled instrumental. Developers who highly valued close interaction and communication between the development team and others involved were called communicative. Developers created their product by quickly building, testing, and revising several prototypes or early versions of the product were named pragmatic. There were 14 instrumental developers in the group, 7 communicative developers, and 3 pragmatic developers. Among the 24 interviewed developers no one with an artistic rationality was found. There are a few possible explanations for this. It could be that there are hardly or no developers with an artistic in the domain of educational design and development; We may not yet understand the essence of the artistic rationality well enough, which has blurred the development of the instruments and analysis; According to Eisner, the most creative decisions about a curriculum are made by teachers, while enacting the curriculum and teachers were not included in this research. Table 1 shows the basic paradigm of each developer and a short project description. The four developers who used several projects to illustrate their remarks are indicated with an *.

3.2 Themes in educational and development practice

In the following sections, several common themes in the technical-professional as well as the socio-professional domain are discussed. The technical-professional domain is ordered along four development activities: analysis, design, evaluation and implementation. The socio-professional domain focuses on the variety of personal backgrounds in a team. Descriptions of activities in the socio-professional domain such as how to get actors committed to the project, or how to negotiate with actors are not included in this chapter.

	Paradigm	**Project description**
SB1	Instrumental	(Re)development of a math method for all classes and types of secondary education
SB2	Communicative	Development of 5 independent lesson materials (4-page leaflets) on cancer, for elementary education
SB3	Communicative	Development of new math method (books, such as lesson, work, and assignment books) for all classes of elementary education
SB4	Pragmatic	Development of new French method for all classes and types of secondary education, consisting of text books, audio tapes, CD, video
CR5	Instrumental	Development of national exam programs for cultural and art education, k7-10
CR6	Communicative	Analysis of current training 'communication', using practice job profiles; development of objectives for renewed training
CR7	Communicative	Development of didactics book for technical domain, to be used in pre- and in-service training
CR8	Instrumental	* Development of curriculum for geography for pre-service training for elementary teachers and several exemplary materials (and other projects)
MM9	Pragmatic	Development and implementation of on-the-job multi media training and development of job aids
MM10	Pragmatic	Development of three simulation programs on gas technology, to be used flexibly in several basic training programs
MM11	Instrumental	Development of a 20-minute video program, as introduction on a topic from the language field
MM12	Instrumental	Development of courseware and teacher guide for practicing understanding reading. The shell of the courseware was to be usable for other topics as well
IT13	Instrumental	Development and implementation of HRD program preparing employees for a new job/function
IT14	Instrumental	Development of user guide to an existing software program as integrated software package
IT15	Instrumental	Development and implementation of 2,5 year HRD-program that prepares employees for their future demands of their jobs

	Paradigm	Project description
IT16	Instrumental	Development and implementation of trainer and trainee materials for an ongoing training program
ET17	Communicative	Development of skills training program to prepare employees for a new task
ET18	Communicative	* HRD development projects
ET19	Communicative	Design and development of skills training program to prepare employees to the new release of software
ET20	Instrumental	Development and implementation of a 18-module self-study program that prepares current and new employees for their job
DE21	Instrumental	Development of interactive computer program with cases for environmental problem solving
DE22	Instrumental	Development of interactive court simulator computer program for law students
DE23	Instrumental	* Development of distance education courses
DE24	Instrumental	* Development of distance education courses

SB = School Book development, CR = curriculum development, MM = development of multimedia, IT = internal HRD, ET = External HRD, DE = development of distance education
* = more than one project used

Table 1. Paradigms and project description per developer

3.2.1 Technical-professional domain

Analysis: Nature of analysis activities

Large projects usually had a clear analysis phase (SB1, SB3, CR5, MM12, IT13, IT16, DE21, DE24). These projects share two characteristics. Firstly, almost all of them are "generic" (Walker, 1990) (except IT13, IT16). Secondly, they all focused on the development of a new product or preparation for a new job. Therefore, developers needed time to figure out what they needed to make or for what function they needed to prepare people.

Developers of smaller projects usually paid less attention to the analysis phase than suggested by development models, even the instrumental developers. Developer MM11 for example needed only one meeting with the client to get her analysis done. This was also supported by developers from other paradigms (CR6, MM10). The three pragmatic developers clearly see evaluation of their first prototype as a kind of analysis. Developer MM9 argued: "*Formative evaluations are more important than a major analysis phase at the beginning of the project. In the analysis phase a developer could talk to people and observe them at work, but they only get a view of the existing situation. It does not say anything about the way in which the product will contribute to improving their work. By formative evaluation you can find that out*".

Design: The shaping of the solution

All developers indicated that they usually start the development process with already an idea of the solution. So, they work "solution-driven", rather than "problem-driven"(Visscher-Voerman & Plomp, 1996) as supposed by ISD-models such as the OKT-model by Plomp, 1982, 1992; IDI-model by the National special media institute, 1971). In some cases, clients may ask for a certain solution when they contact developers, in other cases, developers themselves develop a sense of the solution, as soon as clients have posed them their question. To communicate their solution ideas and other possible solutions to the client, five developers took concrete products from former, similar projects and discussed the pros and cons with the client. These developers view this as a useful way to stimulate the specification of needs and wishes. Pragmatic developers particularly reported that it is never too early to start with an idea for the solution. It is their experience that the most efficient way to get problem specifications clear is to provide the client with ideas or solutions, in the format of a concrete product or prototype.

The findings are congruent with findings in other reconstruction studies from e.g. Rowland and Wilson (1994), and Walker (1990) who found that designers think more in terms of solutions than in terms of problems. Rowland (1992) also found that with experienced developers, the production of possible ideas and solutions usually precedes the formulation of requirements for the solution.

Several developers, especially the instrumental ones, perceive it as a restraint to work solution-driven. Developer IT13, for example, warned that one takes the risk of following that specific solution which may be less suitable than other unexplored options. Developer DE22 stated that one could risk the fact that -during the process- it may become clear what a better solution could have been, but that it may be too late to change the process. Both referred to the ISD-models as "the right" or "the best" way of working. Pragmatic developer MM10 also indicated to never start working on the solution straight away. He argues that is important to think over the consequences for the client and its users before starting the development of the solution.

Design: Media selection

In more than half of the projects discussed, the developer started the project with an idea of the kind of medium for the solution. This is not surprising if one realizes that developers are usually specialized in a certain (medium) field, for example computer simulations, multimedia, school books, on-the-job training, etc, and are approached by the client to make such a kind of product. If developers 'feel' that the client's question is not out of place, they do not pursue the issue of media selection further, but concentrate upon the elaboration of the chosen medium, eventually, by generating several alternatives within the product setting.

This finding is not congruent with ID-models in literature, in which the medium choice is part of the design phase, and often the result of a thorough analysis of the problem.

Design: Starting from target group instead of subject matter

19 Developers expressed that it is essential to take the target group and their jobs as a starting point for the design of the product, rather than the subject matter, because it increases the chance for a useful product. 4 Communicative (SB3, CR7, ET17, ET18) and 1 instrumental developers (DE22) stated explicitly that they distinguish themselves from other colleagues on this point, and think this as an aspect that makes them good developers. They indicated that many of their colleague developers often still use the subject matter as a starting point, which results in products that may be theoretically perfect, but are often too long, and not adaptable to the work situation. The interview data from the other 5 developers show that they also design from a target group perspective although they did not explicitly say so. This attention for user and user setting can be characterized as an implementation anticipating activity.

This finding seems to be congruent with instructional design literature. Many ID-models identify a phase in which the target group and their context, and -in corporate settings- their jobs are analyzed. According to these models, developers should use this analysis information as a starting point for the design of the content and layout of the product.

Evaluation: The use of formative evaluation

All developers conducted one or more formative evaluation activities in their project, to a more or lesser extent. Although only 9 of them used more formal activities such as pilot tests with users in 'real settings' (SB3, SB4, MM9, MM12, MM13, IT14, DE21, DE22, DE23) all developers used informal activities, such as asking feedback from a colleague or from the client. Reasons for conducting an evaluation are first, to get commitment from clients or team, and second, to get a product that is steered to the needs of the client and/or the user. According to 3 developers (CR6, MM9, MM11) it depends on the project and on what has been agreed upon with the client whether or not this 'informal formative evaluation' is done systematically by using a checklist or questionnaire, or not. Developers perceive evaluation activities per se as important. This can be derived from the fact that 20 developers mentioned working in small steps and checking early versions with client and users as one of the most important characteristics of a good development process. They also supposed these activities to have contributed to the success of their project.

It is interesting, however, that more than half of the developers did not use the term 'formative evaluation' to refer to these kinds of activities, but rather spoke of 'informal checks,' 'reviews,' or 'tests.' At least two of them (MM10 and MM11) explicitly mentioned that what they did could not be named formative evaluation. They argued that formative evaluation is money consuming and needs a lot of respondents to enable the developer to derive good and statistical conclusions. Developer DE24 also argued that a good formative evaluation would take at least six months. Apparently, several professional developers still have a negative connotation with the term 'formative evaluation.' They perceive it as an activity that is time-consuming, involves statistics and needs many respondents. The activities the developers do employ, however, have the same purpose as formative evaluation activities: to determine the worth of what has been made/done so far, and to come up with suggestions for improvement. It seems that these -often informal- formative evaluation activities are so integrated in their development activities and part of their professional repertoire, that they do not view them as separate activities anymore.

Developers who conducted one or more formal formative evaluation activities as a separate phase in the development process (e.g. pilot test) indicated that this is often one of the first activities to be omitted or reduced, mostly because of time and money constraints. This finding is consistent with the findings from the surveys of Wedman and Tessmer (1993) or Winer and Vázquez-Abad (1995) and from the study of Van den Akker, Boersma and Nies (1990), who concluded that formative evaluation results are often presented too late, so that improvements cannot be made anymore.

The fact that formative evaluation can help improve the quality of products or processes is also expressed by Maslowski and Visscher (1998), Nieveen (1997), and Tessmer (1994). Apparently, the findings in our study give also reason to believe that formative evaluation can make a difference. It will be interesting to have a closer look at the data to find out how exactly, how often, and why these formative evaluation activities are performed.

Implementation: designing from an implementation perspective

Only four developers were directly involved in the implementation of their product (IT13, IT15, IT16, ET19). Although the others were not, they paid some or considerable attention to the future implementation of the product. They go beyond most of the standard instructional design models that focus only on the product itself. The developers argue that it is important to pay attention to the implementation and prepare it as well as possible, because that will increase the chance that the product will be used after they have left the scene. Activities to prepare the implementation range from informing future users by means of leaflets or meetings (especially in generic situations: SB1, SB3, SB4, CR7, CR8, MM9, DE21, DE22, DE23); organizing train the trainer courses (in specific training situations: ET19, ET20); indicating the factors that improve the implementation (SB1, MM9, DE21, DE22, AE23) or asking several 'critical' persons to participate in the development team (CR6, CR7, ET17, ET18, ET20). One developer (IT16) reported to use formative evaluation activities as a means to prepare and anticipate the implementation. By asking actors to evaluate a product, they get to know it, come to realize its potential, and acknowledge their function to support the product (see in this respect also Van den Akker & Verloop, 1994).

The role of implementation improving activities during the development process is also mentioned in literature by Maloney (1993), Short (1983), Tessmer and Harris (1990) and Van den Akker and Plomp (Van den Akker, 1996; Van den Akker & Plomp, 1994). In traditional ISD models, implementation is, however, still seen as an apart and distinct phase (except Plomp, 1982; Kessels, 1993). The interview data need to be explored further to provide more insights in the why and how of implementation improving activities.

3.2.2 Socio-professional domain

Working in a team

Twenty-two out of the twenty-four interviewed developers worked in a team, consisting either of colleague developers or client and users. Team sizes differed from 2 (e.g. IT15) to more than 10 (e.g. DE21). 9 Developers explicitly mentioned working in a team and having people with different expertise in the development team as a quality criterion, because it increases the chance that the problem and its solution will be considered better and from different perspectives. As one person put it: "*you are never able to think through all perspectives on your own. You need others to challenge you.*"

Professional background

14 Developers had studied design (Educational Science and Technology and/or specific design courses), 4 studied educational sciences. The other 6 had a subject matter background. Two of these subject matter experts (DE22 and DE23) were of the opinion that their more than 10 years of design experience fully compensated for their 'lack' of formal design education. Developers with the formal design background indicated that they need some basic knowledge about the subject matter, to be able to communicate with subject matter experts. They usually gain that in the course of the process, by reading about the subject matter. It was interesting to see that 5 developers with a formal design background (MM12, IT16, ET19, DE23, DE24) mentioned it as an advantage that they are no subject matter experts themselves. They argue that they are better able to design from a target group perspective than subject matter experts, who are bound by the logical structure of the content. In fact, this notion fits findings reported by Saroyan (1993), who studied instructional designers and subject matter experts on a formative evaluation task. Developers MM11 and IT15 took time to master the subject matter by reading books about the subject or by attending a course, because they would be fully responsible for the content of the product, and IT15 also because she would have to teach that content.

3.2.3 Overall structure of the process

Most developers explicitly refute the idea that development is a systematic, linear process. Processes can pass off chaotically, with a lot of flipping back and forth between stages and activities. Without doubt, processes are iterative and cyclic.

It was, however, remarkable to see that the systematic ISD-models still have a major impact on how the process is structured. Fourteen developers worked in an instrumental way. Seven of these instrumental developers work in an organization where standard (instrumental) development procedures are available and -in most cases- have to be used. Two of these developers told that they are held accountable for using this approach. Four of them told that they would like to work in another way, but that they are not able to realize that. For example, developer DE22 explained that he was to make a formal plan for the course before he could develop the course, although he had in mind exactly what to do. Without the plan, the course could not be implemented. He felt this as an impediment.

4. CONCLUSION

This chapter focused on the similarities between developers and on common trends in their work. Some of the similarities refer to specific tactics that developers use, such as examples from previous projects. Most similarities, however, refer to general notions or ideas, such as the perceived importance of evaluation activities or of developing from an implementation perspective. A closer look at the data shows that developers elaborate these notions in their own specific ways. Two strategies that seemed highly potential from a theoretical point of view were also consistently found in most of the projects: formative evaluation and the anticipation of implementation. All developers put great importance to these strategies, although they conducted different concrete activities in the technical-professional as well as the socio-professional domain. The descriptions of the activities conducted, the results, and the reasons for doing it, might inspire other developers to improve their own process and methods.

REFERENCES

Akker, J.J.H. van den, 1988, *Ontwerp en implementatie van natuuronderwijs* [Design and implementation of science education], Swets & Zeitlinger, Amsterdam.

Akker, J.J.H. van den, 1996, Designing from an implementation perspective, In Plomp, Tj. and Ely, D.P. (Eds.), *International encyclopedia of educational technology* (2nd ed.), pp. 290-293, Pergamon, Oxford.

Akker, J.J.H. van den and Plomp, Tj., 1994, Curriculumontwikkelingsonderzoek vanuit implementatieperspectief [Curriculum developmental research from an implementation perspective], In Joha, B.C. and Joosten, F. (Eds.), *Ontwikkelend onderzoek*, (pp. 79-93), VU Uitgeverij, Amsterdam.

Akker, J.J.H. van den, Boersma, K.Th. and Nies, A.C.M., 1990, *Ontwikkelstrategieën in SLO-projecten* [Development strategies in SLO-projects], Dutch National Institute for Curriculum Development, Enschede.

Akker, J.J.H. van den and Verloop, N., 1994, Evaluation approaches and results in curriculum research and development in the Netherlands, *Studies in Educational Evaluation, 20*, 421-436.

Andrews, D.H. and Goodson, L.A., 1991, A comparative analysis of models of instructional design, In Anglin, G.J. (Ed.), *Instructional technology: Past, present and future* (pp. 133-155), Libraries unlimited, Englewood, CO.

Berg, E. van den, 1996, Effects of inservice education on implementation of elementary science, Doctoral dissertation, University of Twente, Enschede.

Edmonds, G., Branch, R. and Mukherjee, P., 1994, A conceptual framework for comparing instructional design models, *Journal of Educational Technology Research and Development, 42*(4), 55-62.

Goel, V. and Pirolli, P., 1992, The structure of design problem spaces, *Cognitive Science, 16*, 395-429.

Goodlad, J.I., 1994, Curriculum as a field of study, In Husén, T. & Postlethwaite, T.N. (Eds.), *The international encyclopedia of education* (pp. 1262-1267), Pergamon, Oxford.

Gustafson, K. and Branch, R., 1997, *Survey of instructional development models* (3rd ed.). ERIC Clearinghouse on Information Resources, Syracuse University, Syracuse, NY.

Holcomb, C., Wedman, J.F. and Tessmer, M., 1996, ID activities and project success: Perceptions of practitioners, *Performance Improvement Quarterly, 9*(1), 49-61.

National Special Media Institute, 1971, What is an IDI?, Michigan State University, East Lansing, MI.

Johnson, J.C., 1990, Selecting ethnographic informants, Qualitative research methods series 22, Sage, Newbury Park, CA.

Kessels, J.W.M., 1993, *Towards design standards for curriculum consistency in corporate education,* Doctoral dissertation, University of Twente, Enschede.

Keursten, P., 1994, *Courseware-ontwikkeling met het oog op implementatie: De docent centraal* [Courseware development from an implementation perspective: A central role for the teacher], Doctoral dissertation, University of Twente, Enschede.

Klimczak, A., 1997, *Instructional design project success factors and indicators: An empirically based model,* Paper presented at AERA meeting, Chicago, March.

Kvale, S., 1996, *Interviews: An introduction to qualitative research interviewing,* Sage, Thousand Oaks, CA.

Maloney, C., 1993, Implementing curriculum: A case study, *Curriculum Perspectives, 13*(3), 23-32.

Maslowski, R. and Visscher, A.J., 1998, *Methoden en technieken voor formatieve evaluatie in sociaal-wetenschappelijke ontwerpsituaties [Methods and techniques for formative evaluation in social scientific design situations],* University of Twente, Enschede.

Miles, M.B. and Huberman, A.M., 1994, Qualitative data analysis: An expanded sourcebook, Sage, Thousand Oaks, CA.

Moonen, J.C.M.M., 1996, Prototyping as a design method, In Plomp, Tj. and Ely, D.P. (Eds.), *International encyclopedia of educational technology* (2nd ed.), pp. 186-190, Pergamon, Cambridge.

Nieveen, N.M., 1997, *Computer support for curriculum developers: A study on the potential of computer support in the domain of formative curriculum evaluation,* Doctoral dissertation, University of Twente, Enschede.

OSF-program, 1993, Een sociaal-wetenschappelijke ontwerpmethodologie [A social-scientific design methodology], University of Twente, Enschede.

Plomp, Tj., 1982, *Onderwijskundige technologie: Enige verkenningen* [Educational technology: An exploration], Inaugural address, University of Twente, Enschede.

Plomp, Tj., 1992, Onderwijskundig ontwerpen: Een inleiding [Educational design: An introduction], In Plomp, Tj., Feteris, A., Pieters, J.M. and Tomic, W. (Eds.), *Ontwerpen van onderwijs en trainingen* (pp. 19-38), Lemma, Utrecht.

Richey, R.C. and Nelson, W.A., 1996, Developmental research, In Jonassen, D. (Ed.), *Handbook of research on educational communications and technology* (pp. 1213-1246), Macmillan, New York.

Roes, M., 1997, *Nascholing op basis van lesvoorbeelden in de context van curriculumvernieuwing* [Inservice education through exemplary lesson materials in the context of curriculum innovation], Doctoral dissertation, University of Twente, Enschede.

Rowland, G., 1992, What do instructional designers actually do? An initial investigation of expert practice, *Performance Improvement Quarterly, 5*(2), 65-86.

Rowland, G. and Wilson, G., 1994, Liminal states in designing, *Performance Improvement Quarterly, 7*(3), 30-45.

Saroyan, A., 1993, Differences in expert practice: A case from formative evaluation, *Instructional Science, 21*, 451-472.

Schön, D.A., 1983, *The reflective practitioner: How professionals think in action,* Basic Books, New York.

Schön, D.A., 1987, *Educating the reflective practitioner: Toward a new design for teaching and learning in the professions,* Jossey-Bass, San Francisco.

Semb, G.B., Ellis, J.A., Fitch, M.A., Parchman, S. and Irick, C., 1995, On-the-job training: Prescriptions and practice, *Performance Improvement Quarterly, 8*(3), 19-37

Short, E.C., 1983, The forms and use of alternative curriculum development strategies: Policy implications, *Curriculum Inquiry, 13*(1), 43-64.

Strauss, A. and Corbin, J., 1994, Grounded theory methodology, In Denzin, N.K. and Lincoln, Y.S. (Eds.), *Handbook of qualitative research,* (pp. 273-285), Sage, Thousand Oaks, CA.

Tessmer, M., 1994, Formative evaluation alternatives, *Performance Improvement Quarterly, 7*(1), 3-18.

Tessmer, M. and Harris, D., 1990, Beyond instructional effectiveness: Key environmental decisions for instructional designers as change agents, *Educational Technology, 30*(7), 16-20.

Tessmer, M. and Wedman, J., 1995, Context-sensitive instructional design models: A response to design research, studies, and criticism, *Performance Improvement Quarterly, 8*(3), 38-54.

Tripp, S.D. and Bichelmeyer, B., 1990, Rapid prototyping: An alternative instructional design strategy, *Educational Technology and Development, 38*(1), 31-44.

Visscher-Voerman, J.I.A. and Plomp, Tj., 1996, Design approaches in training and education, In Plomp, Tj. and Ely, D.P. (Eds.), *International encyclopedia of educational technology* (2nd ed.), p. 22-26, Pergamon, Cambridge.

Walker, D., 1990, *Fundamentals of curriculum*, Harcourt Brace College, Fort Worth.

Wedman, J. and Tessmer, M., 1993, Instructional designers' decisions and priorities: A survey of design practice, *Performance Improvement Quarterly*, *6*(2), 43-57.

Winer, L.R. and Vázquez-Abad, J., 1995, The present and future of ID practice, *Performance Improvement Quarterly*, *8*(3), 55-67.

Zemke, R., 1985, The systems approach: A nice theory but ...., *Training*, October, 103-108.

Chapter 5
A Relational Approach to Curriculum Design

Joseph Kessels
Leiden University

Keywords: Curriculum design, Corporate education, Curriculum consistency, Systematic approach, Relational approach

Abstract

The complex mechanisms in corporate education, where cognitive operations of individual learning intertwine with social processes of an organisational context, demand an extended theory of curriculum design that seeks to explain the existing successes and failures of training systems. The author advocates integrated design procedures that combine a systematic and relational approach. Empirical research findings support this proposition and indicate that relational aspects in curriculum design predict most accurately programme effects.

Effective educational provisions are not constructed, but negotiated. Successful curriculum designers are above all competent social engineers. They skillfully manage the social enterprise of educational decision-making.

1. INTRODUCTION

This chapter addresses the question: What are the driving forces that make a curriculum design successful? The research study that offers the empirical evidence for the proposed answers to that question was carried out in the context of corporate education. Curriculum design focuses primarily on the acquisition of skills and competencies that are sustained by the day to day work environment of the participants in an educational programme. The complex mechanisms in corporate education, where cognitive operations of individual learning intertwine with social processes of an organisational context, demand an extended theory of curriculum design that seeks to explain the existing successes and failures of training systems and predicts the results of new actions.

Corporate education provides intentionally designed learning situations aiming at the mutual effects of individual and organisational behaviour. Therefore, the curriculum design theory needed should not only incorporate indicators for the development of curriculum materials, but also prescribe approaches that relate to the strategic issues of an organisation, to structural feedback mechanisms, as well as to the design of a work environment that inherently holds constructive educational values. The traditional systematic approach to curriculum design, based on rational and logic step by step procedures by a curriculum designer that lead towards the selection and arrangement of content and the choice of learning experiences, needs to be integrated in a relational approach. A relational approach provides activities that encourage the various stakeholders to become involved in the design and implementation process of a programme. As corporate education does not solely focus on the acquisition of competencies of employees, but also on changes in performance and impact on the organisation, the curriculum design should support the social enterprise of the educational decision making process.

Currently, human resource developers as well as managers often regard corporate education as isolated training in a classroom or conference centre. Consequently, the impact of such isolated programmes is very poor. Latham and Crandall (1991), Broad and Newstrom (1992) estimate that only 10% of training resulted in observable behaviour change on the job. A more comprehensive approach to curriculum design, integrating systematic and relational aspects should lead to more successful educational programmes.

This chapter discusses the main characteristics of the systematic and relational approach. The main research question is: does the skilful application of an integrated systematic and relational approach, generate educational programmes that accomplish significantly better results than those programmes with weak design approaches?

The dynamics of the systematic and relational approach have been examined in the design of 31 corporate education programmes. Moreover, the application of these approaches were related to the perceived effects of these programmes. The study was conducted in several phases: On the basis of the review of the literature, first an operationalization of the systematic and the relational approach was drawn up. Propositions on the systematic and relational approach were first tested in an exploratory research phase. Here, 17 programmes were investigated. The aim of the exploratory phase was to detect factors in the process of corporate curriculum design that affect the quality of the outcomes. For this purpose, two sets of contrasting cases were selected: nine successful and eight unsuccessful curricula.

In the second step the findings of the exploratory phase, combined with a literature review, resulted in a blueprint of the design standards that support the systematic and relational approach. The application of this set of design standards was tested against the empirical findings in 14 corporate curricula designed on the basis of the standards. Although the research basis is found in the domain of corporate education, the proposed integrated approach to curriculum design might also evoke a dialogue on improved design approaches in the domain of public education.

2. CURRICULUM CONSISTENCY

In the context of corporate education, we define the term curriculum as: 'the course of action open to an organisation, for influencing the necessary competencies of employees, that contribute to goal-oriented changes in their performance and in their work environment, thus striving for a desired impact on the organisation, by applying planned learning activities and the resulting learning processes' (Kessels, 1993, p. 4). The main concept of the theory presented is the concept of curriculum consistency here considered as one of the attributes that foremost determines the impact of educational programmes. The term 'consistency' serves to describe the contingencies between the constituting elements within a curriculum (the logic relationships between the needs analysis, objectives, learning environment and materials) and the congruencies among the various perceptions of a curriculum (the perceptions of managers, developer, trainers and participants of the main goal and how to achieve this goal). A distinction is made between internal and external curriculum consistency.

The concept of *internal consistency* applies to the logic contingencies between the changes that are needed in the work environment (by means of needs assessment), the necessary competencies of managers and employees to bring about these changes (job analysis and instructional objectives), and the learning situations that facilitate the acquisition of these competencies (instructional strategies, materials and learning environment). Similarily, internal consistency also implies that learning processes should enable employees to acquire competencies that influence their performance, so that the affected work environment has an impact on the organisation.

The concept of *external consistency* refers to the coherence between the perceptions of (senior) managers, developers, supervisors, trainers and trainees (the stakeholders) of what the problem is that has to be resolved and how this will be achieved. As well as being consistent in itself, a curriculum should show external consistency among the stakeholders' perceptions.

The concept of curriculum consistency - the contingencies between its elements and the congruencies between its appearances, as perceived by its stakeholders - is an elaboration of Stake's model for curriculum evaluation (Stake, 1973) and of Kirkpatrick's model for training evaluation (Kirkpatrick, 1975). Throughout this chapter, curriculum consistency, both internal and external, is used as a descriptive framework for quality in corporate education.

3. DESIGN APPROACHES

The main purpose of developing and applying design standards is to improve curriculum consistency. The theory developed here advocates a systematic approach that leads to internal consistency and a relational approach that supports external consistency. The two approaches seem to trigger a powerful combination of systems thinking and social engineering. The integration of a systematic and relational approach in design standards is held responsible for curriculum consistency and subsequently for corporate education of a high standard (Kessels and Harrison, 1998). Therefore, an integrated systematic and relational approach, is to generate educational programmes that accomplish significantly better results than those programmes with weak design approaches.

4. SYSTEMATIC APPROACH

The systematic approach to curriculum design is well known in the literature and follows directly from the work of the prominent American curriculum scholar Ralph W. Tyler. What later became known as the 'Tyler Rationale' (Tyler, 1949), started as a framework to guide the efforts of participating schools in a large curriculum project. The four main questions to be answered are:

- What educational purposes should the school seek to attain?
- What educational experiences can be provided that are likely to attain these purposes?
- How can these educational experiences be effectively organised?
- How can we determine whether these purposes are being attained?

The systematic and analytical approach to curriculum design, as advocated by Tyler has led to design procedures that are still dominant and that heavily rely on needs assessment, task analysis, stating instructional objectives, matching assessment instruments and devising appropriate instructional strategies. Authoritative design procedures that stem from Tyler's rationale are amongst others: Taba's *Curriculum development: Theory and practice* (Taba, 1962), Briggs' *Instructional design: Principles and applications* (Briggs, 1977), Tracey's *Designing training and development systems* (Tracey, 1984), Dick and Carey's *The systematic design of instruction* (Dick & Carey, 1990), Branson and Grow's *Instructional systems development* (Branson & Grow, 1987), Plomp's *Onderwijskundige technologie: enige verkenningen* [Exploring educational technology] (Plomp, 1982), Romiszowski's *Designing instructional systems* (Romiszowski, 1981) and Rothwell and Kazanas's *Mastering the instructional design process. A systematic approach* (Rothwell & Kazanas, 1992). The logic and rational step by step approach, including the iterative use of feedback from formative evaluation, is characteristic of most of these systematic design procedures. The systematic approach implies the logical design sequence of orientation, design, development, evaluation and implementation. This approach, when skilfully applied, leads to a well structured and logically ordered curriculum design with a strong internal consistency. This design on paper is referred to as the formal curriculum.

5. COMPETENCIES FOR A SYSTEMATIC APPROACH

The skilful application of the systematic approach requires specific competencies of the curriculum designer. The American Society for Training and Development, ASTD (McLagan, 1989) and the International Board of Standards for Training, Performance and Instruction, IBSPI (Foshay, Silber & Westgaard, 1986) conducted large-scale research projects on design and development competencies in corporate settings. A selection of competencies from the above mentioned sources has been made to accomplish a systematic approach in curriculum design in the context of corporate education, as advocated in the previous sections:

- Conduct *needs assessment.* Identify ideal and actual performance and performance conditions and determine causes of discrepancies. Employ strategies for analysing individual and organisation behaviour.
- Perform *job and task analysis.* Employ analysis strategies and reporting procedures. Investigate best practices, the inherent cognitive models and attitudinal aspects.
- State *instructional objectives.* Transform job requirements into objectives, so that performance measurement and selection of instructional strategies is facilitated.
- Develop *performance measurements.* Transform needs, performance requirements and objectives into evaluation criteria and appropriate assessment instruments.
- Sequence the performance objectives. Draw a blue-print for the desired *learning environment*, appropriate for achieving the desired changes of performance.
- Specify the *instructional strategies.* Devise instructional interventions to put the blue-print learning environment into action.
- Design *instructional material.* Develop print, audio-visual or electronic-based learner materials, job aids, simulation devices, trainer guides and plans to facilitate the instructional interventions.
- *Evaluate* the educational interventions. Appraise the instructional methods, sequences and materials, and improve.
- *Assess* results. Evaluate learning results, performance improvement and the related impact on the organisation.

However, the systematic and analytical approach to curriculum design, and the resulting formal curriculum is not very often found in reality (Andrews & Goodson 1980). Even when developers apply the prescribed systematic design procedures, programme implementation remains problematic (Kessels, 1993). Apparently, the unilateral, systematic approach does not guarantee success.

6. RELATIONAL APPROACH

The relational approach provides activities that challenge stakeholders to become involved in the design and implementation process and that reveal their perceptions of what the central goal is and how it can be achieved. The assumption is that if the mutual perceptions are made explicit, they can be modified and slowly become compatible. When skilfully applied, the relational approach leads to a strong external consistency: consensus among parties involved on methods of solving the problem, implementing the programme, and creating favourable transfer conditions in the day to day work environment.

It was Walker (1971) who studied the practice of many curriculum designers, and drew our attention to the fact that course design often does not show the step by step approach as advocated in the previous section. Walker identified three basic planning phases: platform, deliberation and design (Walker, 1971, 1990). On the basis of these findings he developed a framework for the process of curriculum planning for which he used the term 'Naturalistic Model'. This model is not a normative model of how curriculum design should take place, but a descriptive model of how it often occurs in reality. In the 'platform-phase' participants talk, discuss and argue about their beliefs, ideas, theories, aims, images and potential procedures concerning the curriculum. When a group achieves clarity and consensus about these constituent elements of the platform, they move into the phase of deliberation. Walker's model specifies that the process of deliberation includes exploring specific conditions, generating alternatives, examining costs and consequences, and selecting a feasible alternative. The platform and deliberative phase involve intensive exchange of ideas and beliefs. Reaching consensus is essential for moving into the 'design-phase'. This phase can become an extremely difficult task, especially when participants hold to their fixed perceptions, or when they feel uncomfortable when the chaos of conflicting ideas, and images can not be resolved in time. When the planning group does reach consensus about the basic principles of the curriculum, they move into the design activities, which include the decision making about specific content, instructional strategies and materials. In this process of curriculum design it is extremely important that participants make their individual beliefs and values explicit as well as their perceptions of the instructional task and their assertions about how to proceed. The importance of Walker's (1971, 1990) deliberative approach is that it recognises the variety of beliefs, aims and images that participants in a project on curriculum design adhere to. This variety of perspectives may frustrate a rational, systematic and linear design process, as proposed by the various sets of design instructions that promote an unilateral internal consistency.

Several other authors have stressed the importance of deliberation in curriculum development. Banathy (1987, p. 93) states that "the process of arriving at better decisions is not a process of optimisation. It is rather, a process of negotiation among those with different points of view and value systems in order to find a satisfying solution." This calls for a participative design where major stakeholders are involved. Banathy (1987) emphasises an iterative and spiralistic design process where the designer may pass several times through the various phases of the design cycle. Recently, design approaches that combine participative deliberation and iterative procedures advocate prototyping as a vehicle for curriculum design. Gentry (1994, p. 160) defines a prototype as "a functional version of an instructional unit usually in an unfinished state, whose effectiveness and efficiency can be tested." It offers users an opportunity to find out what they do not like about the proposed unit, which is often easier than exactly indicating what is needed at an initial state. Prototyping can be regarded as a practical way of organising deliberation among relevant stakeholders.

The relational approach in curriculum design finds support by a continuous concern for implementation. Implementation is conceived as an ongoing activity that starts at the initial stage of the problem solving process (Plomp, 1982). Fullan, who defined implementation as "the process of putting a change into practice" considers clarity (or confusion) about goals and means one of the perennial problems of curriculum change (Fullan, 1991). In his view, two approaches are open to the developer: the fidelity approach that strives for high quality in the developmental stages, and the adaptive approach in which further development must be worked out by individuals and groups who are involved in implementation (Fullan, 1986). In fact, Fullan's fidelity approach bears great similarity with the here described systematic approach, as his adaptive approach to implementation comes near to the advocated deliberative and relational approaches.

Gay (1986) states that curriculum affairs involve issues of power, people, procedures and participation: "Curriculum development is a social enterprise. It is a 'people process' with all the attending potentialities and obstacles associated with humans engaged in social interactions. The interests, values, ideologies, priorities, role functions, and differentiated responsibilities form the contours of the interactional and dynamic context in which curriculum decisions are made. .... Curriculum development is neither a purely rational and scientifically objective, nor a neatly sequentialised and systematic process" (Gay, 1986, pp. 471-472).

7. COMPETENCIES FOR A RELATIONAL APPROACH

Walker's deliberative approach, Fullan's adaptive approach, as well as Gay's emphasis on the development *process* instead of the design *product,* support the elaboration of the relational approach. The early ISD-models emphasised unidirectional application of the systematic approach and the concern for internal consistency prevails (Merrill, Reigeluth & Faust, 1979). Though claimed to be vital for successful programme implementation, external consistency and the related competencies and procedures for a relational approach do not receive much explicit attention in formal instructional systems development models.

As curriculum affairs are mainly activities involving human beings communicating with each other, the relational approach applies to all the contacts between the curriculum developers and relevant stakeholders. Besides senior managers, supervisors, trainers and trainees, other parties may be involved, in particular clients, customers, co-ordinators, sponsors and opinion leaders.

Unlike the systematic approach with its clear and rigorous logic, the relational approach may often seem fuzzy, using informal networks, balancing power and influence, and striving for consensus within the limits of culturally determined feasibility (Duncan & Powers, 1992). Political awareness, cultivating support, developing relationships and gaining visibility seem to be ingredients of this aspect of curriculum design (Warshauer, 1988). Activities that belong to the relational approach are sometimes characterised as "walk and talk the job" (Harrison, 1992).

Many sources[1] offer analyses of the competencies professionals should dispose of when they enter into the relational approach. The most salient competencies are listed below:

- *Communication skills*: listening, observing, interviewing, relating to others, self-expression and exchanging constructive feedback.
- *Project management skills*: leadership and chairperson skills, planning, monitoring and negotiating skills.
- *Consulting skills*: building open collaborative relationships, clarifying mutual expectations and responsibilities, and the ability to influence others and gain commitment.
- *Facilitating change*: encouraging widespread participation in the design and implementation of a project, and dealing with friction and resistance.
- *Experimental flexibility*, self-insight and self-esteem.
- *Ability to create an atmosphere of tact*, trust, politeness, friendliness and stability.

The competencies for a relational approach facilitate the developer's activities in the domain of interpersonal dynamics of decision making about educational planning. The relational approach involves social intervention and skilled communicative interaction. The developer organises meetings and interviews managers, supervisors, employees, potential trainees and trainers. These procedures entail consulting with concerned parties, problem solving, negotiating, reaching a consensus, gaining support, and strategically applying gentle pushes and decisive pulls. The goal of these efforts is to achieve a consensus among parties involved on problem definition, methods of solving the problem, implementing the programme, and creating favourable transfer conditions for learning in the work environment (Kessels & Plomp, 1996).

Several procedures support the relational approach. On the basis of the results from the literature review, the research project investigated project management, rapport-building activities during needs assessment and task analysis, involving line management in the development process, creating similarity between learning situation and work environment, recruiting trainers with practical experience in the subject matter field, and the selection of trainees.

---

[1] Often, the performance oriented corporate educationalist is portrayed as a 'change agent' (Clark, 1991; Phillips & Shaw, 1989; Pont, 1991). The ASTD-research projects (McLagan, 1989), as well the IBSTPI-Standards (Foshay *et al.*, 1986), define several competencies for developers that we consider to make part of the relational approach. Furthermore, literature on consulting and coaching offer a variety of requirements and competencies for developers to adopt in their relational approach (Block, 1981; Lippitt & Lippitt, 1986; Phillips & Shaw, 1989.

8. EMPIRICAL EVIDENCE OF AN INTEGRATED APPROACH

The dynamics of the systematic and relational approach have been examined in the design of 31 corporate education programmes. Moreover, the application of these approaches were related to the perceived effects of these programmes. We reported the research design, the methodological aspects and the results of this study in Kessels (1993), Kessels and Plomp (1997).

8.1 Data collection

Three curriculum experts investigated independently the systematic approach by judging the quality of the applied procedures for needs assessment, task analysis, instructional objectives, evaluation criteria, evaluation instruments, design of instructional strategies and the development of course materials for each programme. The experts investigated the relational approach by judging the quality of the applied procedures for project management, rapport-building activities during needs assessment and task analysis, involving line management in the development process, creating similarity between learning situation and work environment, recruiting trainers with practical experience in the subject matter field, and the selection of trainees.

The effect measures for each programme could be established on the basis of questionnaires and interviews revealing the newly acquired skills, changes in performance and the impact of the programme on the organisation, as perceived by senior managers, supervisors, developers, trainers and trainees.

8.2 Results

In-depth analysis of the programmes revealed a series of interesting phenomena. On the systematic approach, the following observations could be made:

- Reports on training needs assessment and task analysis are mostly absent.
- Instructional objectives of unsuccessful programmes tend to be stated in terms of reproductive knowledge aspects, as the objectives in successful programmes tend to stated in terms of reproductive and productive skills.
- Evaluation criteria are neither stated nor documented.
- Evaluation instruments are restricted to the reaction to learning processes. Very few tests of training results were available. Virtually no instruments were found to assess performance in the workplace and organisational impact.
- In the successful programmes, the designed learning situations show a greater variety of activities than in those that are unsuccessful.
- Most programmes used elaborate training materials. Successful programmes devoted more attention to guidelines for the trainers and coaches.

On the relational approach the following observations could be made:

- Successful programmes show strong involvement by line managers.
- The similarity between learning situation and work environment is extremely strong for the successful programmes and is correspondingly weak for the unsuccessful programmes.
- Practical experience in the subject matter field of the trainer is extensive in the successful programmes and limited in the unsuccessful.
- A majority of the successful programmes emphasised creating favourable conditions for implementation. None of the unsuccessful programmes emphasised this aspect.

The combined appearances of the systematic and relational approach showed the following:

- Successful programmes with a weak systematic approach all exhibited a strong relational approach. The high value for the relational approach seems to compensate for the low value of the systematic approach.
- None of the successful programmes showed a weak relational approach.
- The unsuccessful programmes show a negative correlation between systematic approach and effects (r = -.51), whereas successful programmes have benefited from the systematic approach (r = .62). These figures might indicate that investments in systematic design of the formal curriculum may become counterproductive when the development process does not satisfy the relational approach.

A regression analysis of the main variables led to the following statistical relations:

Systematic Approach to Effects	$r = .64$	$p < .001$	slope = .47
Relational Approach to Effects	r = . *83*	$p < .001$	slope = .54

The Systematic and Relational Approach correlate significantly with Effects. The regression line of the Relational Approach shows a slightly steeper slope, which indicates that the Relational Approach probably predicts most accurately the programme Effects.

9. DISCUSSION

We have probably touched on the limitations of rational curriculum design and its systematic approach. Both for practical reasons and from a theoretical point of view, it is relevant, to find the optimal input for relational and systematic approaches. The results do not question the foundations of internal consistency (the logic contingencies among purpose, objectives, evaluation criteria and instruments, and instructional strategies). Whereas external consistency (the homogeneous curriculum perceptions of the various stakeholders) is viewed as conditional, internal consistency is still considered the driving force behind a curriculum. It might, however, be interesting to investigate curriculum design procedures that are not rigorously rational and strive for logical contingencies in the formal curriculum. If curriculum design were also perceived as professional artistry, additional categories of design principles would be explored, for example:

- creating learning situations that mirror the work environment;
- involving line managers as the prime educators;

- recruiting experienced colleagues as trainers; and
- creating favourable conditions for programme implementation.

Of course, the application of these relational design principles should be applied in a systematic way, but emphasis would primarily be put on the dynamics of the interactional context of curriculum design. The findings of the present study justify the conclusion that in striving for quality in corporate education a relational approach should be a high priority. As a consequence, design standards that strive for an internal, rigid logic, but meanwhile hinder the integration of the actors' interests, values, believes and priorities (external consistency), should be abolished and replaced by mainly strategies focusing on the interpersonal dynamics of educational decision-making. In particular, professional curriculum designers should be alerted not to focus unilaterally on the structured and internally consistent formal curriculum. Curriculum development should be regarded, more than up till now, as a social enterprise. Therefore, developers may elaborate on their management role within that social enterprise of the educational decision-making process. Effective educational provisions are not constructed, but negotiated. Therefore, successful curriculum designers are above all competent social engineers, who skilfully manage the social enterprise of educational decision-making.

REFERENCES

Andrews, D.H. and Goodson, N.D.L., 1980, A comparative analysis of models of instructional design, *Journal of Instructional Development, 4*, 2-16.

Banathy, B.H., 1987, Instructional systems design, In Gagné, R.M. (Ed.), *Educational technology: Foundations* (pp. 85-112), Lawrence Erlbaum, Hillsdale, N.J.

Block, P., 1981, *Flawless consulting: A guide to getting your expertise used,* University Associates, San Diego, CA.

Branson, R.K. and Grow, G., 1987, Instructional systems development, In Gagné, R.M. (Ed.), *Instructional technology: Foundations* (pp. 397-428), Lawrence Erlbaum, Hillsdale, NJ

Briggs, L.J., 1977, *Instructional design: Principles and applications,* Educational Technology Publications, Englewood Cliffs, NY.

Broad, M.L. and Newstrom, J.W., 1992, *Transfer of training. Action-packed strategies to ensure high payoff from training investments,* Addison-Wesley Publishing Company, Reading, MA.

Clark, N., 1991, *Managing personal learning and change. A trainer's guide,* McGraw-Hill, London.

Dick, W. and Carey, L., 1990, *The Systematic Design of Instruction,* Second edition, Scott Foresman, London.

Duncan, J.B. and Powers, E.S., 1992, The politics of intervening in organizations, In Stolovitch, H.D. and Keeps, E.J. (Eds.), *Handbook of human performance technology* (pp. 77-93), Jossey Bass, San Francisco.

Foshay, W., Silber, K. and Westgaard, O., 1986, *Instructional design competencies. The standards,* University of Iowa - The International Board of Standards for Training, Performance and Instruction.

Fullan, M., 1986, Curriculum implementation, In Eraut, M. (Ed.), *The international encyclopedia of educational technology* (pp. 485-491), Pergamon Press, Oxford.

Fullan, M., 1991, *The new meaning of educational change,* Teachers College Press, New York.

Gay, G., 1986, Curriculum development, In M. Eraut (Ed.), *The international encyclopedia of educational technology* (pp. 467-476), Pergamon Press, Oxford.

Gentry, C.G., 1994, *Introduction to instructional development. Process and technique,* Wadsworth, Belmont, CA.

Harrison, R., 1992, *Employee development,* Institute of Personnel Management, London.

Kessels, J.W.M., 1993, *Towards design standards for curriculum consistency in corporate education,* Doctoral dissertation, University of Twente, Enschede.

Kessels, J.W.M. and Plomp, Tj., 1996, Course design, In Plomp, Tj. and Ely, D.P. (Eds.), *The international encyclopedia of educational technology*, 2nd ed. (pp. 143-148), Pergamon Press/Elsevier Science, Oxford.

Kessels, J.W.M. and Plomp, Tj., 1997, The importance of relational aspects in the systems approach, In Dills, C.R. and Romiszowski, A.J. (Eds.), *Instructional development paradigms,* pp. 93-126, Educational Technology Publications, Englewood Cliffs, NJ.

Kessels, J.W.M. and Harrison, R., 1998, External consistency: The key to success in management development programmes?, *Management learning journal for managerial and organizational learning, 29*(1) 39-68.

Kirkpatrick, D.L., 1975, *Evaluating training programs.* A collection of articles from the Journal of the American Society for Training and Development. ASTD, Madison, WI.

Latham, G.P. and Crandall, S.R., 1991, Organizational and social factors, In Morrison, J.E. (Ed.), *Training for performance* (pp. 260-285), John Wiley & Sons, Chichester.

Lippitt, G. and Lippitt, R., 1986, *The consulting process in action,* University Associates, San Diego.

McLagan, P.A., 1989, *The Models. Models for HRD practice,* ASTD, Alexandria, VA.

Merrill, M.D., Reigeluth, C.M. and Faust, G.F., 1979, The instructional quality profile: A curriculum evaluation and design tool, In O'Neil, H.F. (Ed.), *Procedures for instructional systems development* (pp. 165-204), Academic Press, New York.

Phillips, K. and Shaw, P., 1989, *A consultancy approach for trainers,* Gower Publishing, Aldershot Hants.

Plomp, Tj., 1982, *Onderwijskundige technologie: enige verkenningen* [Exploring educational technology], Inaugural address, University of Twente, Enschede.

Pont, T., 1991, *Developing effective training skills,* McGraw-Hill, London.

Romiszowski, A.J., 1981, *Designing instructional systems,* Kogan Page, London.

Rothwell, W.J. and Kazanas, H.C., 1992, *Mastering the instructional design process. A systematic approach,* Jossey-Bass, San Francisco.

Stake, R.E., 1973, The countenance of educational evaluation, In Worthen, B.R. and Sanders, J.R. (Eds.), *Educational evaluation: Theory and practice. Frameworks for planning evaluation studies* (pp. 106-124), Wadsworth, Belmont, CA.

Taba, H., 1962, *Curriculum development: Theory and practice,* Harcourt, New York.

Tracey, W.R., 1984, *Designing training and development systems* (revised edition), American Management Association, New York.

Tyler, R., 1949, *Basic principles of curriculum and instruction,* University of Chicago Press, Chicago.

Walker, D.F., 1971, The process of curriculum development: A naturalistic approach, *School Review, 80,* 51-65.

Walker, D.F., 1990, *Fundamentals of curriculum,* Harcourt Brace Jovanovich, San Diego.

Warshauer, S., 1988, *Inside training and development, Creating effective programs,* University Associates, San Diego.

Chapter 6

Walker's Deliberative Approach in a Small-scale Project: The SPIN-case

Ellen van den Berg
University of Twente

Keywords: Deliberative approach, Curriculum development, Formative evaluation

Abstract

This chapter describes the application of Walker's deliberation approach in a small-scale design project. It highlights the way an individual designer may generate, articulate and test design options and decisions. Two major strategies are discussed. Firstly, the grounding of design options in relevant literature and confront preliminary design decisions with expert opinions. Secondly, the use various formative evaluations in the deliberation process. The main conclusion of this chapter is that Walker's approach is, with some modifications, applicable in a design project in which the individual designer has the ultimate responsibility for the decisions made.

1. INTRODUCTION

This chapter is about applying the deliberative approach to curriculum development in a small-scale project: the SPIN[1]-project. This project aims at designing and implementing an inservice education program that intends to be effective in providing teachers with the competencies and beliefs that enable them to implement a more constructivist approach to elementary science.

Walker (1971, 1990) has formulated the deliberative approach as a way of curriculum design. Its empirical basis is to be found in the intensive study of large-scale curriculum development projects, in which development teams took the responsibility for design decisions. This chapter shows that Walker's deliberative approach is also applicable in a small-scale project, in which *one person* took the final design decisions. In section 2 the deliberative approach is outlined briefly. A general account of the application of this approach in the SPIN-project is given in sections 3 to 5. In section 6, two examples are presented in which deliberation processes in the SPIN project are described in more detail. The conclusions of this chapter are presented in section 7. A

[1] The Dutch acronym SPIN stands for Inservice education, Procedural specifications and the Implementation of Elementary Science. In Dutch SPIN means spider.

summary of the way Walker's approach is applied in the SPIN-case is given in section 8.

2. WALKER'S DELIBERATIVE APPROACH

Walker's deliberative approach has guided the design and development of the SPIN inservice program (Walker, 1990, p. 466-499). In this approach a curriculum design team is engaged in a process of:

- formulating a platform of ideas;
- using these ideas to conceptualize the problem;
- using them to generate promising preliminary versions; and
- revising every subsequent version until it cannot be improved further or some constraint forces a halt.

A *platform of ideas* consists of more formal theories and conceptions, about models and strategies, but also about more tacit professional knowledge of the curriculum developers: common sense and conventional wisdom. A platform provides the terminology and concepts to pose issues, to describe alternatives and to frame arguments. It functions both as a means to conceptualize the educational problem or situation at issue and as a means to generate versions of curriculum plans and materials within the constraints to which the implementation context, the materials, and the process of development are subject. A platform of ideas serves as a factual and as a logical basis for design decisions. It gives a project its coherence and identity. To the outside world, the platform is an emblem signifying commitments that onlookers may or may not share (Walker, 1990, p. 475). For example, a design team that intends to develop curriculum materials in line with a realistic approach to mathematics gives a commitment about what type of curriculum it aims to develop. Advocates of a more formal approach to mathematics education "are likely to disagree" with such a signifying commitment.

Even before a curriculum development project actually starts with activities, the design team has an initial platform of ideas of what the curriculum in question might look like. The designers start with an answer to an educational problem, but it is an answer usually lacking clarity and details. Curriculum development is largely an elaboration of the initial platform of ideas. This process of rational argumentation in which possible actions are considered is called *deliberation* (Walker, 1990, p. 163).

The beginning of a curriculum development project is largely spent on extending, refining and elaborating the platform of ideas. Later on in the project, preliminary versions of materials are planned and shaped through deliberation to the point where more can be learned from the test of a preliminary version than from further deliberation based on ideas only. Cycles of designing, testing, and revising ideas, plans, and materials continue throughout the project. However, as a project progresses its main activities are concentrated on instructional design processes.

3. PLATFORM OF IDEAS IN THE SPIN-PROJECT

3.1 Initial platform of ideas

As in many curriculum development projects, the SPIN-project also started with an initial platform of ideas that functioned as a leading commitment. This commitment was to combine the efforts made by the Dutch National Institute for Curriculum Development in the development of innovative curriculum materials to enhance elementary science reform with inservice education. For years, also in the field of science education, the choice for either inservice education or curriculum materials to strengthen implementation efforts was a subject of, sometimes fierce, debate (Ben-Peretz, 1990; Harlen, 1985). However, thinking of these strategies as being mutually exclusive seems to be rather unproductive, because both efforts combined can enhance the implementation attempts of, in this case, a constructivist approach to elementary science. It was expected that the sum of the efforts of professionally developed curriculum materials and of inservice education is synergetic. Innovative curriculum materials can make a valuable contribution to effective inservice education, but inservice education can also help teachers to get over a barrier to use innovative curriculum materials.

3.2 Conceptualization of the problem

In a constructivist view on learning the emphasis is on active learners, constructing and reconstructing their ideas, taking responsibility for their own learning, and negotiating with others toward purposive ends (Watts, 1994). It is the teacher who has to provide the conditions that make active learning possible and to guide student learning into productive directions.

The implementation of a constructivist approach to elementary science is a complicated endeavor (Van den Berg, 1996; Wallace & Louden, 1992). The main causes for the implementation problems are:

- A constructivist approach requires a fundamental role change of the teacher. In this approach, the teacher is no longer perceived as the transmitter of merely factual information, but as a facilitator of learning processes. Such a role change hits the heart of teachers' educational beliefs and demands for teaching strategies with which the majority of teachers are not familiar (Duit & Treagust, 1995). Moreover, the lack of content-specific background knowledge of elementary teachers is a hindrance to change their role.
- Preparing constructivist elementary science lessons is complex and time-consuming. Developing a mental image of such lessons is more complicated than of lessons that are focused on straightforward content delivery. Moreover, organizing the objects for the students' hands-on activities takes much time.
- The outcomes of individual students in terms of higher order knowledge and skills is essential in a constructivist approach. Having a clear view on this kind of student outcomes is problematic for teachers.

It is not an exaggeration to say that implementing a constructivist approach to elementary science is a complex endeavor. Therefore, an inservice education program, how well designed it may be, can only contribute partly to remove some causes of the implementation problems.

3.3 Elaboration of the platform of ideas

Basically, every curriculum design project is guided by two questions: what should be the aim and content of the curriculum and how may the content be effectively organized. As the aim of the SPIN-project has already been mentioned in the introduction, here the focus will be on the content choice and organization of the content.

3.3.1 Content choice

The content choice started with the selection of a *central orientation* for the SPIN program by its designer. A central orientation identifies the focal point of a program and helps to give its elements coherence. It also functions as a filter for subsequent design decisions. Teacher education programs can be based on different orientations such as an academic orientation, a technical orientation, and a personal orientation (Feiman-Nemser, 1990). An academic orientation focuses on content knowledge, a technical orientation on "teacher actions", and a personal orientation on individual growth.

After a brief literature study in the perspective of the aim of the SPIN inservice program and the intended target group, elementary teachers, it was decided to found the SPIN inservice program on a practice-oriented approach (Van den Berg, 1996, p. 52). This approach focuses on experience in the classroom as a source for teachers' professional development. Significant components are both the craft aspects of teaching and the reflection on classroom practice (Anderson & Mitchener, 1994). Tulder, Roelofs, Veenman and Voeten (1992) indicate that teachers appreciate such an approach and that practice-oriented inservice programs have a positive impact on the implementation of the aims of an inservice program into classroom practice.

A consequence of the choice for a practice-oriented approach as central orientation for the SPIN inservice program implies that the core themes of the program are the things that teachers are expected to do differently in their classroom and the reflection on these new actions. In order to strengthen the reflection in the perspective of a constructivist way of elementary science teaching the metaphor of the *mountain climbing guide* has been incorporated into the SPIN program. This metaphor was contrasted with the *puppet-player* metaphor. This latter metaphor represents a knowledge transmission model in which factual knowledge is delivered from the teacher to the students and assessed in the opposite way. The first metaphor represents a knowledge transformation model in which the teacher's role is perceived as a facilitator of the change process of students' initial cognitive structures.

3.3.2 Organization of the content

Very little is known about how to educate teachers to teach in a constructivist manner (Anderson & Mitchener, 1994; Louden & Wallace, 1994; Van den Berg, 1996). However, there are, although rooted in a behavioristic tradition, empirically based indications about the organization of effective teacher education in order to change teacher behavior. As far as inservice education is concerned, the work of Joyce and Showers (1980, 1988) represent pioneering efforts. They distinguish between the following components of effective inservice education programs: theory, demonstration, practice, feedback, and coaching.

A straightforward application of the components of effective inservice education in the design of the SPIN inservice program is not possible. The Joyce and Showers' inservice programs aim at training teachers to be competent in applying general teaching approaches that are derived from research on effective instruction. In the SPIN inservice program the teaching approaches to be implemented are related to a constructivist way of elementary science teaching. Therefore, the content of Joyce and Showers' components are filled in a way more specifically attuned to a particular subject (elementary science) and to specific teaching approaches that are in line with a constructivist philosophy. For example, in the component demonstration it was showed how teachers may guide a negotiation of meaning process among students.

4. PRELIMINARY VERSIONS OF THE SPIN PROGRAM: DESIGN AND EVALUATION

After the elaboration of the platform of ideas, a preliminary version of the SPIN program was designed. It must be noted that the rational argumentation to come up with this version can not be characterized as a consensus seeking team discussion, as intended in Walker's approach. In the SPIN-case the first phase of "rational argumentation" is better typified as an internal dialogue of the designer after confronting ideas with relevant literature. So, the individual designer seeks to reconcile different and sometimes conflicting options into a consistent set of (preliminary) design decisions. For example, constructivism is portrayed in the literature as a radical break with the current practice in education. On the other hand, implementation literature focuses on a more gradual transformation from current teaching practice to a constructivist approach. In order to moderate those conflicting ideas, and given the fact that the SPIN inservice program is an introductory program, the design decision was made to provide teachers with maximum support in making the first steps towards a constructivist approach to elementary science.

4.1 Preliminary design

The preliminary design of the SPIN inservice program can be summarized as follows. The program starts with the rationale or underlying principles of a constructivist approach to elementary science. Subsequently, the teaching actions that are in line with this approach are addressed. Those actions are demonstrated and practiced, and feedback is provided on practice (micro and classroom teaching). Exemplary curriculum materials play an important role both for demonstration and for practice purposes. The two metaphors mentioned before are used as a means to discuss the underlying beliefs behind a constructivist approach to elementary science. This discussion takes place after the teachers have gained some practical experience with the innovative approach.

4.2 Evaluation of the preliminary design

The preliminary design of the SPIN program was evaluated in the following ways:

- discussions with general specialists in curriculum;
- discussions with science teacher educators and curriculum specialists in elementary science; and
- comparing the ideas of the SPIN program with another Dutch inservice program. This latter program aimed at enhancing the implementation of hands-on science and it showed some of the same intentions as the SPIN program.

A comprehensive overview of all evaluation results are beyond the scope of this chapter, but the six main conclusions are summarized as:

- Start the program with hands-on activities by the participants, because many practicing teachers may not have experiences with these vital activities in elementary science.
- It is better to discuss the underlying principles of a new approach to elementary science (in the SPIN case a constructivist one) later on in a program, after the participants have gained some experience with its practical implications.
- If microteaching lessons will be a part of the SPIN program, they can be based on brief proposals for short lessons. Applying microteaching this way will not absorb a disproportional amount of the program's time. However, following this advice also implies that not the full range of teaching approaches fitting in constructivist science teaching can be practiced.
- The vital importance of practical classroom experience with the new approach has been underpinned as a valuable component of an inservice education program.
- The use of curriculum materials with procedural specifications (accurate how-to-do-it advice) is prudent for teachers who are going to implement a new approach for the first time, because the chance of successful experience will be enlarged. Affinity with the topic to be taught and a relative simple lesson preparation are additional characteristics of curriculum materials that are supportive to initial implementation efforts. For the SPIN inservice program this implies that lesson proposals with procedural specifications have been selected that do not require too much preparation time. Moreover, participants should have a free choice among a variety of lesson proposals.
- Presentation and discussion of traditional and alternative assessment procedures seems to be a good means to address a problem many teachers face: uncertainty about students' outcomes.

(For more details, see: Van den Berg, 1996, p. 55-67).

The design of the SPIN program was revised based on the evaluation results. These results led to a substantial modification of the Joyce and Shower model. It was decided not to start with the theory component. Moreover, a component was added to the design of the SPIN program (alternative and traditional ways of assessment) and for other components valuable suggestions were made. The evaluation results also support the central orientation of the SPIN program: a practice-oriented approach. This approach is perceived as appropriate by all respondents.

5. TWO EXAMPLES OF DELIBERATION IN THE SPIN PROJECT

In this section two examples are presented about deliberation processes that took place during the design of the SPIN program. The first example is about the sequence within the program, and the second is about whether or not to include micro-teaching sessions into the program.

5.1 Joyce and Shower's model of competence-based inservice education

As stated earlier, Joyce and Showers (1988) formulated a model of effective inservice education consisting of the components: theory, demonstration, practice, feedback, and coaching. In the first version of the design of the SPIN program the sequence of Joyce and Showers' model was mainly followed (due to the introductory character of the SPIN program, the coaching component was not taken into consideration).

The position of the "theory" component at the beginning of the inservice program was criticized by one of the curriculum experts. She argued that starting with the theory component results in a sequence that is not in line with the advocated innovation a constructivist approach to elementary science; because the inservice program then starts with a rather traditional way of "telling" the participants what the new approach is. Her advice was: "practice as you preach" and choose another start for the program.

Based on this advice a second version of the design of the SPIN program came into being. This version started with science hands-on activities conducted by the participating teachers in order to give them experience with this important aspect of a constructivist approach to elementary science. Another curriculum expert, however, strongly opposed this design decision. This expert's views on curriculum design are aligned to the instrumental paradigm (see chapter 2 of this book). According to his opinion the model of Joyce and Showers should be followed in the sequence they describe, because that is the way the model is described in the research literature.

For the time being, the designer of the SPIN program decided to value the arguments for changing the initial design to a higher plane. An argument for this decision was that Joyce and Showers' model is rooted in a behavioristic tradition, so its sequence may not fit into a constructivist approach. It was also decided to discuss the issue whether or not to start the SPIN program with the theory behind constructivism with experts in the field of teacher education (including inservice education). After the exchange of arguments, the teacher educators agreed that starting an inservice program with theoretical background of the innovation is not such a sound idea. As an additional reason for starting with a more practical introduction of the inservice program, they stated that such a start would be more motivating for the participants. Moreover, many practicing teachers only have experience with more traditional forms of science education, so starting with an innovative activity gives the participants a sense about what a new way of science teaching might look like.

The deliberations led to the final decision to start the inservice program with hands-on activities in stead of the theory behind a constructivist approach to elementary science. The component theory was addressed later on in the program, after the participants had classroom experience with a more constructivist approach to elementary science.

5.2 To microteach or not to microteach

In the first version of the design of the SPIN program microteaching sessions were planned in which the participants teach each other. This design decision was inspired by Yeany and Padilla's (1986) meta-analysis of effective teacher education in science. This meta-analysis shows a positive effect of microteaching sessions. The rationale behind microteaching sessions is that teachers can practice new skills in a less complicated situation than in real classroom teaching, so they may focus more on those skills. Moreover, after microteaching lessons the participants and inservice educators can provide feedback on a lesson.

All curriculum experts agreed on including microteaching lessons into the SPIN program. Some teacher educators, however, expressed a different opinion. Two of them stated that microteaching lessons are artificial and are, therefore, a waste of time.

Two other teacher educators also formulated some hesitations about microteaching, especially because this activity would be too threatening for the participants. According to their opinion, teachers are afraid to fail in front of their colleagues.

All teacher educators expressed concerns about the amount of time that was dedicated to microteaching initially, they thought that this activity would absorb a disproportional amount of time. One of the teacher educators came up with an idea to shorten the microteaching lessons by using one-leaflet lesson proposals in which suggestions are given for brief lesson activities. The decision was made to follow this suggestion, despite of the fact that not all aspects of a constructivist approach to elementary science could be included in the microteaching lessons. Because there were some serious reasons for not incorperating microteaching lessons, it was also decided to give this issue special attention during the try-out of the SPIN program. After the results in favor of microteaching the try-out, the ultimate decision was made to include microteaching lessons into the SPIN program.

After the revision of the preliminary design of the SPIN program, a first draft was developed and evaluated by means of an expert appraisal and a try-out. The results of these evaluations did not lead to major modifications of the program. However, it must be noted that both the expert appraisal as well as the try-out appeared to be extremely useful in debugging and fine-tuning the program (for details see: Van den Berg, 1996 p. 69-82).

6. CONCLUSIONS

The design problem in the SPIN-case can be characterized as ill-defined, because no firm research based knowledge is available about how to support teacher to change their practice in elementary science. Walker's deliberative approach (1971, 1990) turned out to be appropriate to tackle this ill-defined design problem, because it allows the designer to consider and reconsider design options and decisions from multiple perspectives during the entire process. Moreover all kinds of other considerations, varying from characteristics of the target group, the way to organize the content, and the selection of the media, to the layout of the curriculum materials, may enter the deliberation process. In the earlier stages of the design process attention is focused on examining ideas critically by rational argumentation. Try-outs in practice are characteristic for later stages in the process.

Marsh and Willis (1995) question whether Walker's deliberative approach is applicable in small-scale projects, because the approach finds its empirical justification in rather large-scale curriculum development projects. The design and development of the SPIN inservice program demonstrates that application of the approach in a small-scale project is possible. However, the character of the argumentation process to come to design decisions did not take place within a project team with a shared responsibility. In the SPIN program, one person had the final responsibility and the argumentation took place in discussions with experts in various fields outside the project. Consequently, the deliberation processes in the

SPIN project were not shared decision making processes, but had a more distant and individual character.

In this chapter a brief account is given of the design and development process of the SPIN program. This process may be characterized as a *limited commitment mode control strategy* (Terlouw, Feteris & Martens, 1992). Designers trying to solve complex and ill-defined design problems commit themselves to their preliminary decisions, but this commitment is limited. These decisions are subdued to a control strategy by evaluating and eventually changing them. In this chapter two examples have been given in which this limited commitment to design decisions has been demonstrated. In the first example, the original commitment to start the SPIN program with explaining the theory behind a constructivist approach did not hold, after discussing this design option with various experts. In the second example, the original commitment to include microteaching lessons into the SPIN program did hold, but the way these lessons have been planned to be incorporated into the program was modified based on the discussions. The examples demonstrate also the limited scope of the control strategies. Because the SPIN project, just as many other design projects, lack the time and money for applying sophisticated evaluation methods, a variety of more simple methods has been used. This variety may compensate for the separate weaknesses of each single method (Walker, 1992). The main purpose of the evaluations carried out is to inform the decision-making process during the design and development of an educational program with the purpose of improving the program (Flagg, 1990). The evaluation questions to be answered regarding a set of ideas are not whether it is really true, but whether it is useful (cf. Eisner, 1979). The evaluations carried out in the SPIN project served to justify and clarify design decisions and to improve the SPIN program.

7. FINAL REMARK

On many occasions, the responsibility for the design of a small-scale curriculum project will be in hands of one person, instead of a design team. Also in these circumstances, as the SPIN-case shows, a modified version of Walker's deliberative approach is applicable. The main difference between the original and modified approach may be that the former concentrates on communication whereas the latter is better typified as a confrontation strategy[2].

First, the designer confronts initial ideas with various sources in the research literature in order to warrant the theoretical soundness of the program and to sharpen and refine ideas. This phase ends with the formulation of preliminary design decisions.

Second, those design decisions are discussed with various experts in the field. Based on these discussions, the designer reconsiders the preliminary design decisions and formulates more definite ones. This confrontation is mainly meant to ensure the practical coherency and consistency of the design decisions.

Third, the actual construction of the first version of the program takes place. In a try-out the confrontation with practice comes into being. After this confrontation final revisions are made.

[2] This perspective has been suggested by Robert M. Branch in one of his comments on an earlier draft of this chapter.

The way Walker's approach has been applied in the SPIN-project implies that the individual designer works as a spider in the web. The designer makes decisions and constructs a product by considering various sources of information.

REFERENCES

Anderson, R.D. and Mitchener, C.P., 1994, Research on science teacher education, In D.L. Gabel (Ed.), *Handbook of research on science teaching and learning,* 3-44, Macmillan, New York.

Ben-Peretz, M., 1990, *The teacher-curriculum encounter,* State University of New York Press, Albany.

Berg, E., van den, 1996, *Effects of inservice education on implementation of elementary science,* Doctoral dissertation, University of Twente, Enschede.

Duit, R. and Treagust, D., 1995, Students' conceptions and constructivist teaching, In Fraser, B.J. and Walberg, H.J. (Eds.), *Improving science education,* 46-70, University of Chicago Press, Chicago.

Eisner, E.W., 1979, *The educational imagination,* Macmillan, New York.

Feiman-Nemser, S., 1990, Teacher preparation: Structural and conceptual alternatives, In Houston, W.R. (Ed.), *Handbook of research on teacher education,* 212-233), Macmillan, New York.

Flagg, B.N., 1990, *Formative evaluation for educational technologies,* Lawrence Erlbaum, Hillsdale, NJ.

Harlen, W., 1985, Science education: Primary-school programmes, In Husén, T. and Postlethwaite, T.N. (Eds.), *The international encyclopedia of education,* 4456-4461, Pergamon Press, Oxford.

Joyce, B. and Showers, B., 1980, Improving inservice training: The messages of research, *Educational Leadership, 37,* 379-385.

Joyce, B. and Showers, B., 1988, *Student achievement through staff development,* Longman, New York.

Louden, W. and Wallace, J., 1994. Knowing and teaching science: The constructivist paradox. *International Journal of Science Education, 16,* 649-657.

Marsh, C. and Willis, G., 1995. *Curriculum: Alternative approaches, ongoing issues,* Prentice Hall, Englewood Cliffs, NJ.

Terlouw, C., Feteris, A., Martens, H.L.M., 1992, *Onderwijskundig ontwerpen revisited: een kritische analyse van het OKT-model* [Educational design revisited: A critical analysis of the OKT-model], University of Twente, Enschede.

Tulder, M., van, Roelofs, E., Veenman, S., Voeten, M., 1992, Effecten van nascholing op het onderwijsgedrag van leerkrachten [Effects of inservice education on teachers' behavior], *Pedagogische Studiën, 69,* 99-111.

Walker, D.F., 1971, A study of deliberation in three curriculum projects, *Curriculum Theory Network, 7,* 118-134.

Walker, D.F., 1990, *Fundamentals of curriculum,* Harcourt Brace Jovanovich, San Diego.

Walker, D.F., 1992, Methodological issues in curriculum research, In Jackson, P. (Ed.), *Handbook of research on curriculum,* 98-118, Macmillan, New York.

Wallace, J. and Louden, W., 1992, Science teaching and teachers' knowledge: Prospects for reform of elementary classrooms, *Science Education, 76,* 507-521.

Watts, M. (1994). Constructivism, re-constructivism and task-oriented problem-solving. In P. Fensham, R. Gunstone & R. White (Eds.), *The content of science: A constructivist approach to its teaching and learning* (pp. 39-59). Falmer Press, London.

Yeany, R.H. and Padilla, M.J., 1986, Training science teachers to utilize better teaching strategies. A research synthesis, *Journal of Research in Science Teaching, 23,* 85-95.

Chapter 7

Design and Development of Third Generation Distance Learning Materials: From an Industrial Second Generation Approach Towards Realizing Third Generation Distance Education

Paul Kirschner
Open University

Martin Valcke
University of Gent

Dominique Sluijsmans
Open University

Key words: Distance education, Educational model, Paradigm shift

Abstract

Distance education is going through a paradigm shift from the second to the third generation. In the first generation (correspondence education) teachers were craftsmen who coupled traditional learning materials to self-made personalised lessons that they mailed to their students at a distance. In the second generation new, pedagogically enhanced materials were designed and developed via an industrial model specifically for distance education, but the accent remained on a traditional learning paradigm. In the third generation personal, competence based, interactive materials for learning communities are being designed and developed.

This chapter first outlines the characteristics of the first two generations. It then presents a framework for the design, development and delivery of distance education study materials according to the industrial approach. It concludes with a look at how this will change as wc go to the third generation. In chapter 21 an environment for this new paradigm is worked out..

1. THE FIRST TWO GENERATIONS

Distance education, as all education, is mediated experience where the specific aim is the intentional acquisition of knowledge, skills, attitudes and competencies.

First generation distance education (correspondence education) had at its basis the traditional teacher directed, didactic paradigm. It made use of the traditional schoolbook as primary content material and the teacher as primary pedagogic medium. In essence it was the imitation of the traditional teacher-pupil paradigm over distance. The teacher made use of the same materials that he or she used in traditional face-to-face education, augmented this material with written assignments which were mailed to the students (correspondence, hence the name), carried out a dialogue through this same means and thus allowed students who could not attend school in person to follow a course of study (Nilsen, 1986). In first generation distance education the teacher is a craftsman making 'one-of-a-kind' commodities tailored to specific people or small groups. Their 'articles' are the education that they conceive, define, design, develop, produce and deliver.

The second generation could be called guided independent study and followed a different paradigm, namely a pedagogically integrated, cognitive systems paradigm (Bates, 1995). It made, and in fact still makes use of specially designed and developed learning materials in which the content and the pedagogy are fully integrated into the material. The learning material is, by and large, self-supporting and for the most part does not require the intervention of the teacher (Kirschner & Valcke, 1994; Martens, 1998), although tutors can play an important role after delivery of the materials to students. Materials developed for guided self-study contain explicit and implicit didactic techniques necessary for initiating, facilitating and maintaining relevant study processes. These techniques influence the study process by making it more efficient, more effective and/or more satisfying for the learner. The role of materials in this setting can be summed up as:

- setting of objectives;
- structuring the study process;
- providing information and explanation where relevant or necessary;
- giving examples and/or making things concrete;
- guiding study activities;
- developing study skills;
- enabling self-assessment for the student;
- providing feedback as to the process and products of the study; and
- preparation for formal assessment (Martens, 1998).

Materials are developed in teams according to an industrial approach in which different actors fulfil different roles in the design, development, delivery and student use of the materials for large groups of learners.

The third generation could be called interactive or networked multimedia education and follows a more flexible, personal, competence based, interactive (constructivist) paradigm for learning within learning communities (Bates, 1995). It is an open, flexible, collaborative information and communication technology based form of education. After a short discussion of the second generation approach, this third generation will be handled in depth.

2. THE PROCESS OF DESIGN, DEVELOPMENT AND PRODUCTION IN THE SECOND GENERATION

2.1 Phases and products

Materials for second generation distance education are:

- designed and developed in multidisciplinary course teams;
- the direct result of explicit design;
- characterised by their explicit structure, intrinsic dialogue style and use of different forms of assessment as feedback mechanism; and
- based upon the most functional and necessary medium.

The approach used to design is not algorithmic, but is rather heuristic in nature. In this heuristic approach there are *no* absolute criteria that can be used or applied when designing independent study materials (Gastkemper, 1993). The approach is based upon an analysis of the chosen objectives, on the pedagogic requirements for the intended population, on a systematic choice of available media and on the limiting conditions of the situation of either the institute or the student or both (Willis, 1998). When applying this approach a number of factors are considered, weighed and eventually balanced. These are the:

- representation of the goals to be achieved;
- characteristics of students who will be studying the materials, both cognitive (knowledge, skills, competencies) and attitudinal (intention, needs) in nature;
- educational functions/didactic concepts to facilitate the achievement of objectives within the context of the student characteristics;
- characteristics of all media available for the student;
- intended setting or study situation; and
- practical considerations and limitations.

Six phases are involved in the design and development of distance study materials. For a more detailed discussion of the phases and the products, see Kirschner (1996).

Phases	Products
Pre-planning	Global description
Planning	Detail plan and team formation
Development	Draft materials
	Formative evaluation
	Final materials
	Items for examination
Production	Study materials
Delivery	Study arrangement:
	Materials
	Support
	Assessment
	Summative evaluation
Revision	Revision plan

Table 1. Six phases and products in the design and development of distance study materials

2.2 Actors

The industrial approach is not only characterised by the use of distinct phases and products during design and development, but also by the use of interdisciplinary teams. The major actors with their accompanying roles are given in *Table 2*.

Actor	Role
Project chair	Specialist in the global content area Responsible for: Basic content design based upon predetermined objectives Recruiting and evaluating the developers Evaluation of content and cohesiveness of contributions Project management
Material developers	Experts in specific content areas Recruited from different segments of society depending upon the specific expertise required (academia, industry, government et cetera) Responsible for the development of the specific assessment (procedures)
Educational technologist[1]	Designs and develops, together with the project team (chair), the proper pedagogic approach including the media mix and student support Develops the functional specifications for the materials Instructs the developers Manages the development of the electronic media Assesses and makes suggestions on the developed materials Designs, sets up and carries out developmental testing of the materials Carries out staff development with respect to the design and development of the arrangements
Logistician	Day to day co-ordination, planning and administration of: Contracts Rights to materials Work planning of the different specialists Budgets Production capacity Deadlines
Editor	Language and didactic editing
Other roles	Programmers for the actual development of computer programmes Illustrators for design and illustration Illustration researchers for finding and acquiring illustrations Information technology managers for installation and maintenance of hardware and software Desk top publishing personnel for the introduction of the texts in electronic form And so forth

Table 2. Actors and their roles

In the next section, the shift from the second to the third generation is outlined.

[1] The educational technologist systematically looks for and applies relevant theoretical and practical knowledge from the fields of educational technology, psychology, education, instructional technology, curriculum technology, communication sciences, mass media, et cetera as well professionally applying tried and true empirical methods and procedures in the solution of educational problems.

3. A PARADIGM SHIFT FROM THE SECOND TO THE THIRD GENERATION

As stated in the previous section, the second generation approach was predominantly supply-driven. All students got the same book, with the same examples, exercises, et cetera. This is in sharp conflict with the need for more demand-driven education. The era of supply driven education is considered to be ineffective and inefficient. The need for tailoring courses to the specific needs of student is a world-wide issue (Kirschner & Valcke, 1994). Many of the latest information technologies can comprise the platform of a future learning economy, where learning experiences are pulled by demand rather than driven by supply. Adaptive learning systems answer the need for demand driven education. These systems should:

- be designed for an open environment of components and modular services;
- support synchronous and asynchronous interaction;
- permit frequent upgrading and reuse of courseware and on-line services;
- optimise use of the network, data centre/server, and desktop;
- balance consideration for learner interests and competencies with educational objectives and cost;
- be adapted to varied and changing user needs, as well as diverse contexts;
- be reusable and easily integrated with other systems; and
- demonstrate ways to reduce the burden on learners and educators without compromising educational quality.

In addition, proposals for tools, systems and infrastructure are strongly encouraged if they can incorporate instructional paradigms with proven effectiveness, address the training and self-development requirements of the work force, and employ innovative net-centric, web-based business models for the delivery of education and training (Morris, 1997). The five key concepts that characterise this paradigm shift are:

- competency-based education;
- continuous assessment and evaluation;
- co-operative and collaborative learning;
- flexible learning materials; and
- internet-based study environments.

4. THE ORIGINS OF THE KEY CONCEPTS IN THE NEW DISTANCE EDUCATION-PARADIGM

In professional contexts a large and still growing concern about the performance levels of new recruits and existing staff can be observed (Boyatzis, 1982; De Snoeck, 1997). Job descriptions are presented in terms of task profiles, human resource management is based on skills management approaches and the literature on professional training clearly reflects this focus on 'competencies'. In educational contexts there is also a real call for competency-based education although the concept of competencies is not always used. International reports or the evaluation reports on the quality of university studies reflect this in their plea for more skills-oriented education, more real-life orientation, less emphasis on declarative knowledge, et cetera. In higher professional education in the Netherlands, competency profiles are agreed upon at the national level and institutes translate the profiles into their curricula (Van Vilsteren, Hummel, Kirschner, & Wigman, 1997). In the new paradigm of the Open University of the Netherlands the student will be informed about the curriculum throughout a so-called competence-catalogue. A competence-map coupled to an 'intaker' thereupon helps students evaluate their strengths and weaknesses in those competencies. The student then follows an individual learning path to develop the desired competencies. This personal learning path will be documented in a portfolio.

Whereas assessment and evaluation were traditionally seen as control processes at the end of the study, there is a growing tendency to integrate assessment and evaluation into all phases of the study. Prior knowledge tests help both student and institute determine entrance levels, intake tests help fit programmes to learner needs and characteristics, and progress tests help students to check mastery. Moreover, responsibility for assessment and evaluation is handed over to the student when self- and peer-assessment is applied. The availability of information and communication-technology (ICT) reinforces this by offering on-line and interactive assessment and evaluation environments. An important shift with respect to assessment is the growing role of the labour market in influencing the way curricula and their goals are revised (Pilot, 1997). The form of assessment determines whether the student has achieved the knowledge, skills or competencies required for his or her chosen field and the scope of the field will influence, in turn, the content of the assessment (Moerkerke & Terlouw, 1998). By integrating forms of assessment, like self-, peer- and co-assessment in the curriculum, students develop into competent persons and lifelong learners who continuously reflect on their behaviour and their learning. These alternative forms of assessment should be a part of a process of change towards a student-centred approach. This change requires a shift in emphasis from norm-referenced to criterion-referenced assessment, from purely summative to formative and summative evaluation, from external to internal control and from the assessment of product to the assessment of products and processes. Dochy (1992) and Moerkerke (1996) give a good overview of these trends and offer a large number of examples.

Distance education in general suffers from its stress on the individual acquisition of knowledge and skills. This questions the validity of the study programmes because certain goals are difficult to achieve in an individual context. Examples abound such as negotiation skills, chairing a meeting, monitoring a discussion, interviewing decision-makers. It is therefore no wonder that curriculum evaluation at the university level (the so-called 'visitation' reports) are critical about skills development in general and collaborative work in particular. Collaborative learning as a guiding didactic principle has strongly affected the use and position of ICT resulting in a strong change in emphasis on ICT use in which students, educational staff and support staff work together. Such collaboration has a dual function. It supports the use of effective discursive learning methods (make explicit, discuss, reason, and reflect, convince) while allowing for the acquisition of essential social and communication skills (Dillenbourg et. al, 1995; Mirande, Riemersma & Veen, 1997). Moreover, research on achievement has also found co-operation to promote greater intrinsic motivation to learn, more frequent use of cognitive processes such as reconceptualisation, higher-level reasoning, metacognition, cognitive elaboration, and networking, and greater long-term maintenance of the skills learned (Johnson & Johnson, 1993).

Traditional education builds on a massive amount of ready-made materials and standard curricula (default programmes). This reflects a supply-driven paradigm of education that is being replaced by a demand-driven one (Kirschner & Valcke, 1994) in which study activities and resources are adapted to the needs and requirements of the students. Not only are the curricula more in line with what students want or need, the materials are also more in line with specific student demands or characteristics. The reader sees the relationship of this flexibility of learning materials with the need for other forms of assessment (Valcke & Vuist, 1995; De Wolf & Valcke, 1996; Valcke & Van Dam-Mieras, 1998). Bates (1994) summarises as follows:

> Learners will interact with their desk-top or portable workstations in a variety of ways, determined by the nature of the learning task, and their preferred style of learning in the work situation. These preferred styles will vary considerably, both within a single person, depending on the task, and, for the same task, between different individuals.
> The learning context will need to encompass the following:
> - working alone, interacting with learning material (locally or remotely);
> - working collaboratively with fellow workers at different remote sites, either synchronously or asynchronously: both these modes are likely to be multimedia;
> - as 'apprentice' or 'student', working with a more experienced worker, supervisor, or instructor; and
> - as 'instructor' or 'supervisor' for other less experienced colleagues.
>
> The same person may find themselves [*sic*] in each of these roles within a single working day. Learners will also need to be able to work from home, or from a work-site, or while in transit. They will need the following:
> - access to information (searching, downloading) from multiple sources in multiple formats;
> - selection, storage and re-ordering/re-creation of information;
> - direct communication with instructors, colleagues, and other learners;
> - incorporation of accessed/re-worked material into work documents; and
> - sharing and manipulation of information/documents/projects with others.
>
> Learners will need to access, combine, create and transmit audio, video, text, and data as necessary. If we take this as the design requirement, there is then a need to build systems that support this form of learning, both for formal and informal learning.

This leads to the final key concept, namely internet-based study environments. The concurrent development of computer and communications technology have led to the introduction of innovative methods (Gibbs, 1992), a growing interest in resource-based learning (Exley & Gibbs, 1994; Hawkridge, 1996), and substantial investment in the development of Computer Based Learning materials (Slater, 1996). The internet has arrived on the educational scene, making a range of difficult to achieve learning concepts, such as distributed learning (Tait, 1997), possible.

The shift from an industrial economy to an information economy is creating more adults who require lifelong learning. They will change jobs frequently during their working life, and will require additional education while not being able to 'attend' school in the traditional sense. In addition, there will be a significant increase in the number of 'traditional' students who wish or need to participate in higher education.

This presents the educational community with a tough challenge (if we assume that governments will not increase funding to match the increased need). One way to meet this challenge is to develop distributed learning environments (Dede, 1996; 1997). A distributed learning environment is an approach to education and training that is intended to be learner-centred, enabling both synchronous and asynchronous interaction through the integration of pedagogically-appropriate technologies. The model is based on blending suitable educational technologies with aspects of campus-based delivery, open learning systems and distance education. The approach gives instructors the flexibility to customise learning environments to meet the needs of diverse student populations, while aiming to provide both high quality and cost-effective learning opportunities. Learning environments promote the use of the internet (www) to help students find, evaluate and process information, solve problems, communicate ideas, work collaboratively, and learn how to learn. This distributed learning and teaching will result in:

- the creation of educational paradigms that use computer networks to create learner-centred, project based education that is free from constraints of time and place, offer a variety of learning styles and promote collaboration;
- improved communication and increased collaboration among all members of the community of learners;
- learning activities that improve student skills in assessing information, use them to solve problems, communicate ideas and learn how to learn; and
- new partnerships between education and the private sector.

Further research will result in knowledge resources that support the design, development, management, and assessment of innovative learning systems.

All of this has not, for example, escaped the Dutch context where a multitude of governmental and private initiatives have been initiated to push the use of ICT and distributed learning in education. Two recent examples at the higher education level are the initiation of a subsidy programme for educational innovation based upon the findings of the Commission for Quality and Study-ability (Wijnen, Wolfhagen, De Bie, Brouwer, Ruijter & Vos, 1992) and the programme ICT and higher professional education (Mirande et al., 1997). The integrated use of internet is a pre-dominant issue in these programmes.

5. DESIGN AND DEVELOPMENT OF STUDY MATERIALS IN THIRD GENERATION EDUCATION

Table 1 presents the six phases and products in the design and development of study materials and *Table 2* the role of the educational technologist in second generation distance education. These two aspects are merged into one table for third generation distance education (see *Table 3*).

Phases	Products	Role of the educational technologist
Pre-planning	Global description of: • Competencies • Domains • Functions • Interactions/transactions • Actors	Helps determine and validate competencies Helps define actors and pedagogic functions
Planning	Team formation Functional specifications of: • Competence map • Tasks and criteria • Modes of interaction • Domain maps • Links between elements	Designs pedagogic environment: • Creates and validates competency map • Defines pedagogic model used • Defines task form • Models assessment criteria
Development	Draft arrangements Trial version for evaluation Final version Criteria for certification	Helps develop arrangement with an accent on pedagogic aspects: • Instructs team • Evaluates the quality and usability of the arrangements • Designs/supervises development and filling of electronic environment • Sets-up, carries out and reports on beta testing
Production	None	None: The educational technologist has no specific role in the production process
Delivery	Electronic delivery of: • Arrangements • Support • Co-operative learning facilities • Summative evaluation	None
Revision	Continuous revision (updating and upgrading)	Monitors pedagogic upgrading of the environment and the arrangements

Table 3. Six phases and products in the design and development of third generation distance study materials and the role of the educational technologist therein

Current practices in the Open University of the Netherlands are characterised by the development of an Electronic Learning Environment (ELO), in which all key elements as described above are integrated. The next section illustrates the model of this innovative environment, which is not going to be designed for distance education, but for higher education in general.

6. THE ELECTRONIC LEARNING ENVIRONMENT (ELO)

From a technical perspective, the ELO is an internet based, multimedia computer environment for learners, staff, administrators, logisticians, et cetera in the context of designing, developing, carrying out, supporting and assessing study processes. It is both a study and a work environment. The staff of the Open University of the Netherlands logs onto the environment and is automatically authorised for those tasks, roles and responsibilities that are in line with their job profile. Students log onto the environment and are automatically authorised to study and work within their specific study environment in line with their goals and profiles. This environment is currently under development. Parts already exist in the existing StudieNet (intranet) environment at www.ouh.nl. Moreover, a number of large scale projects such as International Law help to check the validity of the ELO-design and are the first experimental implementation fields for 1998. *Figure 1* shows the position of 'objects' in relation to one another in view of designing the basic structure and functions of the new ELO.

The figure shows the vital position of a 'map' with competencies as starting point for all further processes and products. The intake process defines what the - default and eventually ad hoc developed - curriculum of the student is planned to be. The building blocks of the curriculum (sometimes clustered as curriculum units) are 'study tasks'. Study tasks are linked to operational objectives and assessment-objects. Students tackle tasks by working on the task prescriptions, using on-line and off-line resources, entering collaborative learning environments. They can decide whether or not to take tests, to send in assignments or other products that result from (independent or individual) task-activities. They can check their portfolio, etcetera.

As to the assessment and evaluation approach in the ELO, the figure shows a variety of approaches used. The triangles show assessment and evaluation approaches with a variety of functions:

- The intake assessment plays a major role in the negotiation process to determine the specific instantiated curriculum for a specific students, building upon their preferences, prior knowledge, professional orientation, et cetera.
- At the curriculum level students can assess their own prior knowledge and skills. At the task level they can take formative tests to monitor their progress. In relation to a specific knowledge domain students can take tests that assess the mastery of specific knowledge or skill elements in that domain. These assessment types are not the basis for the certification of the mastery of competencies. They are intended to help students to guarantee that they master the underlying principles, techniques, skills, and declarative knowledge elements.
- At the task level task an assessment object (trapezoid) is presented. This object assesses mastery of a competency. At the Open University of the Netherlands, task execution - actual student performance - is the assessment object.

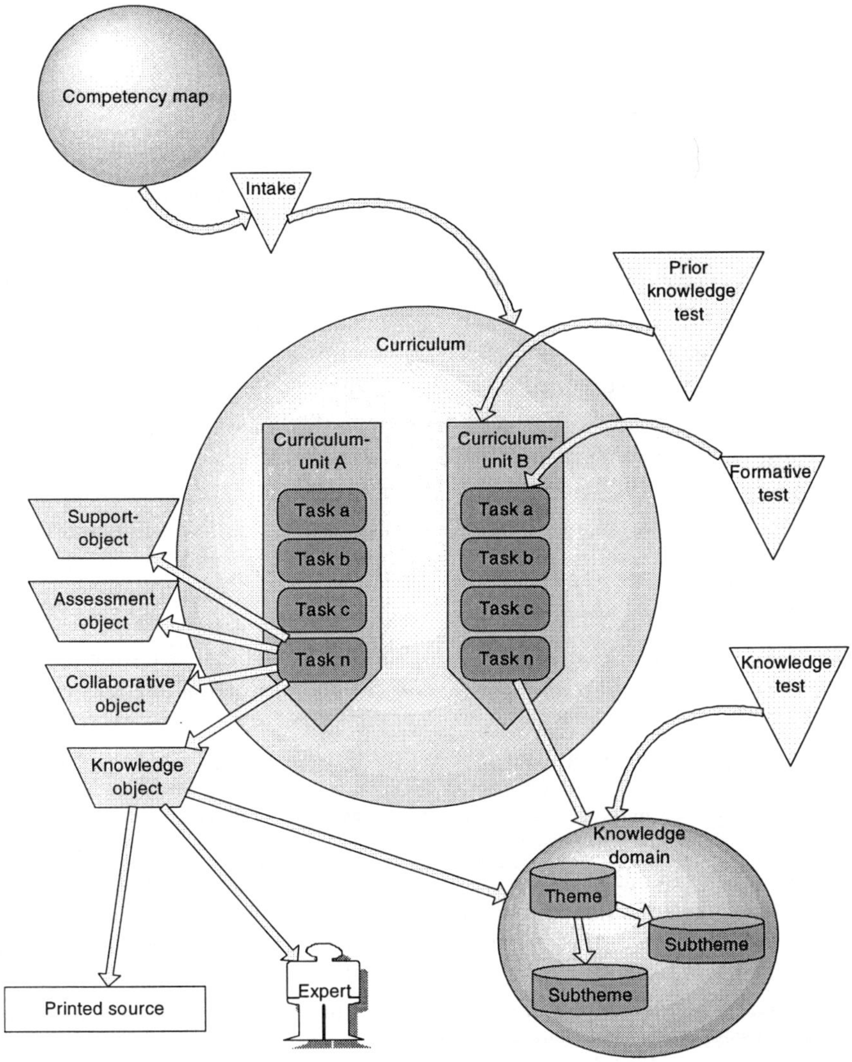

Figure 1. An object model of the ELO

The student in the ELO-environment has access to his/her own individual portfolio or dossier. This portfolio contains the formal agreements between the student and the educational institute as to the curriculum to be studied and the outcomes of the assessment and evaluation cycles, such as final scores and the product outcomes.

Valcke, Kirschner and Bos elaborate on this environment for the new paradigm in chapter 21 in this book.

7. CONCLUSION

In third generation education resources and services are made available beyond the boundaries of any single discipline. Information technology makes these resources more reusable and links instructors and students (Morris, 1997). Morris says that the best way to avoid a mismatch between supply and demand is to deploy technology in response to clearly articulated needs. Through a flexible, learner-centred curriculum, professional growth will be facilitated in distance education while students are challenged to develop decision-making, problem-solving, and critical thinking skills.

Tools for developing instructional content and courseware will need to operate across different platforms with other tools used to build and manage learning systems. In other chapters in this book new electronic learning environments and tools are discussed. These new environments and the materials within them correspond with the mission of third generation education as it develops from staid, supply-driven learning aimed at content specific learning into flexible, demand-driven learning that encourages life-long learning.

REFERENCES

Bates, A.W., 1994, *Educational multi-media in a networked society,* World Conference on Educational Multimedia, Ottman and Tomek, Vancouver.

Bates, A.W., 1995, *Technology, open learning and distance education,* Routledge, London.

Boyatzis, R.E., 1982, *The competent manager. A model for effective performance,* John Wiley & Sons, New York.

Dede, C., 1996, Distance learning--distributed learning: Making the transformation, *Learning and Leading with Technology, 23,* 7, 25-30.

Dede, C., 1997, Distributed learning: How new technologies promise a richer educational experience connection, *New England's Journal of Higher Education and Economic Development, 12,* 2, 12-16.

Dillenbourg, P., Baker, M., Blaye, A. and O'Malley, C., 1995, The evolution of research on collaborative learning, In Reimann, P. and Spada, H. (Eds.), *Learning in humans and machines. Towards an interdisciplinary learning science,* 189- 211, Pergamon, London.

Dochy, F.J.R.C., 1992, *Assessment of prior knowledge as a determinant for future learning,* Lemma, Utrecht.

Exley, K. and Gibbs, G., 1994, Course design for resource based learning, Centre for Staff Development, Oxford Brookes University, Oxford.

Gibbs, G. (Ed.), 1992, *Improving the quality of student learning,* Educational and Technical Services, Bristol.

Hawkridge, D., 1996, The next educational technology in higher education, *Innovations in Education and Training International, 33,* 1, 5-12.

Gastkemper, F., 1993, Development of media, methods and technology: central issues in higher distance education, In Vocht, C. de and Henderikx, P. (Eds.), *Flexible responses in higher education. Strategies and scenarios for the use of open and distance education in mainstream higher education,* 13th-14th December 1993, Brussels, Belgium, 129-137.

Johnson, D. and Johnson, R., 1993, What we know about cooperative learning at the college level, Cooperative Learning , 13(3); also http://www2.emc.maricopa.edu/innovation/CCL/ whatweknow.html.

Kirschner, P.A., 1996, *Design, development and delivery of distance study materials: An industrial approach to distance education,* In Coetzee, J.H. and Smith, T.G. (Eds.), Proceedings International Conference on Education and Change, University of South Africa: Pretoria.

Kirschner, P. and Valcke, M., 1994, From supply-driven to demand-driven education: new conceptions and the role of ICT therein, *Computer in Human Services, 10,* 31-53.

Martens, R.L., 1998, *The use and effects of embedded support devices in independent learning,* Lemma, Utrecht.

Mirande, M., Riemersma, J. and Veen, W., 1997, *De digitale leeromgeving,*[The digital learning environment], Wolters Noordhoff, Groningen.

Moerkerke, G., 1996, *Assessment for flexible learning,* Lemma, Utrecht.

Moerkerke, G. and Terlouw, C., 1998, Herontwerp van toetsing [Redesigning assessment], *Tijdschrift voor Hoger Onderwijs en Management, 1*, 19-25.

Morris, R., 1997, *Adaptive learning systems,* National Institute of Standards and Technology, http://www.atp.nist.gov/atp/97wp-lt.htm.

Nilsen, E., 1986, On the definition of correspondence education, *Epistolodidaktika*, 1986/1, 3-30.

Pilot, A., 1997, *Curricula voor een op informatie gerichte maatschappij* [Curricula for an information society], Paper presented at the Consortium COO HBO and SUNCOO conference, December 11-12, 1997, Utrecht.

Slater, J., 1996, The impact of TLTP, *Active Learning*, 4, 3-6.

Snoeck, J. de, 1997, Zoeken naar de receptuur: competentiemanagement [Searching for a recipe: Competency manager], In N.V. Bekaert S.A., *Gids voor Personeelsmanagement, 76*(11), 40-45.

Tait, B., 1997, Object orientation in educational software, *Innovations in Education and Training International, 34,* 3, 167-173.

Valcke, M. and Vuist, G., 1995, A model-based design approach for the flexibilisation of courses, In Lockwood, F. (Ed.), *Open and distance learning today,* pp. 185-196, Routledge, London.

Valcke, M. and Dam-Mieras, M.C.E. van, 1998, Designing flexible learning material resources in natural sciences in natural sciences to support tailor-made courses, In (Eds.), Proceedings Bite-Conference "Bringing Information Technology to Education", pp. 472-484, Universiteit Maastricht, Maastricht.

Vilsteren, P. van, Hummel, H., Kirschner, P. and Wigman, M., 1997, *Het verwerven van competenties bij de Open Universiteit* [The acquisition of competencies at the Open University of the Netherlands], Open University of the Netherlands (Center for Educational Technology and Expertise), Heerlen.

Wijnen, W.H.F.W., Wolfhagen, H.A.P., Bie, D. de, Brouwer, O.G., Ruijter, C.T.A. and Vos, P., 1992, Te doen of niet te doen? Advies over de studeerbaarheid van onderwijsprogramma's in het hoger onderwijs [To do or not to do? Advice on approproateness of study of educational programmes in higher education], Ministerie van Onderwijs, Cultuur & Wetenschappen, Zoetermeer.

Willis, B., 1998, *Instructional development for distance education,* University of Idaho, http://www.uidaho.edu/evo/dist3.html.

Wolf, H. de and Valcke, M., 1996, Onderwijsfuncties en multimedia - Competentiegericht onderwijs en het ontwikkelen van flexibele leermaterialen [Educational functions and multimedia - Competency based education and the development of flexible learning materials], In Wolf, H.C. de (Ed.), *Het onvermijdelijke samengaan van onderwijs en multimedia*. LUC, Diepenbeek.

Chapter 8

The Design and Prototyping of Digital Learning Material: Some New Perspectives

Jef Moonen
University of Twente

Key words: Design, Prototyping, Digital learning material

Abstract

Stimulated by the growth of multimedia, the World Wide Web, and the convergence of communication and information technologies, the nature and potential of digital learning materials are rapidly changing. Design methods and in particular prototyping activities have to adapt to the new situation. New design approaches are not only needed in the area of digital learning material. From a more general perspective rational and relational design approaches can be examined in the context of their contribution to design strategies. In order to cope with the new situation, a new design approach called 'the 3-space design strategy' is presented and applied to the design and prototyping of digital learning material. Furthermore an example of a design project about a media-selection support system is discussed from the perspective of the 3-space design strategy.

1. INTRODUCTION

Learning material exists in three basic forms: printed material including graphs and pictures, analogue audiovisual material, and digital material involving computer-accessible components. Slides, film, audiotape, videotape, and radio and television broadcast form the basic ingredients for audiovisual learning material. Digital learning material is based on or relies on digital components. Examples are educational computer programs, multimedia programs on CD-ROM, and WWW documents or systems accessible through the Internet.

A main characteristic of traditional printed learning material is its organization of the *content* and the *structure* of the material in chapters, subchapters, paragraphs, etc., and the logical flow and build-up throughout those chapters. The visual aspects of printed learning material have also received considerable attention, related for example to the choice of fonts; use of graphics, colors, and pictures; and so on. The same characteristics of structure and visualization also apply to digital learning materials.

Also, as interaction is a main additional feature of digital learning materials, the *user-interface* has become an important design focus as well.

Furthermore, and in particular for learning materials with high hyperlinking potential and a modular structure, the size of and relations between the different "to be linked" modules, are also of importance. These issues relate to the *granularity* of "learning objects". This aspect is comparable to the choice an author makes when he or she divides a traditional text into chapters, subchapters, and paragraphs. However, the hyperlinking facility within digital learning material creates a new navigation potential not present in traditional text, video and broadcasting. Besides opportunities to navigate freely through an information base and to consult additional information, users have a strong need for navigation support via a guiding structure. Finding an optimal combination of both features (free *navigation* and guiding *structure*) is an important problem to be addressed in the design of digital learning material.

Although content, structure, interface, and granularity are key characteristics of digital learning materials, their nature and potential are rapidly changing because of the growth and wide availability of multimedia, the World Wide Web (WWW), and the convergence of communication and information technologies. This evolution raises a number of questions: (a) Are existing design and production methods for digital learning materials appropriate for practical application when dealing with the new forms? (b) Do we need a new design and production strategy to cope with these new forms and does the 3-space design strategy, introduced in this chapter, provide a solution? and (c) What is the specific role of prototyping in the 3-space design strategy and how does prototyping apply to the design and production of new forms of digital learning material?

This chapter addresses these questions by first, in Section 2, giving a short reflection on existing design methods, classifying them as rational or relational. In this section it is also argued that traditional rational design approaches for digital learning material do not appropriately address the specific needs that occur during the design-and-production process in practice given the new nature and potential of technology. In Section 3 a new design approach, the 3-space design strategy, is introduced. Sections 4 and 5 illustrate the 3-space design strategy in practice with particular attention to the role of prototyping; and indicate implications of the strategy for the design of digital learning materials reflecting the new possibilities of technology.

2. GENERAL DESIGN STRATEGIES: RATIONAL VERSUS RELATIONAL

General design strategies for software development (Conger, 1994) and therefore for digital learning materials, have evolved over time. They, in turn, can be seen in terms of two broad categories of general design methodologies for the social sciences. The first category is characterized by the design approach described by Simon (1969) and the second category is characterized by the design approach described by Schön (1983). The essence of the design approach advocated by Simon refers to a decomposition of the problem situation into subproblems, after which the solution of each subproblem, being downscaled in complexity, can be designed in an easier way. Characteristic for his approach is a strong end-means connection. A basic assumption thereby is that the problem situation is well defined and that there is a clear agreement about the goals of the project. This could be called a "rational" approach.

Design methods have been traditionally based on a rational-decomposition approach. Often however, and certainly in the social sciences, a rational-decomposition approach is not appropriate. In addition, the changing context of the situation and the (potential) influence of major stakeholders in the process will often create fuzzy situations and uncertainty. To be able to execute design processes in such situations, Schön, in contrast to Simon, suggests an approach that is based upon 'reflection-in-action', whereby means and ends are used iteratively as basic elements of the situation, not assuming prior agreement about ends. This could be called a "relational" approach, as it emphasizes the gradually evolving relationship between product and intended user.

A relational approach fits with the conclusion of Sphorer (1998), reporting on eight years of research by the Apple Advance Technology Group, whereby more than hundred different tools for software designing and authoring were used. He concludes from these experiences that three lessons stand out: (a) *users first*; "users had to be involved from two perspectives: cognitive fit (for usability) and social fit (for dissemination)", (b) *complexity kills*; successful tools were ones that had a single key innovation that users could readily see the value of and learn to incorporate into their daily practice, especially when the innovation could be fit into an incremental improvement to an existing product'", (c) *cognitive fit is easier to attain than social fit*; the time constant in social fit is typically much longer than the time constant for cognitive fit" (p. 131).

Social fit relates to the perspectives of Dourish (1995) who is concerned about the static nature of the current systems-design methodologies and argues that "the design process does not end with the delivery of a system to some community of users. Instead, it continues as they use and adapt the system" (p. 44). And more recently, in his book about Web Site Engineering, Powell (1998) states that "software becomes less and less useful unless it is changed over time. Consequently, software should be designed with change in mind" (p. 37). Such ideas result in a strong tendency to relate the software design process to the context in which the finished product has to be used (Beyer & Holtzblatt, 1998; Moran, 1994).

The question is: Does a rational decomposition approach to software design allow for an adequate contextual analysis, or is a relational approach more appropriate? Depending on the specific situation at hand both approaches are useful. As soon as (parts of) the goals of the project are agreed upon and operationalised, a rational and systematic approach is the most effective. However, when such agreement is not within reach, a reflection-in-action approach seems to be more appropriate. This implies that it is a wise choice to combine, in a global design strategy, the strengths of both approaches. For the context of educational and training design Kessels (1993) presents such an approach when he makes a distinction between a systematic design approach (leading to internal consistency) and a relational design approach (leading to external consistency)[1]. Kessels' suggestion to a design approach that integrates both a rational and relational approach can also be applied to the design of digital learning material. Typically however, software design methods so far have assumed a more-or-less "waterfall" approach (Humphrey, 1989) consistent with Simons' problem-decomposition conceptions.

In considering how to integrate both rational and relational approaches into a design approach, the following characteristics of educational-design activities, and certainly the design and production of digital learning material, need to be taken into account: (a) there is often *uncertainty* about how to proceed, (b) there is a major influence of the *context,* and (c) many of the design activities have to do with *adaptation* (Moonen, 1998). Uncertainty is mainly caused by the complexity of the situation (many actors, many factors, unclear and almost constantly changing interactions), and as a consequence the lack of comprehensive theories about the underlying processes. Design and production are often highly based on adaptation because of the evolutionary developments in technology which result in the need for new versions of previously used software systems. And finally, the influence of the context in which the design has to be realized is very significant because of the many interests involved in the adaptation of digital learning resources and the related dominance or weakness of stakeholders in the decision process about the adaptation. In the following section, these characteristics underlie a new design approach for digital learning materials, based on both relational and rational perspectives.

3. COMBINING RATIONAL AND RELATIONAL APPROACHES: THE 3-SPACE DESIGN STRATEGY

How could a combination of the two global methodological approaches (rational and relational) work out in practice? A strategy could be to begin with the major characteristics of digital learning material and identify an approach balancing rational and relational aspects that tries to overcome the problems caused by each of the characteristics. In this Section, the three characteristics of uncertainty, context, and adaptation, mentioned in the previous paragraph, are further developed as basic concepts for such an approach (Section 3.1), followed by a description of such a new strategy, the *3-space design strategy* (Section 3.2).

3.1 Basics: Cognitive flexibility, prototyping, half products

The characteristics at the end of Section 2 lead to the following suggestions with respect to an educational design strategy: (a) reduce the uncertainty given a specific situation, (b) deliberately create opportunities for the stakeholders to influence the design process and, (c) focus on adaptation of already existing materials or half product. To translate such suggestions into a practical design and development strategy,

[1] See also chapter 5 of Kessels in this book

integrating both rational and relational perspectives, these three general suggestions are expanded in the following sections.

Reducing uncertainty: Cognitive flexibility and pragmatism

Uncertainty about how to proceed with design activities is less obvious when dealing with the design of technical systems. When dealing with a technical system, more-or-less established procedures exist indicating the major directions of the design process. It would be of help if a comparable procedure could be followed for social-science design activities. To reduce the uncertainty in a social-design process a kind of "temporary agreement" between the stakeholders about the design process and the final product can be established early in the design process. Such a temporary agreement is a replacement of the established procedures available when designing a technical system. A temporary agreement should be based upon theoretical considerations, if available, but should also take into account the context and the interests of stakeholders. The temporary agreement has to be negotiated between the stakeholders and should represent a consensus between them. It is based upon what seems to be working in practice, should be global and simple, and should also incorporate a huge amount of pragmatism. Professional practice and examples can be helpful as a basis for such temporary agreement. Such an approach is in agreement with the cognitive flexibility theory of Spiro, Feltovich, Jacobson and Courson (1992) and with the writings of Dills and Romiszowski (1997) when the latter mention that "instructional design is, and will be, practice based on multiple paradigms" (p. xii).

Opportunities for contextual influence: Prototyping

In the evolution of many design schemes and strategies available for software design there has been a tendency to incorporate the user more explicitly in the design process (Schach, 1993). Prototyping creates opportunities for such activities. Prototyping is the process of creating an early version of the final product (Moonen, 1996). That version could be incomplete in functionality and usability characteristics, or lack the performance of the final system. The primary objectives of a prototype are to provide users with a working example of the proposed system and to help them to identify, define, and revise the specifications more precisely.

Applying various cycles of prototyping during the design process also facilitates possibilities to better adjust the temporarily reached agreement to the needs of the eventual users. Those interests were originally addressed by the stakeholders in the negotiations coming to the temporary agreement at the beginning of the project. However those stakeholders may have emphasized aspects which do not adequately reflect the opinions of end users not represented earlier. In addition a prototyping environment creates opportunities to adapt components of already existing products.

Adaptability: Half products

Crucial in the design of digital learning resources is coming to consensus or temporary agreement at the beginning of the design process. To ensure that the temporary agreement is still valid after the design process has started, the prototyping activity creates opportunities to check how well the agreed consensus applies to a wider variety of stakeholders and users. However, given practical circumstances that check can only be applied to a limited number of users. The ultimate check about the appropriateness of a product has to be done by the real end users of the product. It is therefore recommended that a product available after each cycle of the prototyping activity has the format of a half product, (in principle) adaptable to the final specifications of each individual end user. In such an approach, the consensus about the product, first described in the temporary agreement in the beginning of the design process, culminates into a final individual agreement about the product by each end user[2]. Of course the kind of product that is designed will determine how feasible it is to make extensive adaptation possibilities available. Because of the growing integration of information-and-communication technology aspects in many digital learning products, possibilities for adaptation activities of half products are growing, and therefore should be incorporated in a global design process and strategy.

3.2 A new approach: The 3-space design strategy

In order to deal with the basics described above, a new design strategy called the 3-space design strategy has been developed within the context of a multidisciplinary research project on the nature of design in the social sciences (Moonen, 1998), emphasizing three kind of activities or activity spaces, each to be executed in a different phase of a project: (a) a consensus space, (b) a task space, and (c) an implementation space. Within each space the activities combine a rational and a relational approach, following main design characteristics related to the central ideas of Simon and Schön. Each space is focussed on specific activities to be performed in that space, starting from well-described inputs and resulting in well described outputs. In the following sections the strategy is described in more detail.

Visual representation of the generic approach

The three constituting elements of the design strategy are called "activity spaces" and not "phases" to avoid the suggestion that these three elements *have* to be executed in a linear order. Within each space the build-up is straightforward: input-process-output. Often however, a linear order of execution between spaces (but not necessarily within spaces) will be the common practice. Therefore each space has an input and an output, the output of the previous space being the input for the next. But, it is conceivable that in a particular situation and after the consensus space activities have led to the input for the task space, the consensus space continue to coexists with the task space, for instance through a steering committee that accompanies the project. In such a situation, the temporary agreement from the consensus space is not only (potentially) changing in the task space because of the comments received from users participating in prototyping activities, but also because the output of the consensus space is being reviewed by the original stakeholders or steering committee. Clearly such parallel activity will create confusion in the task space. But it often happens. *Figure 1* gives a visual representation of the generic structure of each of the three activity spaces.

---

[2] The rapid prototyping method described by Collis and de Boer (chapter 20 of this book) illustrates this approach.

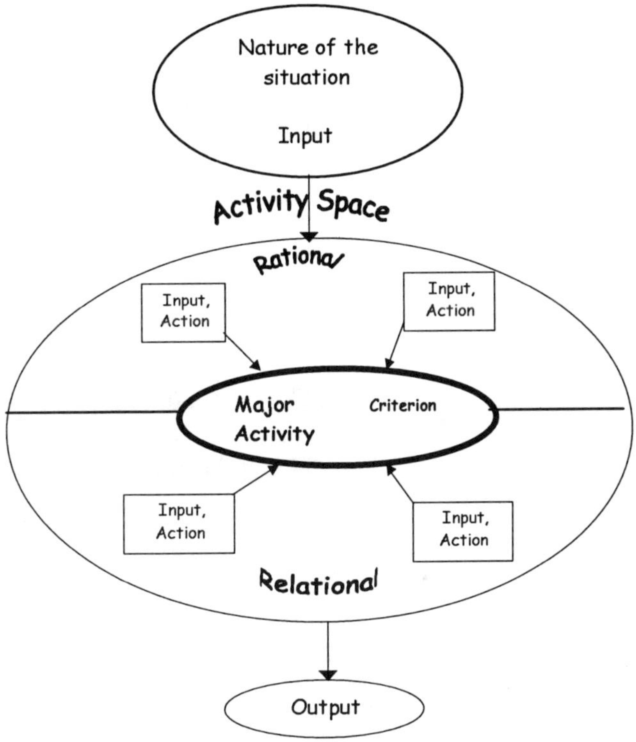

Figure 1. The generic approach for each space in the 3-space design strategy

Consensus space

In the consensus space the main objective is to move from an unstructured situation and very global ideas and specifications towards a structured design problem and functional specifications. The inputs of the process are the design problem, already-available solution ideas for the problem, and global specifications of the projected product. To support the process of reaching a temporary agreement and consensus about the product and about how to proceed, the main activity within this consensus space is that of balancing existing theory and decomposition of the problem (the rational perspective) with the influence of context, concerns, interests, and professional practice (the relational perspective). A process of social constructivism should take place, whereby stakeholders construct a framework, discuss, and evolve towards temporary agreement about potential solutions. Such potential solutions should consider the adaptation of existing resources and end up into a simplification of the problem space by the participating partners, resulting in a working consensus for the design process and product. The net result of the consensus space activities are initial functional specifications for the product. *Figure 2* gives a visual representation of the consensus space.

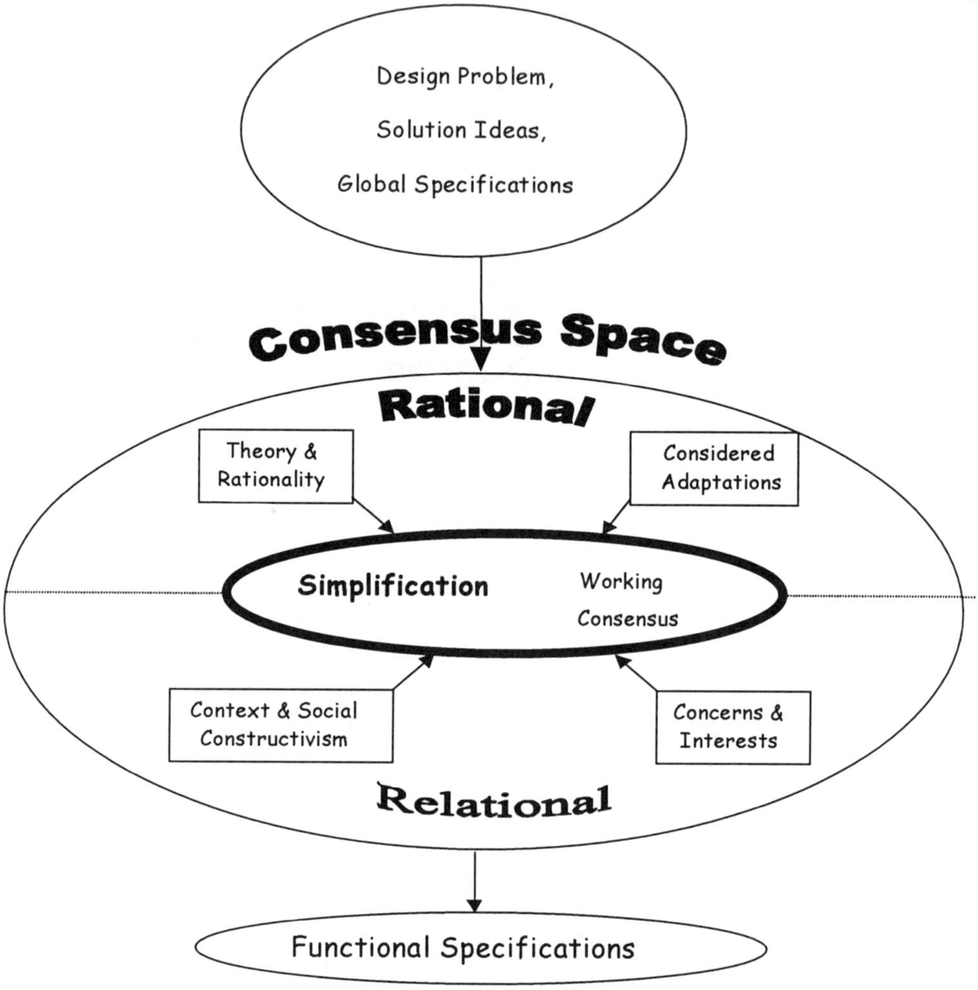

Figure 2. The consensus space of the 3-space design strategy

Task space

In the task space the main objective is to iteratively formulate technical specifications and to construct, using prototyping, a series of half products. Inputs for the task space are the initial functional specifications that resulted from the consensus space. The main activity within this task space is a balancing activity combining decomposition of tasks based upon the functional specifications, and through prototyping activity, reflection-in-action confronting the partial results with the actual opinions of representatives of projected end users. This process leads to an adaptation of the functional specifications. Rapid prototyping and formative evaluation form iterative design activities and lead to various versions of a half product. The final half product, should be "internally" acceptable, meaning that it satisfies the functional and technical specifications evolved during the task space. The final half product will have the format of an adaptable product. *Figure 3* gives a visual representation of the task space.

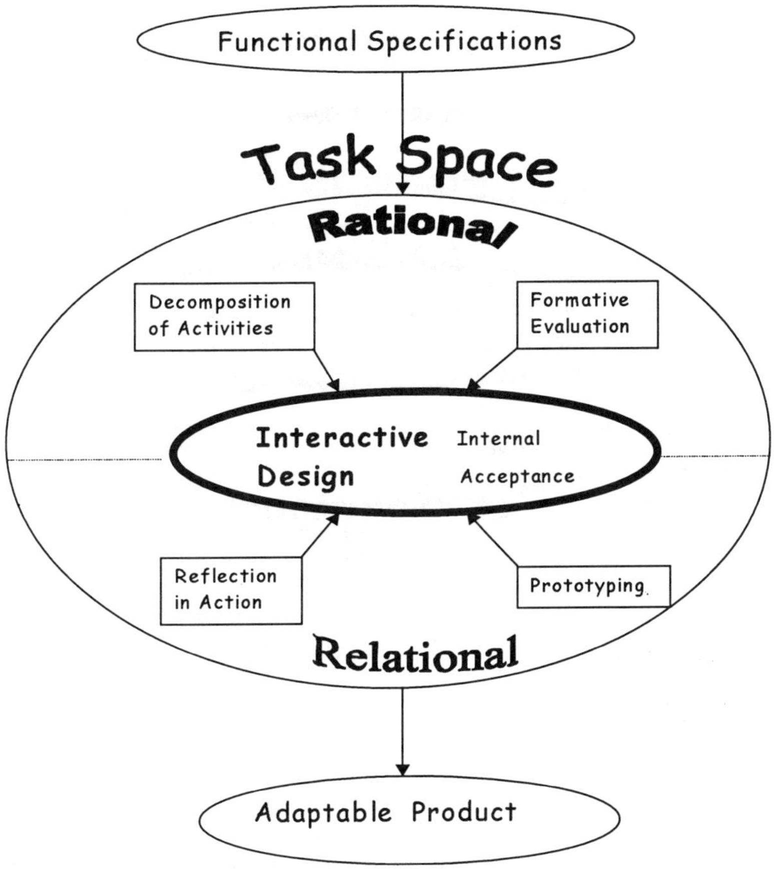

Figure 3. The task space of the 3-space design strategy

Implementation space

In the implementation space the main objective is to confront the actual end users with the final half product, from now on called the adaptable product, and create an opportunity for the end users to adapt the product to their own individualized specifications. Using tools made available, either with available technologies (for instance WWW editors in the case of a WWW resource) or already integrated into the adaptable product, the end users transform that adaptable product into a product that fits their specific needs. At the end of this process the product is "externally" acceptable to the extent that the end user feels satisfied with the product. *Figure 4* gives a visual representation of the implementation space.

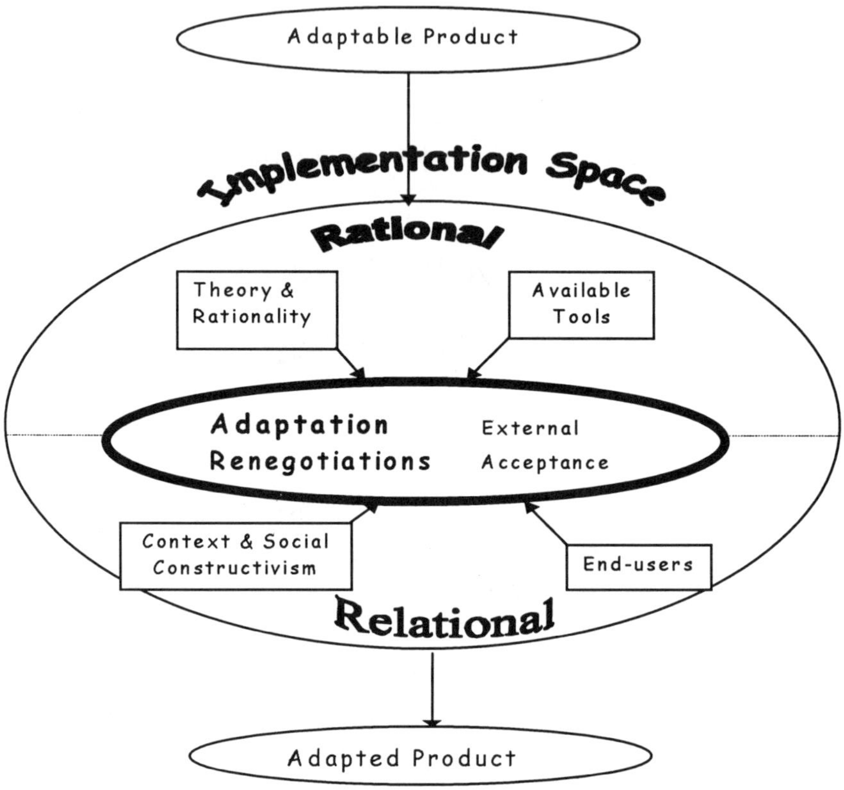

Figure 4. The implementation space of the 3-space design strategy

4. THE 3-SPACE DESIGN STRATEGY IN PRACTICE

Many products have already made provisions for user-adaptation activities in the implementation space. For example, many of the products produced by *Microsoft* have features that allow the end users to adapt the product to their own personal preferences, such as, for instance, in a typical word processing package. In such a package fonts can be changed, the lay-out of the screen can be changed, colors can be changed, the depth of applicability of many of the available tools (such as for instance a spelling checker) can be adapted to individual needs, and so on. In addition, the user also has the opportunity to choose items out of many menus in order to work in an already pre-specified format (a letter format for instance, or a report format, etc.), and to work in an environment that fits his or her particular needs Another example is the possibility for an instructor to download preselected video segments through the WWW and thus compose an individualized WWW-based learning resource by integrating these segments into a WWW-based course-support environment. Also becoming familiar in practice is the process of instructors or students selecting modules, articles, or chapters of a book from a publisher's database or from a variety of WWW sites and linking them together via a WWW page so that in a way they composes a book that perfectly fits their needs. This is "publishing on demand" and is already commonplace in the form of WWW-based bookmark lists (Collis, 1998). The *MyYahoo* tools used by millions of persons worldwide, give the users the possibility to tailor their own WWW sites, both in terms of interface and layout choices and also in terms of dynamic content (http://my.yahoo.com). Given the general tendency in society whereby consumers are becoming producers of tailored products, and supported by the evolution of information and communications technology, this third space in the design cycle will gain in importance in the future.

As a more-extensive example, a case based on a recent PhD study (Verwijs, 1998) can be cited. The study was about realizing a specific example of digital learning material: a media-selection support system, to be used by instructors employed in the Fire-Fighting Training Center of the Royal Dutch Airforce.

The general design approach in this study was basically a rational approach, but with considerable emphasis on different iterations of prototyping and end-user formative evaluation. In this context several prototypes were developed: (a) a paper-based prototype, (b) a so-called shallow software prototype whereby the chosen functionalities of the product were available in computer-screen format but those screens were not yet automatically interconnected and/or information about the topic being discussed were still missing or incomplete, and (c) finally an extended prototype that was functioning as was originally planned and specified.

The major conclusion of the evaluation of the final product was that some parts of the support system that was developed (features which automated activities that the instructors were already doing and thereby reduced their task load) were highly appreciated by the instructors, but that the main part of the support system, in which the selection of media was handled, did not really appeal to them. The final conclusion of the project was that it was unlikely that the instructors were going to use the media-selection support system in their daily practice.

Upon this experience, the question becomes: What went wrong? The chosen design strategy, including prototyping, was followed as it should have been. However the chosen approach could not turn the product in something that the instructors perceived as useful. Analyzing the situation in terms of the constructs of the 3-space design strategy, there were three major problems that were not solved in an appropriate way.

- The first, and maybe the most severe problem, occurred in the beginning of the design process, in the consensus space. The study was originally planned together with the management of the training centers of the Royal Airforce. From those discussions the assumption arose that, given the fast evolution of media and ICT, instructors needed a support tool to keep informed about new media. In that negotiation phase of the project however, there was no contact with the eventual target group. Consequently, the problem as perceived by the management was not acknowledged by the eventual users group.

- The second problem occurred when the selection of the Fire-Fighters Training Center was made. Although there was a general willingness by the target group to participate in the experiments, it was extremely difficult to fit the formative evaluation of the prototypes into the time schedule of the instructors. Eventually there were some contact activities, such as a walkthrough by the instructors through the shallow prototype from which useful comments were gathered and improvements were made towards the extended prototype, but generally speaking, extensive contacts with the target group had to be reduced to a minimum level.
- The third problem was created by the misconception that a general media-selection support system *could be* constructed. From the experience of this study, it became very clear that for at least substantial parts of a media-selection support system, these parts should be very closely related to the specific circumstances of the eventual user. This meant that those parts had to be customized to the specific context and needs of that particular user. In addition, the fast evolution of media/ICT created an extra problem in terms of actuality of the information put into the media-selection support system. This could eventually be overcome by including a connection to specific addresses on the WWW, to be downloaded each time that this was necessary. The product however, remained conceived as a stand-alone system, and the issue of obtaining up-to-date and adequate information was never dealt with. It was concluded that the projected final media-selection support system could not be more than an adaptable product, to be finalized by each individual end user, given his/her context and circumstances. How to do this, without using the WWW, was not worked out.

Based upon this experience and in relation to the 3-space design strategy described in section 5, the project should have been executed in the following way:

- In the consensus space of the project, real representatives of the target group should have participated in the discussion. The result of the consensus space activities should be functional specifications that should have taken into account practical contextual variables brought forward by the target group.
- In the task space of the project, the active participation of the representatives of the target group should have been secured so that the further negotiation of the functional specifications from the consensus space could have taken place
- In the implementation space, the end users should have the opportunity to adapt the product, to fit their particular circumstances. During the task space, tools and strategies (such as using the WWW) should have been developed in order to make end-user adaptations possible.

5. IMPLICATIONS

The 3-space design strategy has a number of implications, particularly with regard to the distinction between internal and external acceptability and the prototyping process. These are reflected upon in this final section of the chapter.

5.1 Internal versus external acceptability

In previous design methodologies based upon a rational approach, it was implicitly assumed that when a product was internally acceptable, it would also be externally acceptable. In an ideal situation this should indeed be the case. However, in his dissertation research about adaptable courseware, De Vries (1997) concluded that having a functional and usable product (which means that the product is made according to the requirements and functional specifications and therefore being internally accepted), does not guarantee or imply that such product will also be used by the target group, meaning being externally accepted. This situation (of high internal acceptance and low external acceptance) can happen in practice because of at least two reasons that can be explained in terms of the constructs of the 3-space design strategy. In the consensus space, when the functional specifications are formulated, these specifications are based upon a consensus reached by parties who participate in the negotiation. Often and given realistic circumstances, end users will not or not fully be represented in those discussions, nor can complete contextual circumstances sufficiently be taken into account. It is therefore doubtful, but at the same time unavoidable, that the consensus reached will satisfy all potential future end users. A second reason is that there is often a time gap between the formulation of the consensus and initial functional specifications and the availability of the product. In that time period many things may have changed. Two of the most-typical changes are: (a) the ideas that started the product design have changed, and (b) the technology that is incorporated in the product has evolved. Given these, it is reasonable to expect that, certainly in the area of digital learning materials, an internally acceptable product if only based on initial consensus will not necessarily be acceptable to end users (that is externally acceptable). Consequently it is difficult, if not impossible, to come up with "final" functional specifications as a guiding formula for a product production.

How to deal with this dilemma? A solution is to add an additional 'personalized' design activity in the implementation space whereby a sort of "final" specification, fitting the specific circumstances and context of that particular end user, can be (implicitly) formulated and applied. Because of the context and the individual aspects involved, that activity should preferably be done by the end users themselves. Tools for this final adaptation should be as easy to use as bookmark tools in a WWW browser in order to optimize the chance of user acceptance. The *TeleTOP* WWW-based course-support system is such a tool[3].

5.2. Prototyping and specifications

As has been emphasized, it is often a problem to formulate in the consensus space detailed specifications that will give final directions to programmers to produce a software product. One approach to deal with such problems is to produce software in a non-sequential way using iterative prototyping (Crinnion, 1991) or rapid development (McConnell, 1996) as methods. The same applies to digital learning material.

[3] See Collis and de Boer, chapter 20 of this book

With the advent of more powerful (prototyping) tools, these trends are even becoming stronger. Jonassen (1991) has noted that prototyping causes a paradigm shift in design activities. Because of the availability of prototyping methods and accompanying tools, less emphasis is given to the explicit working out of specifications and, as a consequence, of instructional goals, task and content analysis, and instructional strategies. Functional specifications are often not worked out into finalized technical specifications. During and through the prototyping activities in the task space, formative-evaluation activities evolve into new rounds of negotiation with stakeholders, now represented by more "typical" users of the product. This again supports the tendency for less-"final" specifications at the end of the consensus space. Because of the availability of prototyping, the rational approach often shifts towards a relational approach. However, sometimes this shift is not primarily caused by the gradual resolution of real issues concerning uncertainty about the situation, as described in section 3, but is mainly caused by the simple fact that prototyping seems to make it easier not to think ahead. In that respect the availability of prototyping has its potential positive and negative effects.

What are the consequences for the production of digital learning material? In section 1 the following characteristics of digital learning material were mentioned: content, structure, user-interface, and granularity of learning objects. How should prototyping activities relate to these characteristics, given the interests of the stakeholders? Which of these characteristics of digital learning materials should be focused upon by prototyping activities in the task space? Which characteristics should be left open for adaptation in the implementation space, given the general atmosphere for more personalized and individualized products supported by a socially fueled feeling that consumers should also become producers?

While producing a digital learning product, the *content* is probably stable, but also can be augmented continuously by new resources available via the WWW. The technology used however is shifting almost permanently and new technologies become available regularly, for instance the Internet, virtual reality, and intelligent agents. The final user, instructor or learner, should have the tools to augment and extend content, via at least the additional of links to additional material or notes that add comment to the material.

Depending upon the technology used for the digital learning material, there are de facto a limited number of graphical *user-interface* (GUI) standards (Jansen, 1998), in particular for materials that are displayed through a computer screen. Given the strong movement towards standardization of the graphical user interface of computer-related products, the basic aspects of the user interface of a product are not generally the object of prototyping activities. There is much (although not complete) evidence (Mayhew, 1992) in the ergonomics and human-computer interaction fields about what works and what is less acceptable in user-interface design. The producers of digital learning material should take into account general guidelines about user-interface design derived from that evidence. However, it is very well conceivable that the specific "look and feel" of a product, in terms of how the graphics, frames, and color patterns of a user-interface fit together, is object of prototyping activities and formative evaluation. The crucial question is therefore what should be considered as "basic" aspects of a user-interface and what specific features makes a screen particularly conducive to learning?

To what extent should the end user be able to tailor the user interface to his or her own preferences?

A typical activity of a teacher is to *structure* a learning environment, using a specific didactical approach and learning material to support that approach. The chosen structure is also supportive for the students' learning activities. In the case of digital learning material, the structure of the material not only consists of the kinds of elements (chapters, subchapters, or more general 'resources') that are available, but also includes the navigation possibilities between hyperlinked resource modules. The structure of digital learning material is probably a dominant factor for individual teachers and learners when they select such material.

In order to optimize the chances for digital learning material to be chosen by potential teachers and learners, a carefully executed prototyping activity particularly focused on navigation has to be executed in the task space. Given the huge number of potential individual preferences with respect to didactical approaches and learning styles, it will not be possible to respond to every potential comment given during a prototyping trial. Therefore, it is essential that during the prototyping activities a balance should be found between the original objectives and specifications of the learning material with respect to structure, navigation, and hyperlinking, and the individual comments and suggestions of the test group. Also, in the implementation space, the end user should have the tools to reorganize the structure of material, through the addition of links or new materials.

The educational world is rapidly moving towards a substantial increase of flexible learning situations at the program or institutional level which in return means more attention to the *granularity* of learning materials. Flexible learning situations often imply flexibility in the learning materials supporting a course or curriculum. This issue is also important in the context of the exchange of courses between different organizations in order to fit their particular curricula and institutional frameworks. Potential adaptation of digital learning material in this respect is thereby essential. A crucial aspect thereby is the granularity of the learning object. What is the size of a "unit of learning material" (ULM)? Is it comparable to a chapter of a book, or is a videotape of 30 minutes about a subject a good example, or is it a whole course that should be the unit to be exchanged without alterations? Or is the ULM much smaller, for instance and in terms of a textual representation, is it just one paragraph of 5 to 30 lines describing one particular aspect of an issue? Is an ULM identical to a WWW document without its external hyperlinks? There are many possibilities.

The granularity of (components of) digital learning material will determine how easy or difficult it is to make selections of the content of that material or restructure its architecture, navigation, and hyperlinking. Finding out what is a feasible size for an ULM and how manipulation of ULMs by end users can be supported by appropriate tools should be a major prototyping activity in the task space.

In summary: Prototyping in the task space of the 3-space design strategy has to concentrate on how to structure the material, how to divide the material in appropriate learning objects (ULMs) and how to navigate through those learning objects in a hyperlinked pattern. In addition, the prototyping in the task space should also concentrate on how to create a visual appearance of the material that fits with the intentions and style of the producers.

6. CONCLUSION

This chapter started by introducing a number of questions: (a) Are existing design and production methods for digital learning materials appropriate for practical application when dealing with new forms of digital learning material? (b) Do we need a new design and production strategy to cope with these new forms and does the 3-space design strategy provide a solution? and (c) What is the specific role of prototyping in the 3-space design strategy and how does prototyping apply to the design and production of new forms of digital learning material? The analysis in this chapter provides some answers to these questions.

Given the developments in technologies and societal urges for more-modular educational structures, it is proposed that to realize digital learning material under these new circumstances one should follow a route that can be described in terms of a 3-space design strategy consisting of a consensus space, a task space, and an implementation space. In such a design strategy negotiating, prototyping, and final adjustment of an adaptable product by the end user are main characteristics of the approach. Therefore a design approach integrating rational and relational perspectives, made concrete through a series of prototyping and revision iterations, is necessary.

Based upon such a design strategy, prototyping activities in the task space should be concentrated on optimizing the structure and navigation facilities of the digital learning material, and the granularity of the modules concerned, and within a framework of basic guidelines, upon the user-interface.

REFERENCES

Beyer, H. and Holtzblatt, K., 1998, *Contextual design,* Morgan Kaufmann Publishing, San Francisco.

Collis, B., 1998, *Teleware: Instrumentation of Telelearning,* Inaugural address, University of Twente, Enschede.

Conger, S.A., 1994, *The new software engineering,* Wadsworth Publishing Company, Belmont, CA.

Crinnion, J., 1991, *Evolutionary systems development,* Pitman Publishing, London.

Dills, C. and Romiszowski, A. (Eds.), 1997, *Instructional Developments Paradigms,* Educational Technology Publications, Englewood Cliffs, NJ.

Dourish, P., 1995, Developing a reflective model of collaborative systems, *ACM Transactions on Computer-Human Interaction, 2*(1), p. 40-63.

Humphrey, W.S., 1989, *Managing the software process,* Addison-Wesley, Reading, MA.

Jansen, B., 1998, The graphical user interface: An introduction, *SigCHI Bulletin, 30*(2), 22-26.

Jonassen, D., 1991, Objectivism versus constructivism: Do we need a new philosophical paradigm? *Educational Technology Research and Development, 39*(3), 5-14.

Kessels, J., 1993, *Towards design standards for curriculum consistency in corporate education,* Doctoral dissertation, University of Twente, Enschede.

Mayhew, D., 1992, *Principles and guidelines in software user interface design,* Prentice-Hall, London.

McConnell, S., 1996, *Rapid development,* Microsoft Press, Redmond, WA.

Moran, T.P. (Ed.), 1994, Context in design, *Human-Computer Interaction, 9*(1).

Moonen, J., 1996, Prototyping as a design method, In Plomp, Tj. and Ely, D. (Eds.), 1996, *International encyclopedia of educational technology,* Second edition, (pp. 186-190), Pergamon, Oxford.

Moonen, J., 1998, *A design methodology for Social Sciences: Version 0.2,* Paper presented at the Social Science Design Conference, University of Twente, 27-28 March, 1998.

Powell, T.A., 1998, *Web site engineering,* Prentice Hall, Upple Saddle River, NJ.

Schach, S.R., 1993, *Software engineering,* Irwin, Boston.

Schön, D.A., 1983, *The reflective practioner,* Basic Books, New York.

Simon, H.A., 1969, *The science of the artificial,* The MIT Press, Cambridge, MA.

Spiro, R.J., Feltovich, P.J., Jacobson, M.J. and Courson, R.L., 1992, Cognitive flexibility, constructivism, and hypertext: Random access instruction for advanced knowledge acquisition in ill-structured domains, In Duffy, T.M. and Jonassen, D.H. (Eds.), *Constructivism and the technology of instruction: A conversation* (pp. 23-48), Lawrence Erlbaum Associates, Hillsdale, NJ.

Spohrer, J., 1998, ATG educational research, The authoring tools thread, *SIGCHI Bulletin, 30*(2), 124-133.

Verwijs, C., 1998, *A mix of core and complementary media: New perspectives in media-decision making,* Doctoral dissertation, University of Twente, Enschede.

Vries, S. de, 1997, *Aanpasbare courseware: ontwerp en gebruik binnen het Voortgezet Onderwijs* (Adaptable courseware: design and use within secondary education), Doctoral dissertation,University of Twente, Enschede.

Chapter 9

Supporting Instructional Design with an Information System

Renate Limbach
Ton de Jong
Jules Pieters
University of Twente

Gordon Rowland
Ithaca College

Keywords: Instructional design, Information systems, Discovery learning, Constructivistic learning environments.

Abstract

In this chapter the emphasis is on the support of instructional design for constructivistic learning environments. Until now there is little known about the way designers should be supported in the design of constructivistic learning environments. Usually designers apply an instructional design model that was originally meant for expository environments. However, we argue that for several reasons these instructional design models do not include characteristics of constructivistic learning environments. For instance learning goals are very prominent in these design models whereas in constructivistic learning environments activities are the main determinants of achieving knowledge development. In addition, designers usually do use the techniques prescribed by these models but they hardly follow the predetermined order of design activities. An alternative design approach is proposed in which the emphasis is not on the order of design activities, but on the support designers can be delivered during a design process that is personally assembled. For delivering this support, we developed an information system.

1. INTRODUCTION

Instructional dcsign models for expository teaching claim that designing instruction is a rational and systematic process, that leads from goals through means to ends (e.g. Dick & Carey, 1990; Merrill, 1991; Rothwell & Kazanas, 1992). This process encompasses analyzing, defining goals and objectives, designing test items, strategy and materials, and developing evaluation. Development of instructional theory and also everyday design practice stress the enhancement of traditional instructional design models (Richey, 1995). Still in 1991, Gustafson could conceive little substantive change in the conceptual framework of instructional design models (Tessmer & Wedman, 1995). But a couple of years later, researchers reported considerable differences between normative theoretical approaches of designing instructional systems and the descriptive nature of the designing instruction in practice (De Hoog, De Jong & De Vries, 1994; Pieters & Bergman, 1995; Rowland, 1992; Tessmer & Wedman, 1995). Tessmer and Wedman (1995) report that the gap between theory and practice indicates a need for descriptive rather than prescriptive models. Also, they argue on the basis of empirical evidence that "designers shape the ID process to accommodate the contextual factors surrounding a given project by minimizing, deleting, or delaying ID tasks." And, "The design process is actually more holistic than linear; several tasks may be carried out at once. (...) It may also be helpful to think of instructional design as concurrent with development (rather than preceding development)" (Tessmer & Wedman, 1995, p. 40). In this chapter we concentrate on learning in constructivistic learning environments, operationalized by us through environments for discovery learning. Discovery learning is a highly self-directed and constructive way of learning (De Jong & Van Joolingen, 1998) with the purpose to acquire deep, flexible, and transferable knowledge (De Jong et al., 1998). In discovery learning, learners are invited to construct their knowledge by extracting the relevant information from the discovery learning environment by stating hypotheses, doing experiments and drawing conclusions, in short by acting as a scientist. Though this may sound like a contradiction, also this new way of learning calls for instructional activities. The designer is expected to develop sufficient structure in the learning environment to support the learner in the process of discovery. In a metaphorical sense the designer could be compared with an architect of a recreation-ground.

Until now, there is little known about the way designers should be supported in designing discovery type of constructivistic learning environments. As noted before, designers could apply an instructional design model. For at least two reasons, however, we assume that instructional design models for expository teaching are not highly appropriate to be applied in this new and seemingly paradoxical task, designing instruction for non-instructive environments. The first reason has to do with the purpose of these models: they are more suited for the expository mode of teaching. The second reason is more general, they do not fit into the design process as it is carried out in practice. In the following section we will discuss each reason.

1.1 Instructional design models for expository teaching

By paraphrasing authors referred to above, one may argue that existing models of instructional design are tailored to an expository mode of teaching. Main elements in these models refer to defining goals and designing instructional solutions that are intended to move learners into the direction of the goals by dissipating degrees of freedom (see Jonassen, 1991, for a discussion of objectivistic and constructivistic approaches in education). Whereas in the expository mode of teaching the decisions on the context and structure of the instruction presented are almost fully determined by someone (or something) else than the learner, in discovery learning learners are invited to take the initiative. For the design of discovery learning environments, an intriguing design issue concerns the balance between allowing freedom and giving direction to the learners.

Another issue is related to the important role of the goals and subgoals which learners are required to achieve in models meant for expository teaching (see also Jonassen, 1991). In discovery learning the (sub)goals are not explicitly defined. More emphasis is put on the development of tasks carried out by the learner. Learners are expected to be increasingly responsible for their own knowledge construction process, and the designer does not need to specify in detail the learning path from start-state to end-state. In most existing instructional design models, on the contrary, determining detailed (sub)goals is pivotal and further design activities are consecutive. This leads to the idea of automating the process of instructional design (e.g. Merrill, Li & Jones, 1990).

1.2 The alleged mismatch between the instructional design model and the design process in practice

A second, more general reason why we assume that these models are not suited for designing discovery learning environments is that these models are not based on the actual process of designing (e.g. Gros, Elen, Kerres, van Merriënboer & Spector, 1997; Holcomb, Wedman & Tessmer, 1996). Perez and Emery (1995) put it as follows: "... they are models describing what designers should think and do rather than models of what designers actually think and do" (p. 81). To put it positively, recent holistic, synthetic approaches "... place more emphasis on understanding and accommodating the entire design context. These models will embody situational adaptation, product visioning, contextual analysis, and concurrent design" (Tessmer & Wedman, 1995, p. 46).

The main reason for the mismatch can be the underestimation of the complexity of the design process (e.g. Tripp & Bichelmeyer, 1990). Existent models assume the process of designing to be logical and determined. Not only this is very doubtful, practice shows that the design of instruction is a complex ill-structured problem-solving task (e.g. Rowland, 1992). Almost all models have a linear (waterfall) structure (even when intermediate regressions are allowed, these models basically remain linear), but research shows that the design process may not be linear at all (e.g. De Hoog et al., 1994; Jonassen et al., 1997; Richey, 1995). Tripp and Bichelmeyer (1990) state that "... any design methodology which acknowledges the complexity of the situation may be more efficient because it anticipates and short-circuits the kinds of problems designers typically encounter" (p. 34).

Another reason for the mismatch between the design model and the design process in practice has to do with a focus on the learner rather than on the users of a model. While in the instructional design models the target learner has a central place (for instance by taking the level of prior knowledge into consideration), the user (or a group of users) of an instructional design model is seldom taken into account. However, studies (e.g. Gayeski, 1990; Greeno, Korpi, Jackson III, & Michalchik, 1990; Perez & Emery, 1995; Pieters & Bergman, 1995; Rowland, 1992) show that there are considerable individual differences between novice and experts designers. Though still a lot is unknown about the design process of instructional design, recently the importance of gaining knowledge about the design process has clearly emerged. Rowland (1992), for example, drew the following conclusion: "It has become increasingly apparent that in our literature we have abundant information on what authors / designers say they do, or say others should do, but little idea of what expert designers actually do themselves" (Rowland, 1992, p. 65). Studies by Greeno et al. (1990), and Goel and Pirolli (1992) intended especially to contribute to cognitive psychological research of (instructional) design processes. The motive for the study by Perez and Emery (1995) was the lack of knowledge about cognitive processes which are required for instructional design. The objectives of the study by Rowland (1992) and by Pieters and Bergman (1995) concerned the practical implications. A disadvantage of these studies is that they provide a description of the design process of expository environments and that the knowledge acquired by the studies has not been used as a concrete input of providing support for designing instruction.

2. SUPPORTING INSTRUCTIONAL DESIGN WITH AN INFORMATION SYSTEM

As stated above designing instruction is a complex, ill-structured, problem solving task. Perez and Emery (1995) add two conclusions to this. First of all they state, because of "it is not known what combinations of knowledge, skills, talents, and / or cognitive processes are required" (...) "the nature of instructional design as a domain is not well understood" (Perez & Emery, 1995, p. 80). Secondly, they conclude that the "use of these models has been problematic primarily because they require that the designer derive general principles of instruction from general principles of learning and apply them to specific tasks or domains. This is no easy task" (Perez & Emery, 1995, p. 81). In the situation of designing discovery learning environments the instructional design task is even more difficult: the changes needed in designing discovery environments as compared to expository teaching are not simple. They require the most difficult change of all - changing the designer's mind.

To conclude, there seems to be a need for a flexible support measure (adapted to the individual design process and user characteristics), which assists the designer during the ill-structured task (and therefore in addition to the more logical aspects of the task it should also support more creative aspects of the task (e.g. Richey, 1995)), and which consists of specific support to acquire a new skill for developing discovery learning environments. The emphasis should not be on the order of design activities, but on the quality of the activities of a situated, holistic and personally assembled design process. Especially for flexible interactive environments, and according to the current technological possibilities, we should reconsider the traditional way of presenting instructional design models, consisting of fixed design phases. A computer tool, as a way to support designers, seems a valuable option for more flexible designing. When we look at the influences of computers as an intermediate support tool on the problem solving behavior of the user in general, there are generally two important aspects, namely amplification and transformation. "In the case of the former, the job is finished faster and more accurately, while in the second, the changes due to amplification are expected to lead to a transformation of the structure of the activity required to perform the task" (Karasavvidis, 1999, p. 53). For designing instruction, the emphasis for providing support with a tool is on amplification; to facilitate the design which makes the development of the design process more efficient. But for a designer the pedagogical changes needed in designing discovery environments as compared to expository teaching are not simple and a change of the structure of the activity is a necessity.

Since designers need specific instructional information for this pedagogical change, the main purpose of the support should be to provide the designer with knowledge, implications, and possibilities about the instructional aspects of the design of a discovery learning environment. The role of knowledge is important in this respect. A design could be seen as the construction of instantiated knowledge and the design process as a process of knowledge acquisition (Tripp & Bichelmeyer, 1990). Experts apply knowledge intensive techniques (e.g. Anderson, 1987; Royer, Cisero & Carlo, 1993) in carrying out cognition tasks, e.g., design tasks. In this way an other type of tool also seems a promising way to support designers with knowledge for the purpose of the instructional design discovery learning environments, by information systems. Information systems are interrelated components that collect, process, and disseminate information to support problem solving, decision making, and analysis (adapted to Laudon & Laudon, 1996).

As far as we know, the use of this type of information systems has seldom been applied in the context of on-line support for designers of instructional design. Support mostly exists in the form of on-line help to provide support on the operational aspects of the design in stead of the pedagogical aspects. De Vries and De Jong (1997) stated that "information systems are not longer only used for fact finding, but are considered a promising tool for knowledge transfer in educational contexts as well as in specific contexts" (De Vries & De Jong, 1997, p. 3). As stated before, information systems seem to be a good candidate for supporting the designer: they provide the user with knowledge about the instructional aspects of the design of a discovery learning environment, and they are constructive; they actively engage designers in designing constructive learning environments, and, subsequently, creating knowledge that reflects their comprehension and conception of the design process, and certain aspects of it. The use of an information system puts through the constructivistic thoughts about support (provide freedom, but when necessary, guidance). When providing support by an information system, designers are expected to take more initiative and therefore the information system should provide them with the knowledge by which they are able to construct and adapt their own knowledge base. In addition, the psychological ideas behind information systems are based on current psychological ideas of human learning processes (e.g., schema theory and associative networks), in contradiction with the psychological ideas behind expository environments. Merrill et al. (1990) state most models "are based on the psychology of the 50's and 60's; they are analytical, not synthetic; they are component rather than model or schema oriented; and their application requires considerable effort" (Merrill et al., 1990, p. 8).

In the following section a more specific view of the above mentioned ideas is described, that is, an information system for supporting the instructional design of discovery learning environments, which is part of an authoring system for simulation based learning environments.

2.1 An example of an information system for the support of the design of a discovery learning environment

In the next sections we outline a few of our considerations for creating an information system. During the design in an authoring system, designers should have continuous access to the information system, both directly and in a context sensitive way, from the activity they are performing. In this example, the authoring system (called SIMQUEST) is an environment for designing and creating simulation-based learning environments (De Jong et al., 1998). It enables designers to create computer-based simulation learning environments, including simulations and instructional support for learners. In SIMQUEST, designers have access to instruments for balancing freedom and support for the learner. It is not necessary to predefine the goals. In addition, SIMQUEST is a flexible authoring environment which allows every possible sequence of design[1].

The content of the information system available in the authoring environment SIMQUEST

The information is particularly meant to provide the designer with information needed for an effective design of a discovery learning environment. The next question therefore was how the content of such an information system should look to support the designer in this educational discovery challenge. To determine the topics needed in the information system we performed two studies. The first study investigated the questions designers may have during designing of a discovery learning environment and analyzed the terminology they used (De Jong, et al., 1996). *Figure 1* shows an overview of some results of the questions designers put forward after they had inspected screendumps of the authoring environment SIMQUEST (see for more information about this study De Jong et al., 1996). The purpose of the second study was to gain insight into the design process of discovery learning environments itself (see for more information about this study Limbach, De Jong & Pieters, 1998b).

Based on the results of these studies, topics were selected which played an important role during the instructional design and therefore should be discussed in the information system. First of all the studies showed that novice designers have little insight into the learning processes involved in discovery learning and the ways learner support could trigger these processes. As stated above one of the main challenges for a designer is to find a (new) balance between sufficient freedom for learners to regulate their own learning process and guidance of the learning process. The designers in the study felt a need for control and as a result produced environments that were very structured and followed fixed sequences of predetermined steps. In addition, designers need help with the pedagogical change towards more freedom for the learner. The information system therefore, contains concrete ideas of what kind of control could be taken into consideration (for instance the issue of control of the sequence, the availability of instructional support, or aspects like the time learners could/should use a simulation). In discovery learning the (sub)goals are less clear beforehand. The acquisition of a particular discovery learning process (for instance generating a

[1] Descriptions of this authoring system can be found in De Jong et al. (1998) and in Chapter 18 of this book.

Questions about the environment			
What are the restrictions of the environment? * for the designer * for the learner	What are the restrictions for designing simulations?	Is it easy to adapt/change the material?	Are there applications available?
What is the advantage of this environment above other programs?	Are there possibilities to change the simulation e.g., to show different sides of a figure?	Are there templates available e.g., graphic arts?	Is there a different type of help for the designer and for the learner?
Can I give a visual explanation?	Can I adapt the assignments to the answers of other assignments?	Can I make individual learning paths?	Does the environment provide help to the learner?
Questions about instruction			
General	What are the instructional goals of the environment?	Which subject is appropriate?	What is the target-population?
	What are the experiences of other designers?	Is it possible for the learner to work at home?	What is the place of the environment in the curriculum?
	Are there different working forms possible?	Is it appropriate to let learners make their own environment as a research?	Do the learners need any books or other hardware?
	What is the role of the teacher when the learners are working with it?		
Discovery learning	Do learners need guidance or can you let them discover? How can I support the learners' discovery process?	Is there a balance between the learners challenge to discover and the distraction of thoughts by e.g. assignments?	Do learners need an introduction of discovery learning?
Skills	Which skills will be learned?	How to learn specific skills in the environment?	
Learning result	Are the learning results better than learning with a book?	Are the learning results faster than learning with a book?	
Time	How much time do I need before I understand the environment?	How much time do I have to spend to design an environment?	

Figure 1. Questions designers put forward after they had inspected screendumps of the authoring environment.

hypothesis or interpretation of a graph) could also be a goal on its own. The same issue plays a role with measuring the results. The designers in the studies handled goals and the way the results should be measured comparable to the way they normally do in an expository learning environment. However, for discovery learning a more appropriate assessment method is, for instance, a test which asks learners to make qualitative predictions (Swaak & De Jong, 1996). The way the above mentioned topics were filled in by the designer depended on several other aspects namely, the domain-, learner-, context- characteristics, the way the content is presented and the way this all would be implemented in the curriculum. The design process depends on the results of the requirements set and the constraints experienced by the designers. All the above mentioned topics were therefore discussed in the information system.

The structure of the information system

The structure in an information system could follow a hierarchy (especially meant for search activities) or hyperlinks (especially meant for browse activities). De Vries and De Jong (1997) studied the structure of the information (hierarchy, network or mixed) for both ways of use (search and browse). We concluded, based on their results, that for our purpose (both ways of use, browse and search, should be possible) a mixed structure of hierarchy and network seemed the best structure. The use of a mixed structure also prevents Binder's (1989) fear that on the one hand designers get lost in the network of relations, but on the other hand are not too bound by a hierarchy.

The resulting information system was built up as a matrix with 'topics' on one axis, and on the other four categories of information: 'what is', 'example', 'background' and 'design consideration' (in Chapter 18 of this Volume a description of these information categories is described). *Figure 2* presents the information system available in the authoring environment SIMQUEST.

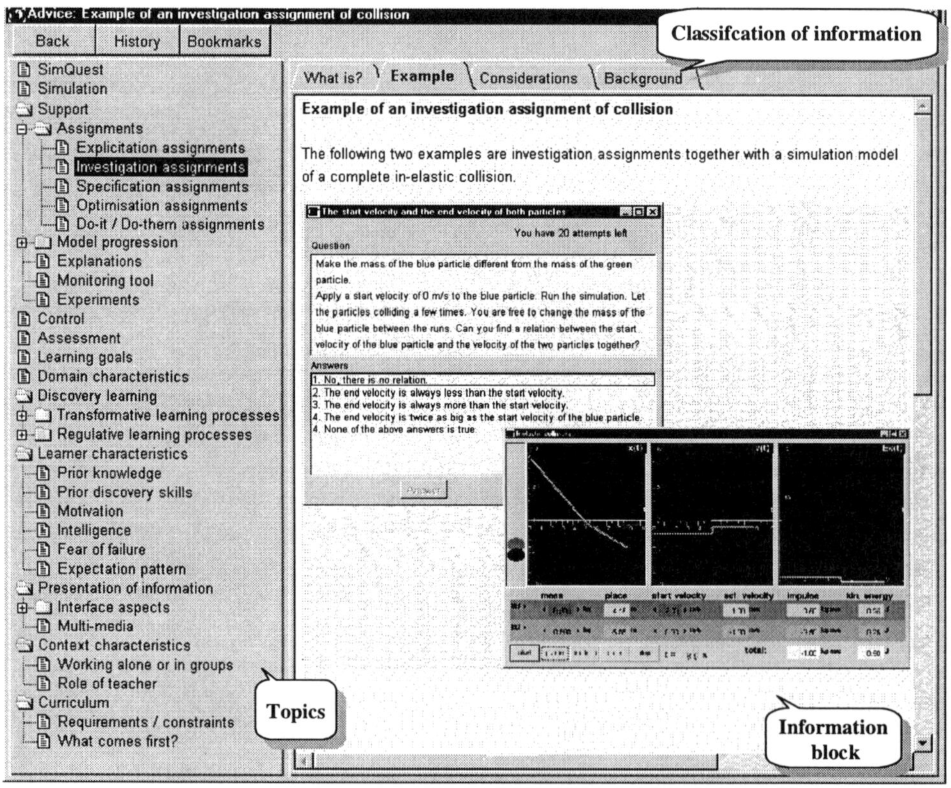

Figure 2. The information system available in the authoring environment SIMQUEST (example on collission)

So, for each (sub)topic (12 topics and thirty-nine subtopics in total) designers can ask for information in each of these categories. Especially the presence of examples (e.g. Catrambone, 1996) and analogies (e.g. Venville & Treagust, 1996) for enabling the transfer of information to their own situation seems to be important for designers, as was confirmed in the first mentioned study (De Jong et al., 1996).

2.2 An example of the flexibility of an information system

Above we suggested that an information system with special characteristics may have several benefits. In addition to context sensitive support, another major benefit of such an information system is its flexibility. Through its flexibility, an information system can easily be adapted to other, but comparable circumstances. For instance to a more extreme constructivistic point of view where the learner's knowing is seen as designing. Designing and the creation of artifacts - if seen as equivalent to learning - suggest learning activities that are open-ended, ill-structured problem-solving tasks. In these, the nature of 'discovery' learning becomes one of unpredictable outcomes as well as unique learning paths, and the structure of the environment is more difficult to prescribe. Essentially, a first question is what sort of features should an environment have in order to support learning if we consider learning to be this truly open-ended inquiry. A second question is how designers can be supported in this completely open-ended task. Here we offer preliminary responses to each:

- In terms of the learning environment itself, a first implication is the capability of significant growth and revision.
- Second, engagement with the information system would likely occur as an on-going conversation or dialogue rather than creation followed by use of a design product. This would imply a move from a guided and semi-controlled interaction in a designer-defined environment to a more open, constructive, reflective, and evolving exploration.
- Third, and related to the previous point, the roles of designer and learner might change significantly. For example, the role currently seen as that of the designer might shift to a partnership between teacher and learner, and in this sense even lead to an inseparability of environments for designing and learning.

Responding to the second question concerning supporting designers, a shift in concept of what is to be designed seems essential, from structured tutorials, predefined learning pathways, and resource definition, to environments that enable learning activity. Underlying this is a shift in perception of how learning will occur, from 'compliance' (that is, following a predetermined path) to learner control, self-determined (sub)goals, and obtaining guidance on request. Also accompanying this would be a shift in conception of function/role, from designer → teacher → learner (a simple linear chain of causal relations, that is, one taking the role of affecting behavior of another) to designer, teacher, and learner collaboration. If models are employed, these shifts imply models that are evolutionary and continual rather than fixed and one time. In this particular example a cycle of implementation → evaluation → revision → implementation would continue indefinitely, with a heavy reliance on feedback and the learner being seen as a collaborative partner in designing and in his/her own learning.

In this example with a learner as a designer, the information system provides support (this instead of an instructional design model). The information system should be adapted for this constructionistic purpose; emphasizing the process (building knowledge of the domain by designing) instead of the product (an optimal design and implementation of a discovery learning environment). In this particular case, especially, the relation between domain, simulation, and the underlying mathematical model, setting goals and self-assessment should be emphasized and less attention can be paid to topics like the context, learner characteristics, the way the information should be presented. In addition, a designer (learner in this example) still has to create an environment. In this situation the above presented topics are sufficient and should be tackled by the learner as a designer.

3. CONCLUSION

In this chapter we discussed the support of instructional design for discovery learning environments. The emphasis was on the consequences of an instructional design model for designers in the case of individual knowledge construction process of learners. An alternative design approach was proposed in which the emphasis is not on the order of design activities, as is often determined by a design model, but on the support designers can be delivered during a design process that is personally assembled. For delivering this support, we developed an information system. The information put into this system was based on knowledge acquired in practical design sessions. We support the benefit of using an information system and a possible way of presenting the information (see also Limbach, De Jong & Pieters, 1998a). Further research should decide the value of the presented benefits. In this Volume, in the second part of Chapter 18 the authoring environment SIMQUEST is presented and in addition to the information system other cognitive tools for providing support to the designer (operational and methodological support) are discussed.

REFERENCES

Anderson, J.R., 1987, Skill acquisition: compilation of weak-method problem solutions, *Psychological Review, 94,* 192-210.

Binder, C., 1989, Hypertext design issues, *Performance Improvement Quarterly, 2(3),* 16-33.

Catrambone, R., 1996, Generalizing solution procedures learned from examples, *Journal of experience psychology: learning and memory, and cognition, 22,* 1020-1031.

Dick, W. and Carey, L., 1990, *The systematic design of instruction (3rd ed.),* Harper Collins, New York.

Gayeski, D.M., 1990, Are you ready to automate design?, *Training & Development Journal, 44(1),* 61-62.

Goel, V. and Pirolli, P., 1992, The structure of design problem spaces, *Cognitive Science, 16,* 395-429.

Greeno, J.G., Korpi, M.K., Jackson III, D.N. and Michalchik, V.S., 1990, *Processes and knowledge in designing instruction,* Internal paper, Stanford University.

Gros, B., Elen, J., Kerres, M., van Merriënboer, J. and Spector, M., 1997, Instructional design and the authoring of multimedia and hypermedia systems: does a marriage make sense?, *Educational Technology, 1,* 48-56.

Holcomb, C., Wedman, J.F. and Tessmer, M., 1996, ID activities and project success: perceptions of practitioners, *Performance Improvement Quarterly, 9(1),* 49-61.

Hoog, R. de, Jong, T. de and Vries, F. de, 1994, Constraint driven software design: An escape from the waterfall model, *Performance Improvement Quarterly, 7(3),* 48-63.

Jonassen, D.H., 1991, Objectivism versus constructivism: do we need a new philosophical paradigm?, *Educational Technology Research & Development, 39(3),* 5-14.

Jonassen, D.H., Hennon, R.J., Ondrusek, A., Samouilova, M., Spaulding, K.L., Yueh, H.P., Nouri, V., DiRocco, M. and Birdwell, D., 1997, Certainty, determinism, and predictability in theories of instructional design: Lessons from Science, *Educational Technology, 37(1),* 27-34.

Jong, T. de and Joolingen, W.R. van, 1998, Scientific discovery learning with computer simulations of conceptual domains, *Review of Educational Research, 68,* 179-201.

Jong, T. de, Joolingen, W.R. van, Swaak, J., Veermans, K., Limbach, R., King, S. and Gureghian, D., 1998, Self-directed learning in simulation-based discovery environments, *Journal of Computer Assisted Learning, 14,* 235-246.

Jong, T. de, Kuyper, M., Hoog, R. de, Gellevij, M., Limbach, R., Veldhuizen, B. van, Wieberdink, B. and Horst, M. ter, 1996, Author support in SIMQUEST, In Jong, T. de (Ed.), *Designing integrated computer simulation environments for discovery learning: First project progress report* (pp. 97-139), SERVIVE project (ET 1020), University of Twente, Enschede.

Karasavvidis, I., 1999, *Learning to solve correlational problems: a study of the social and material distribution of cognition,* Doctoral dissertation, University Twente, Enschede.

Laudon, K.C. and Laudon, J.P., 1996, *Management information systems: Organization and technology (4th ed.),* Prentice Hall, New York.

Limbach, R., Jong, T. de and Pieters, J., 1998a, Supporting instructional design with an information system: the SIMQUEST advice tool, In Jong, T. de and Joolingen, W. van (Eds.), *SIMQUEST: An authoring system for integrated simulation learning environments* (pp. 75-95), SERVIVE deliverables D7.2 and D8, University of Twente, Enschede.

Limbach, R., Jong, T. de and Pieters, J., 1998b, The design of instructional simulation-based discovery learning environments: An exploratory study, In Jong, T. de (Ed.), *Evaluation of the SIMQUEST authoring system and SIMQUEST applications* (pp. 17-40), SERVIVE deliverable D6, University of Twente, Enschede.

Merrill, M.D., 1991, Constructivism and instructional design, *Educational Technology, 31(5),* 18-23.

Merrill, M.D., Li, Z. and Jones, M.K., 1990, Limitations of first generation instructional design, *Educational Technology, 30(1),* 7-11.

Perez, R.S. and Emery, C.D., 1995, Designer thinking: how novices and experts think about instructional design, *Performance Improvement Quarterly, 8(3),* 80-95.

Pieters, J.M. and Bergman, R., 1995, The empirical basis of designing instruction: What practice can contribute to theory, *Performance Improvement Quarterly, 8(3),* 118-129.

Richey, R.C., 1995, Trends in instructional design: Emerging theory-based models, *Performance Improvement Quarterly, 8(3),* 97-111.

Rothwell, W.J. and Kazanas, H.C., 1992, *Mastering the instructional design process: A systematic approach,* Jossey-Bass, San Francisco.

Rowland, G., 1992, What do instructional designers actually do? An empirical investigation of expert practice, *Performance Improvement Quarterly, 5(2),* 65-86.

Royer, J.M., Cisero, C.A. and Carlo, M.S., 1993, Techniques and procedures for assessing cognitive skills, *Review of Educational Research, 63(2),* 201-243.

Swaak, J. and Jong, T. de, 1996, Measuring intuitive knowledge in science: the development of the WHAT-IF test, *Studies in Educational Evaluation, 22,* 341-362.

Tessmer, M. and Wedman, J., 1995, Context-sensitive instructional design models: A response to design research, studies, and criticism, *Performance Improvement Quarterly, 8(3),* 38-54.

Tripp, S.T. and Bichelmeyer, B., 1990, Rapid prototyping: An alternative instructional design strategy, *Educational Technology Research and Development, 38(1),* 31-44.

Venville, G.J. and Treagust, D.F., 1996, The role of analogies in promoting conceptual change in biology, *Instructional Science, 24,* 295-320.

Vries, E. de and Jong, T. de, 1997, Using information systems while performing complex tasks an example from architectural design, *International Journal of Human-Computer Studies, 46,* 31-54.

Chapter 10
Prototyping to Reach Product Quality

Nienke Nieveen
University of Twente

Keywords: Prototyping, Quality criteria, Formative evaluation

Abstract

This chapter provides a framework for product quality consisting of the following three criteria: validity, practicality and effectiveness, and provides insight into the applicability of the framework in various domains of educational product development.

In order to reach product quality, the prototyping approach is seen and understood as a suitable approach. This chapter discusses three significant characteristics of a prototyping approach: extensive use of prototypes, high degree of iteration and the role of formative evaluation, and the paramount importance of user involvement.

The chapter illustrates the way the prototyping approach has been instrumental in developing a computer support system for instructional developers. During the prototyping process, the framework assisted in deciding the focus of each prototype and enhanced the transparency of the entire process.

1. INTRODUCTION

Few people would argue the importance of developing high quality products for the educational setting. Nevertheless, quality criteria remain often implicit, which complicates not only the discussions about the quality of a product, but also the way high quality products could be developed. By providing a framework for product quality and elaborating on an approach to develop such products, this chapter intends to provide a better basis for such discussions.

The chapter starts with a framework for product quality criteria (section 2). The framework is developed from the perspective of developing learning materials (as a sub set of the wide variety of educational products). The section also provides insight into the applicability of the framework in another educational product development domain: the domain of developing computer support systems for instructional developers.

In order to reach product quality, the prototyping approach is seen and understood as a suitable approach. Section 3 discusses this approach in general, whereas section 4 illustrates the way the prototyping approach has been instrumental in developing a computer support system for instructional developers.

The chapter closes with some concluding remarks on the connections between product quality criteria and the prototyping approach (section 5).

2. PRODUCT QUALITY CRITERIA

In general, the wide array of educational products plays many important roles in education. For instance, exemplary learning materials (as a sub set of all possible educational products) can be of great importance especially during initial stages of implementing a new curriculum (cf. van den Akker, 1988; Ball & Cohen, 1996), as such materials can:

- provide theoretical background information on the meaning of the change;
- demonstrate the practical meaning of the change at stake;
- give potential users the opportunity to experiment with the materials to get insight into the consequences of the change for their daily practice; and
- stimulate discussions about the educational changes among teachers who are using the materials.

To fulfill these functions, learning materials must be of good quality. Although few people would argue the importance of good quality products, people tend to be somewhat vague about what they really mean with the term. Sometimes they point at the requirement that the product should work well; while at other times they mean that product use should lead to better results. In order to make the concept "quality" more transparent, it was related to a typology of curriculum representations, resulting in a framework with three quality criteria: validity, practicality, and effectiveness. Section 2.1 discusses this framework and section 2.2 provides insight in its applicability to other educational products besides learning materials.

2.1 Product quality framework

The typology of curriculum representations starts from the notion that each product may be manifested in more and less concrete forms. In the field of curriculum development usually six curriculum representations are distinguished (Goodlad, Klein & Tye, 1979; adapted by van den Akker, 1988, 1990). *Table 1* provides an overview of these representations.

Curriculum representations	
Ideal	Reflects the original assumptions, visions and intentions that are laid down in a curriculum document.
Formal	Reflects the concrete curriculum documents such as student materials and teacher guides. In some studies the term '*Intended curriculum*' is used, which refers to a combination of the ideal and formal curriculum.
Perceived	Represents the curriculum as interpreted by its users (especially teachers).
Operational	Reflects the actual instructional process as it was realized (also often referred to as curriculum-in-action or the enacted curriculum).
Experiential	Reflects the curriculum as the students experience it.
Attained	Represents the learning results of the students.

Table 1. Typology with six curriculum representations

This typology has proven to be an aid in understanding the relationships and discrepancies between different representations of the curriculum in practice. In this section the typology is used to clarify the notion of product quality.

First, as far as good quality material is concerned, the material itself (the intended curriculum) must be well considered. The components of the material should be based on state-of-the- art knowledge (content validity) and all components should be consistently linked to each other (construct validity). If the product meets these requirements it is considered to be *valid.*

A second characteristic of high-quality materials is that teachers (and other experts) consider the materials to be usable and that it is easy for teachers and students to use the materials in a way that is largely compatible with the developers' intentions. This means that consistency should exist between the intended and perceived curriculum and the intended and operational curriculum. If both consistencies are in place, we call these materials *practical.*

A third characteristic of high quality materials is that students appreciate the learning program and that desired learning takes place. With such *effective* materials, consistency exists between the intended and experiential curriculum and the intended and attained curriculum.

Table 2 illustrates the links between the three quality aspects (validity, practicality and effectiveness) and the curriculum representations (see also Nieveen, 1997).

	Quality aspects		
	Validity	Practicality	Effectiveness
Represen-tations	Intended (ideal + formal): • State-of-the-art • Internally consistent	Consistency between • Intended ↔ Perceived • Intended ↔ Operational	Consistency between • Intended ↔ Experiential • Intended ↔ Attained

Table 2. Linking quality aspects and curriculum representations

So far, we have been referring to the quality of educational products from the perspective of developing learning materials. However, we consider the three quality aspects (validity, practicality and effectiveness) also to be applicable to a much wider array of educational products.

2.2 Product quality in CASCADE study

To clarify the applicability of the three quality criteria to other educational products besides the learning materials, the CASCADE study is introduced. The CASCADE study aimed at developing a high-quality computer support system (called CASCADE) that assists instructional developers during their formative evaluation efforts (Nieveen, 1997) To study the quality of CASCADE during its development, the same three quality criteria (discussed in section 2.1) were used in the following manner:

- *Validity*. The computer support system CASCADE should include 'state-of-the-art knowledge' and should be internally consistent.
- *Practicality*. CASCADE should meet the needs, wishes and contextual constraints of the members of the target group (the instructional developers).
- *Effectiveness*. CASCADE should impact the formative evaluation efforts of the target group.

Applying the three criteria to the CASCADE system enabled the developers of the support system to become clear about the system's quality requirements. This transparency supported the communication between the developers of the system, the stakeholders and target group.

Interaction between the developers of CASCADE, the target group, and the stakeholders appeared to be of paramount importance in the CASCADE study as it was difficult for all of these groups to provide detailed specifications for the CASCADE system in an early stage of the project. Not only were the type(s) of support that the target group needed largely unknown, it was also difficult to decide which aspects of formative evaluation should be included in the support system and which characteristics the user interface should have. In such cases, when there are many uncertainties, several authors (such as Goodrum, Dorsey & Schwen, 1993; Shneiderman, 1992; Tessmer, 1994) recommend allowing the specifications to evolve using a prototyping approach.

Applying a prototyping approach during the development of the CASCADE system together with holding to the three quality aspects proved to be very powerful in the CASCADE study. In the remainder of this chapter general characteristics of the prototyping approach are considered (section 3), and the way this approach was applied in the CASCADE study is described (section 4).

3. PROTOTYPING

A prototyping approach may enable the process of discovering specifications and their adequacy. In this section the following significant characteristics of a prototyping approach are discussed: extensive use of prototypes (3.1); high degree of iteration (3.2); and participation of representatives of the target group (3.3).

3.1 Extensive use of prototypes

According to Smith (1991, p. 42), a prototype is a "preliminary version or a model of all or a part of a system before full commitment is made to develop it." In this definition, the term 'develop' refers to the construction of the final product. Prototypes are all products that are designed before the final product will be constructed and fully implemented in practice. According to Smith (1991), prototypes may be used in two ways. On the one hand, a prototype may be continually refined (based on reflections of developers on the prototype and formative evaluation results) and evolve towards a final deliverable. Smith refers to this approach with the term *evolutionary* prototyping. On the other hand, developers may design *throw-away* prototypes. This kind of prototype clarifies possible consequences of specific design ideas. After being evaluated, a throw-away prototype will be discarded and its evaluation results are taken into account in a next prototype. This will continue until all uncertainties are covered and the final product can be constructed.

In software engineering, a variety of prototypes is being utilized, such as: scenarios, paper-based prototypes, and working (or computer-based) prototypes. A *scenario* is a "fairly open-ended narrative description of typical and critical situations that prospective users participate in, and which the design work may transform" (Carroll, 1994, p. 65). Scenarios may be used to make the tentative design specifications more concrete. This makes it easier to communicate the potentials of a system with the target group about. With respect to *paper-based prototypes*, Rettig (1994) contends that the value of these prototypes is often underestimated: "I see this [paper-based prototype] as a wonderfully simple and effective tool that has somehow failed to come into general use in the software community" (p. 22). One way to create a paper-based prototype is to make a pile of papers representing all screens that may appear during the use of the system. Users may 'walk through' the screens to get an idea of the intentions of the computer system. Paper-based prototypes focus the attention of the user more on content and overall structure than on appearance (Rettig, 1994). Further, a *working (or computer-based) prototype* may help in making the design considerations even more concrete, as these prototypes may be actually used by members of the target group. A working prototype need not be completed or operable on the target hardware/software platform (Monk, Wright, Haber & Davenport, 1993). Finally, after the computer-based prototype is tested and revised (several times), and all stakeholders are satisfied with the program, a final deliverable may be developed. In the case of an evolutionary prototyping approach, the final deliverable is another subsequent version of the working prototype. However, this is not always possible, for instance, if the hardware used for prototyping purposes and the hardware used by the target group are incompatible, then the final version of the working prototype will be 'thrown away' and a new program will have to be developed based on the last prototype. In the final deliverable, the entire system is complete and operates as an integrated whole on the target hardware.

3.2 High degree of iteration

Compared to the provision of abstract specifications, the use of a series of concrete prototypes may provide a better foundation for identifying the requirements of the support system in interaction with members of the target group, experts, and other groups. Formative evaluation of prototypes is a crucial part of each prototyping process. It gives participants of the evaluation as well as the developer of the prototype insight in the potentials of the support system and its desired characteristics. Results of the formative evaluation may lead to revision of the prototype and adaptation of the specifications that underlie the support system. In this way, each prototyping cycle represents the evolution of intentions of the system. Based on several such cycles, the computer system may evolve towards a high-quality final deliverable.

Especially in development projects that aim at innovative and complex products, with few experiences from which to draw, a prototyping approach appears to be appropriate. According to Tripp and Bichelmeyer (1990), in these situations "little formal research is needed to begin a project, and much information can be gathered from research conducted as learners [or users, in case of the CASCADE study] use the prototype" (p. 40). According to Moonen (1996, p. 189), the prototyping approach may even cause a shift in the development activities, where "less emphasis is given to the writing of the specifications [at the beginning of a project] and more emphasis is given to the direct visualization on a screen of (parts of) the projected product, eventually ending in 'visual' specifications as a starting point for the next phase of the development process."

To stimulate an iterative approach, de Hoog, de Jong and de Vries (1994) propose a methodology that should facilitate nonlinear development. They suggest decomposition of the product into several components that may be built separately. Developing the components of a product is called 'local development', whereas the care about the whole product is called 'global development'. According to this methodology, the key components of the product will be developed in an incremental way. Each subsequent version of the component should show an improvement compared to the former state and should reflect the quality requirements of that particular component (as far as they are explicit). In addition, each version of the component should be consistent with the other components that are under development at the same time. Different components may be in different states of development.

Another important characteristic of iterative development approaches is the notion of 'think big, but start small'. By first trying to develop a small part of the final system, one keeps the development process manageable and one can learn from failures and apply successes when developing the next parts.

3.3 Participation of representatives of the target group

When applying a prototyping approach, the participation of members of the target group is of paramount importance. Monk et al. (1993, p. 5) stress that "involving users in design does not mean having endless meetings where some abstract specifications are discussed with union representatives or management. It requires access to people typical of those who will actually use the system, not their representatives or management." Goodrum et al. (1993) agree with this by pointing out that "as collaborators their [the participants'] responsibilities extend beyond their traditional role where one might interview them about their needs, have them sign-off on a list of specifications, or have them test the product after it has been designed and developed" (p. 14). More user involvement may lead to (cf. Moonen, 1996; Shneiderman, 1992):

- more accurate information about the tasks which will be supported;
- more intensive discussions about the requirements of the system;
- better opportunities to negotiate and justify design ideas;
- increase of user commitment and ownership of the final deliverable;
- stimulation of the professional development of the participants; and
- changes in the work environment of the participants.

In addition, developers have a strong tendency to 'design for themselves' (cf. Monk et al., 1993). Because developers of a support system may have different backgrounds than those of the target group and have a logical understanding of the overall support system, this may lead to problems. A user-centered approach with involvement of end users during the development process may prevent developers of a support system from being naive at this point.

4. PROTOTYPING APPROACH IN THE CASCADE STUDY

The prototyping stage in the CASCADE study was guided by the question: What are characteristics of a high quality computer support system for formative evaluation of lesson materials within the SLO context? This section describes the prototyping stage as it has unfolded, including the types and foci of the prototypes (4.1) and the formative evaluation of each prototype (4.2).

4.1 Type and focus of the prototypes of CASCADE

As was mentioned in the former section, the development process of a software product could be made more transparent, when decomposing the software product into several key components (de Hoog et al., 1994). The CASCADE system was decomposed into three of such key components. The first component was the *content* of the system, referring to the conceptualization of the formative evaluation of lesson materials. The *support* included in the system in order to assist the target group in performing a formative evaluation of the lesson materials they are developing was viewed as the second key component of CASCADE. The third key component was the *user interface* that should assure that the support is accessible for its users. Each of these components could contribute to or obstruct the overall effectiveness of the support system. Therefore, the three quality aspects (validity, practicality and effectiveness) discussed in section 2 were considered to be requirements for each key component of CASCADE (content, support and user interface). In addition, the whole system should show coherence between its key components.

To thoroughly address the importance of quality, five subsequent prototypes have been developed. The first prototype (*preliminary computer-based prototype*) was a throw-away prototype. This prototype was designed in order to get feeling for the notion of computer-based support. Based on the formative evaluation results, the decision was made to develop a *paper-based prototype* first, in order to illustrate the potentials of computer-support without the need to put much effort into actually programming the support system. After its formative evaluation, the paper-based prototype was discarded and a *first computer-based prototype* was constructed, based on the design ideas collected. Based on the promising results of the formative evaluation of the first computer-based prototype, the decision was made to continue building on it until *the second computer-based prototype* was ready. So far, an Apple Macintosh-based prototyping tool called 'HyperCard' (together with its programming language, 'HyperTalk') was used in order to design the computer-based prototypes of CASCADE. Based on the formative evaluation results of the latter prototype, it was decided to rebuild the prototype within the Delphi programming environment in order to obtain a *final version* that would be compatible to the Windows-platform used by the intended target group. A full description of the prototypes and accompanying formative evaluation efforts is presented in Nieveen (1997) and Nieveen and van den Akker (1999).

Although each prototype forms an integrated whole of the key components (content, support, and user interface) each individual prototype had its own central focus. The first four prototypes and their formative evaluation activities concerned the validity and practicality of the key components of the system (content, support and/or user interface), whereas the major attention of the evaluation of the final version was on its practicality and effectiveness. The primary attention of each prototype is illustrated in *Table 3*.

		prelim comp. based	**paper-based**		**first computer-based**		**second computer-based**		**final version**	
		Users (n=5)	experts (n=3)	users (n=5)	experts (n=2)	users (n=4)	experts (n=6)	users (n=4)	users (n=17)	users (n=4)
Validity	content		✓ ea							
	support						✓ ea			
	interface				✓ ea					
Practi-cality	content					✓ ce		✓ ce	✓ ws	✓ to
	support	✓ ce		✓ wt		✓ ce	✓ ea	✓ ce	✓ ws	✓ to
	interface	✓ ce		✓ wt	✓ ea	✓ ce		✓ ce	✓ ws	✓ to
Effec-tiveness	entire system								✓ ws	✓ to

Note:
prelim comp. based = preliminary computer-based prototype
✓ = primary attention of prototype and formative evaluation
ce = cooperative evaluation; wt = walk through; ea = expert appraisal; ws = workshop; to = try-out

Table 3. Focus of design and formative evaluation of the prototypes

4.2 Formative evaluation of prototypes of CASCADE

As is indicated in *Table 3*, in the formative evaluation of the prototypes two main types of respondents participated: users and experts. To give a central role to the members of the target group during the discussion about the growing ideas and prototypes, a user group was recruited. The group consisted of five representative developers from the target group (after the first computer-based prototype one user left the group because of other priorities). Besides the user group, various experts reviewed the prototypes. The experts who participated in the formative evaluation of the user interface of the first computer-based prototype are scholars in the field of interface design. All other experts were selected based on their expertise in the field of curriculum development and, more specifically, formative evaluation.

As is illustrated in *Table 3*, five major formative evaluation approaches were chosen: cooperative evaluation, walkthrough, expert appraisal, workshop and try-out. The first approach, *cooperative evaluation*, was selected to gain insight into the practicality of the preliminary, first and second computer-based prototype and to generate revision ideas for improving the practicality.

Monk et al. (1993) explain that by using this method users (one-by-one) work through a set of tasks with the system. The tasks must be representative of the work that the product will support. The tasks are selected by the designer of the prototype because free use of the prototype is often not possible with a partial product. While the users work with the prototype, they are encouraged to work aloud (explain to the designer what they are doing) and to ask questions about aspects which are not clear (cf. Dumas & Redish, 1994; van der Geest, 1991). The designer of the prototype may also ask the users questions and looks for unexpected behavior and comments from the user. Because this evaluation method is focused on prototypical products, it may be possible that users will encounter some [major] problems during their task performance (for instance because of bugs). If this happens, the designer of the product will assist the user in order to prevent any major delays. The need for this type of cooperation provides an indication that revision decisions may be required.

During the cooperative evaluation of CASCADE, all problems and comments on the prototype were audio-taped and transcribed afterwards. Based on review of these raw data revision decisions for a next prototype of the support system were generated. The key component 'content' was not yet very elaborated in the preliminary computer-based prototype. Therefore, the content of CASCADE was not taken into account during the cooperative evaluation of this prototype. During the cooperative evaluation of the first and second computer-based prototype, the practicality of all key components of CASCADE was taken into consideration.

The practicality of the support and user interface of the paper-based prototype was determined by a *walkthrough*. During a walkthrough, members of the user group go from one paper screen to the other while the designer of the prototype takes the role of the computer. For instance, when a user 'pushes' a button to go to another screen, the designer shows the user this screen (which is another paper on the pile).

To guide the user's movements through the paper-based prototype of CASCADE, the prototype was presented to the user in a format that is comparable to the use of scenarios (cf. Carroll, 1994). Each scenario started with a short story about a fictitious person who uses the prototype for a specific purpose, followed by a series of screens which should give the participants an idea of the events that happened when this fictitious person consulted the system. Based on the scenario, members of the user group 'walked through' the prototype. They were asked to work aloud, which made it possible to gain insight into their problems and comments concerning the support and user interface. During a debriefing, a checklist was used to assure that all users answered all questions concerning the paper-based prototype. All problems and comments were audio-taped and transcribed. Based on interpretation of the raw data, revision decisions were generated.

Various experts judged the validity and practicality of the key components of the prototypes during *expert appraisals*. Dumas and Redish (1994) call this type of expert review 'heuristic evaluation'. Experts are asked to examine the prototype and look for problems that violate some of the general principles (heuristics) of the domain of their expertise.

Preceding the review of CASCADE, the purpose of the support system was explained to the user interface experts. Subsequently, the experts were asked to perform predetermined tasks with the system (comparable to the cooperative evaluation with the members of the user group) in order to get to know the support system. Then the experts were invited to review the user interface in light of a checklist with user interface guidelines. The expert appraisal sessions were audio-taped and the raw data were transcribed and interpreted; based on those results, the user interface of CASCADE was changed.

Three experts in the field of curriculum development and formative curriculum evaluation reviewed the validity of the content of the paper-based prototype. They were asked to write their comments on the outline of the content. In a face-to-face meeting the reactions were discussed. Six experts (with expertise in the same field) judged the validity and practicality of the support of the second computer-based prototype. Again, the experts were asked to write their comments on a print-out of the support screens of the prototype. All experts' comments were interpreted and used as grounds for the generation of revision decisions.

In order to reach a wider group of developers of the target group (n=17), a *workshop* was organized. The main focus of the workshop was on learning to prepare a formative evaluation plan with CASCADE. The main part of the workshop consisted of a two hours' training session with hands-on experiences based on a booklet with a 'guided tour'. By means of an assignment, the participants were invited to develop a formative evaluation plan for a case from their own practice (which they had been asked to bring to the workshop). A built-in log function in CASCADE saved the actions of the participants and time spent on each screen. The workshop leader and an assistant only interfered with the process when the participants raised problems or comments and made notes of their observations.

To study the practicality and effectiveness of CASCADE in real practice, four *try-outs* were planned within projects that started up and conducted a formative evaluation of lesson materials for primary or secondary education. The project members agreed to use CASCADE as much as possible during their formative evaluation activities. The evaluator of CASCADE was allowed to intervene with the process only when problems were brought to notice on the initiative of the participants. The participants were invited to make notes in a logbook about the components of CASCADE they used; judgments about the practicality of the components; satisfaction with the process; the time needed to complete a task; support they received from others; and impact of the evaluation. With the help of a log tool, the computer kept track of the use of CASCADE. Interviews were planned after the participants had finished a clear-cut part of the formative evaluation.

5. CONCLUDING REMARKS

Developing high quality products is a complex process, especially where a new product is concerned and when the target group does not have clear expectations. The combination of postponing the writing of detailed specifications and the development of subsequent prototypes give the target group, the stakeholders and the developers of an innovative product the opportunity to gain a better understanding of the consequences of certain design decisions. The use of a series of concrete prototypes gives a more immediate sense of the proposed system, its potentials and its desirable characteristics.

The framework for product quality criteria presented in this chapter assists in deciding on the focus of each subsequent prototype. It makes clear that often times there is a shift in the developers' primary attention with regard to their prototype. During the design and development of the first few prototypes, developers seem to be especially interested in the validity and practicality of their prototype. As soon as they have become more certain about the characteristics of their prototype, their interest shifts towards the effectiveness of the prototype.

Linking the quality requirements (validity, practicality and effectiveness) to key components of a product (in case of CASCADE: content, support and user-interface) enhances the transparency of the prototyping process. Although a risk of decomposition may seem that it can increase the chance for inconsistency in the design decisions of the entire product, the design of an integral prototype serves the consistency between the key components of the system. The design of a prototype makes clear the fact that ideas with respect to one component need to fit in with the entire system and changes in one component often have consequences for the other key components.

Formative evaluation activities lead to revisions of each prototype and adaptations of the conceptual framework underlying the product. Based on several such cycles, the prototype evolves towards a better final version and the insights into important specifications become more fine-tuned. User involvement is of paramount importance during the prototyping stage. Especially end users can help refine the product and enhance the quality of each subsequent prototype.

REFERENCES

Akker, J.J.H. van den, 1988, *Ontwerp en implementatie van natuuronderwijs* [Design and implementation of science education], Swets & Zeitlinger, Amsterdam.

Akker, J.J.H. van den, 1990, Het gebruik van curricula [The use of curricula], In *Onderwijskundig lexicon II*, E3100-1/12, Samsom, Alphen aan de Rijn.

Ball, D.L. and Cohen, D.K., 1996, Reform by the book: What is-or might be the role of curriculum materials in teacher learning and instructional reform?, *Educational Researcher, 25*(9), 6-14.

Carroll, J.M., 1994, Designing scenarios for human action, *Performance Improvement Quarterly, 7*(3), 64-75.

Dumas, J.S. and Redish, J.C., 1994, *A practical guide to usability testing*, Ablex, Norwood, NJ.

Geest, T.M. van der, 1991, *Tools for teaching writing as a process: Design, development, implementation and evaluation of computer-assisted writing instruction*, Doctoral dissertation, University of Twente, Enschede.

Goodlad, J.I., Klein, M.F. and Tye, K.A., 1979, The domains of curriculum and their study, In Goodlad, J.I. and Associates (Eds.), *Curriculum Inquiry: The study of curriculum practice* (pp. 43-76), McGraw-Hill, New York.

Goodrum, L.T., Dorsey, D.A. and Schwen, T.M., 1993, Defining and building an enriched learning and information environment, *Educational Technology, 18*(11), 10-20.

Hoog, R. de, Jong, T. de and Vries, F. de, 1994, Constraint-driven software design: An escape from the waterfall model, *Performance Improvement Quarterly, 7*(3), 48-63.

Monk, A., Wright, P., Haber, J. and Davenport, L., 1993, *Improving your human-computer interface: A practical technique*, Prentice Hall, New York.

Nieveen, N.M., 1997, *Computer support for curriculum developers: A study on the potential of computer support in the domain of formative curriculum development*, Doctoral dissertation, University of Twente, Enschede.

Nieveen, N.M., Akker, J.J.H. van den, 1999, Exploring the potential of a computer tool for instructional developers, *Educational Technology Research & Development, 47*(3), 77-98.

Rettig, M., 1994, Prototyping for tiny fingers, *Communications of the ACM, 37*(4), 21-27.

Shneiderman, B., 1992, *Designing the user interface: Strategies for effective human-computer interaction*, Addison-Wesley, Reading, MA.

Smith, M.F., 1991, *Software prototyping: Adoption, practice and management*, McGraw-Hill, London.

Tessmer, M., 1994, Formative evaluation alternatives, *Performance Improvement Quarterly, 7*(1), 3-18.

Tripp, S.D. and Bichelmeyer, B., 1990, Rapid prototyping: An alternative instructional design strategy, *Educational Technology Research & Development, 38*(1), 31-44.

Chapter 11

The Potential of Formative Evaluation in Program Design Models

Ralf Maslowski
Adrie Visscher
University of Twente

Keywords: Formative evaluation, Design models

Abstract

This contribution discusses the nature and role of formative evaluation in educational design and development. It consists of two sections. In the first section the role of evaluation in educational design, and the importance of conducting evaluations during design and development practices is discussed. Furthermore, limitations of formative evaluation for improving educational programs are identified. In the second section, the nature of formative evaluation in four models for educational design, i.e. the planning-by-objectives model, the prototyping model, the deliberative model and the artistic model, is examined in detail.

1. THE ROLE OF FORMATIVE EVALUATION IN DESIGN AND DEVELOPMENT

With the advent of large scale educational projects in the 1960s, designers began to see program design as a continuous process of tryout and development for improvement. In the United States, Cronbach (1963) made a plea for the inclusion of evaluation in program design: "Evaluation, used to improve the course while it is still fluid, contributes more to improvement of education than evaluation used to appraise a product already placed on the market" (p. 236). Cronbach's 'evaluation for improvement' was later labeled 'formative evaluation' by Scriven (1967). Scriven used the term 'formative' to distinguish between the evaluation of a program[1] during

[1] Following Scriven (1991a) a program is conceived here as the general effort that marshals staff and projects toward defined goals. In the field of education it may concern a school policy, a national educational policy, an instructional program or, for instance, a curriculum.

its developmental stages and the processes involved in determining the effectiveness of a program after its completion, which he labeled 'summative evaluation'.

Summative evaluations are performed by, or commissioned by persons or institutions who intend to use the evaluation findings to determine the impacts, outcomes or results of a particular program. Summative findings may, for instance, be used to decide whether or not to purchase a new teaching package, or to decide whether an innovation project should be continued or not (Scriven, 1991b).

Formative evaluation, on the other hand, is intended to support the process of improvement, and normally commissioned or done by, and delivered to, someone who can make improvements. It is an integral part of the development process as it provides continual feedback to assist in planning and then producing a particular program. Formative evaluation is, as Flagg (1990) denotes, performed "to inform the decision-making process during the design, production, and implementation ... of an educational program with the purpose of improving the program" (p. 23).

Although formative and summative evaluations differ in their purpose, both are concerned with determining the value, or more specifically, the merit and worth of a program. Hereby, merit refers to the 'intrinsic quality' or 'context-free value' of a program, whereas the worth of a program is concerned with its 'extrinsic quality' or 'context-determined value'. Stufflebeam and Shinkfield (1985) illustrate both concepts by means of evaluating a program for preparing history teachers. If such a program produces teachers who confidently and effectively teach others about history, it is considered to be meritorious, but it need not be worthy. For example, if the society has no need for history teachers, because there are already more good ones than can be employed, then continuation at the same level of expansion of this program may not be a worthwhile expenditure of scarce resources.

1.1 Formative evaluation in relation to the improvement of educational programs

As stated before, formative evaluation is aimed at determining the strengths and weaknesses of an educational program, in order to generate suggestions for improvement. However, formative evaluation in itself is an insufficient base for improvement recommendations (Scriven, 1991b). Conceptually, a distinction can be drawn between the *description* (specification of aspects or features), the *evaluation* (assessment of worth or merit), and the *diagnosis* (labeling) of a program (*Figure 1*). In the description phase information on a program is collected and analyzed. However, these results are not classified as satisfactory or unsatisfactory at this stage. This is subject to the formative evaluation phase, which results in the identification of a program's shortcomings. These shortcomings still have to be linked to each other, interpreted and matched to patterns of disorder. A diagnosis aims at this, finding the best 'label' for the observed deficits.

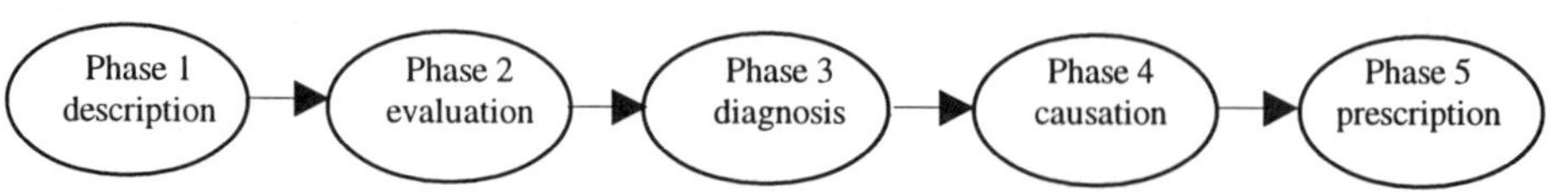

Figure 1. Five phases in the process of design improvement

Although the diagnosis of a program's deficits can include knowledge of its underlying cause there may be no obvious explanation, and *causation* of the problem needs to be investigated further. Insight in the cause of the program's deficits is necessary for identifying a remediation (*prescription*). The fact that a program has its shortcomings, nevertheless, does not imply that improvement is possible and feasible. Besides information on what caused the program's deficit, it is necessary to have a remedy at one's disposal. Furthermore, the severity of a detected problem is related to the quality of the program as a whole; it may, for instance, be relatively small in comparison to the overall quality. The benefits of the revision may not equal out the costs, or remediation of the problem may cause side effects that are as severe as the detected problem itself.

From a conceptual point of view, formative evaluation is a necessary but insufficient condition for generating recommendations for improvement. In addition to the evaluation of the program, improvement requires pre-evaluative (i.e. description) as well as post-evaluative (i.e. diagnosis, causation and prescription) actions. Therefore, a formative evaluation ideally provides information on the seriousness and cause of the problem, information on levers for improvement, and indicates the expected costs of remediation in addition to the mere specification of the program's deficits (Tessmer, 1993). Stated otherwise, formative research should be conducted in such a way that information is provided that directs or strengthens the diagnosis, causation and revision of observed shortcomings in the design.

1.2 Areas for special attention in formative evaluation

In the literature the influential contribution of rigorous formative evaluation to the improvement of educational programs is often stressed (cf. Baker & Alkin, 1973; Nathenson & Henderson, 1980). Despite these indications of favorable effects, however, evaluation does not always lead to improvements in the design at hand, even when improvement is possible and worthwhile. This may have a number of reasons. If a designer, for instance, does not know of any concrete remediation for the deficits that have been detected, or in case the funds for remediation are too restricted, then the findings of a formative evaluation will probably not result in adequate improvements of the design. It is, therefore, not only important to plan and conduct formative evaluations during the design process, but also to ensure that the necessary expertise and sufficient resources to make improvements are available as well.

A second reason for only partial improvements often comes from the fact that designers nearly always depend on informants or users for information on the practicality or effectiveness of a program under development. For instance, if a formative evaluation of an instructional program is largely based on data provided by teachers, students, and other stakeholders, it will be hard to take notice of the program's most important shortcomings, because some of them may simply not have been observed by the stakeholders. As the designer is primarily concerned with correcting the flaws that came up in the evaluation, this will probably result in correcting the actual deficits only partially. Therefore, designers ideally will select a group of diverse informants and users, so the number of shortcomings that get unnoticed will be minimized.

Another frequent problem is that the effectiveness of a program is not solely dependent on the rigor of the design phase (including the evaluation of the program), but on the quality of the implementation phase as well. A high quality program that has not been implemented in the way it was intended, will probably loose much of its value. For example, a textbook that has been developed to stimulate pupils' self-motivation, but eventually proves to be used for direct instruction, will at best reach its goal partially. The value of a program may also change as a consequence of changing user needs, or as a result of alternatives for the program that become available. Therefore, it is important that after the program has been implemented, its value will be determined regularly to check its practicality and effectiveness (cf. Thiagarajan, 1991).

2. THE NATURE AND FUNCTION OF FORMATIVE EVALUATION IN DIFFERENT DESIGN MODELS

The nature of formative evaluation in program development is largely dependent on the design model that is adopted by the developers (cf. Maslowski & Visscher, 1997). In this section, the nature and function of formative evaluation in four such models, as specified by Visscher and Schulten (1997), is discussed[2]. This section focuses on the *potential* role of formative evaluation in each model instead of on its actual role, thereby referring to the aspects of formative evaluation presented in the previous section.

2.1 The planning by objectives model

The planning by objectives model reflects the systematic, rationalistic approach to program development, which was advocated by Bobbitt (1924) and Tyler (1949). In this model the formulation of the program's objectives lays the foundation for subsequent development actions. After the intended outcomes have been formulated on the basis of a preliminary problem analysis, the means to reach these ends are specified.

Formative evaluation is important in this model. It essentially serves two different functions. First, crucial to the planning by objectives model is formulating a large number of intermediate objectives by means of decomposition. For each of these objectives actions are sought that may bridge the difference between the actual and the desired state. This requires that the developer, after each delineated part of the design process, formally evaluates whether a specified objective has been met, before he can move on to subsequent actions.

Second, some objectives may be reached through alternative actions. Designers therefore often explore alternatives. After various strategies have been developed partially or fully, it is evaluated which one serves the objectives best, or seems to be most promising. If the tentative general objectives are decomposed into precise

[2] The four design models are examined briefly here. For a more elaborate discussion of the models, the reader is referred to the chapter 2 in this book.

intermediate objectives, measurable criteria and standards for comparing the alternative actions can be deduced easily from these objectives. However, because alternatives are often evaluated on the basis of a set of criteria, this may lead to ambiguous results. For instance, alternative A may score better on one of the criteria than alternative B, however at the same time it may score much lower on other criteria. In such cases computational methods, like the so-called 'weight and sum' method, may be used for evaluating alternatives (cf. Scriven, 1991b) on more than one criterion.

2.2 The protoyping model

In the prototyping model, after a succinct statement of needs and objectives, a global outline of a program is constructed or specific parts of a program are developed in detail, and thereafter checked with teachers and students (cf. Tripp & Bichelmeyer, 1990). It is assumed that the needs of teachers and students, the contents to be covered and the objectives to be reached, cannot be determined fully in advance. Prototyping is used when design specifications are not clear, as in the development of an innovative program, or in case stakeholders are unable to formulate their needs precisely (Gray & Black, 1994).

In this model preliminary versions of the program are evaluated in cooperation with teachers and students to further specify the design requirements and to determine the usefulness and effectiveness of the program. The development of the program, therefore, is an iterative process in which the users of the program play an important role. Evaluations with users form the basis for further refinements in the program, or even for discarding the prototype and developing a new prototype that better reflects their needs. However, relying on user satisfaction as a main indicator for improvement may result in a program that is practical and feasible, but may be only partially effective. Users' knowledge of other alternatives -which may be better and cheaper- as well as improvement possibilities may be too limited to judge a prototype profoundly.

As in the planning by objectives model, formative evaluation plays a central role in the process of developing a program, however, in the prototyping model it is more oriented towards a program's extrinsic quality (its worth for one or more individuals or groups) than its intrinsic quality (its merit, i.e. its context independent value which may be determined on the basis of professional standards). For that reason, the user's satisfaction with the practicality and effectiveness is regarded to be the most important dimension when evaluating a program. Besides the comments from users during the evaluation, the designer will use his professional standards to determine the merit of the program while refining it. Although these professional standards are hardly ever explicated, they may be regarded as an evaluation basis as well. Both designers and external evaluators should be involved in the evaluation of the intrinsic and extrinsic quality of the program.

2.3 The deliberative model

In the deliberative model communication and negotiation between stakeholders in the design process (e.g. developers, clients and users) is considered crucial. Their exchange of ideas is meant to reach consensus on the problem to be solved, the desired outcome and the way of accomplishing that. The process includes a continuous deliberation of means and ends, possibilities and alternatives (Schwab, 1970). Agreement between participants serves as a shared frame of reference and legitimizes the decisions taken. The deliberative model encompasses the formulation of ideas, discussing, verifying and adapting these ideas, and using them to generate promising versions of materials, assessing the merits of promising early material versions, and revising them until they cannot be improved further (Walker, 1990).

In the early tentative design phases ideas are formulated and verified with practice in an informal way. At that stage the program to be developed is not at hand yet, and verification is especially meant to refine understanding, discussion, ideas and deliberation. If this process leads to consensus on the desired outcomes, a more formal formative evaluation may be carried out to determine to which extent the preliminary program meets the norms as agreed upon by the stakeholders. In the deliberative model the quality of the program is conceived as a matter of extrinsic quality which is likely to be determined by those involved in the design process. The designer who coordinates the process of deliberation will have to make sure that the program on which all stakeholders agree, still meets the professional standards (intrinsic quality). The evaluation dimension may be labeled 'the degree of fit with the design criteria stated'. If this is not the case, or if the stakeholders cannot reach consensus on certain aspects of the program, external evaluators may be involved to overcome the deadlock in the deliberative process. Although in practice usually only one program is designed and evaluated, if possible, a few program alternatives may be developed to find out which one, according to the stakeholders, meets the standards best.

2.4 The artistic model

In the artistic model program development is considered to be highly complex because, as Eisner (1979) denotes, social reality is subjectively perceived and socially constructed. Eisner stresses that the practice of design often is not as rational as assumed in the literature. Although design skills to a certain degree can be taught and learned by experience, according to this approach they cannot be reduced to a simple design heuristic that can be followed and that works under all circumstances. On the contrary, program design is an uncertain endeavor, in which the developer is confronted with a unique case that falls outside the categories of existing theory and technique, and in which he has to deal with different -and sometimes conflicting- perceptions of stakeholders. It requires practical insight and artistic skills to deal with this situation. Developers have to bring in their own values, ideas and use their creativity in shaping the solution.

Because of the uncertainty about the problem and the intended outcomes the developer 'communicates with the situation' (cf. Schön, 1983). While framing, classifying, acting, experimenting and steering, he attempts to grasp and solve the problem. In this ongoing process the developer informally judges his actions continuously, appreciating what is educationally significant and what is not. There are no predetermined evaluation dimensions or standards; the developer's tacit quality standards determine his judgement and follow-up actions. Formative evaluation in the artistic model is essentially interwoven with and integrated in the development actions, as reflection on the actions undertaken by the developer.

3. CONCLUDING REMARKS

Formative evaluation is conceived as an integral part of the design process which provides continuous feedback to inform the design, production and implementation of an educational program with the purpose of improving it. Studies in the field of educational design have shown that formative evaluation may have considerable benefits for the program that is being developed. From a conceptual point of view formative evaluation is a necessary, but insufficient precondition for program improvement, for improvement also requires the successful execution of pre-evaluative and post-evaluative actions.

The role that a formative evaluation plays in the design and development of an educational program, depends to a large degree on the design model that is being adopted. In this article four design models and their potential evaluation role have been characterized. First, the *planning by objectives model*, in which formative evaluation is primarily concerned with whether intermediate objectives of the program design are met by action alternatives, and which alternative action (probably) serves the objectives best. The *prototyping model* has been characterized, in which evaluation serves the role of (further) specifying the design requirements, and determining the user's satisfaction with the program's practicality and effectiveness. The role of evaluation in the *deliberative model* was described as to determine to which extent the preliminary program meets the design criteria as agreed upon by stakeholders, and as a means to overcome the possible deadlock in the negotiations. Finally, the *artistic model* was described, where evaluation plays the role of an informal reflection on (the results of) a designer's attempts to grasp and solve the problem at hand and to base follow-up actions upon.

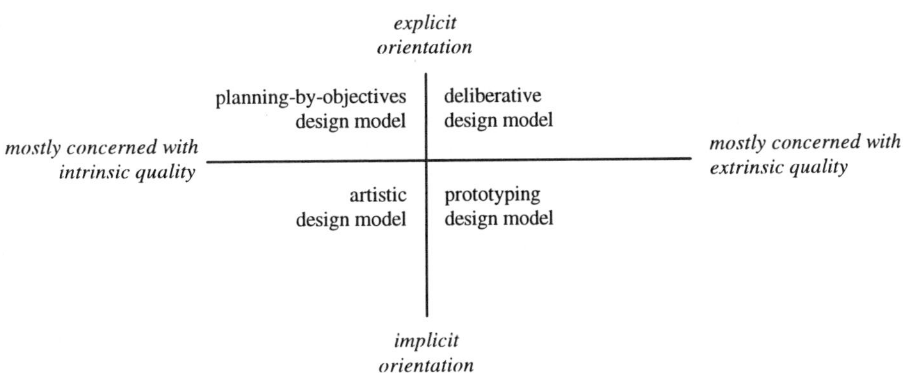

Figure 2. Design models according to their orientation and quality concern

These four design models may be classified according to their primary concern with the intrinsic or extrinsic quality of the program, as well as their orientation towards taking either explicit or implicit decisions during the design process (*Figure 2*). As was highlighted in the previous paragraph, the planning-by-objectives model and the artistic model are primarily concerned with the program's intrinsic quality, whereas the prototyping model and the deliberative model are mostly concerned with the extrinsic quality of the program. Furthermore, in the planning-by-objectives model (by taking the rational means-ends approach) and the deliberative model (by reaching consensus on actions to be taken through discourse), the actions to be taken are made as explicit and clear as possible. In the artistic and the prototyping model, on the other hand, much of the decisions are taken by the designers in a less accountable and more implicit manner.

REFERENCES

Baker, E.L. and Alkin, U.C., 1973, Formative evaluation of instructional development, *AV Communication Review, 21* (4), 389-418.

Bobbitt, F., 1924, *How to make a curriculum,* Houghton Mifflin, Boston, MA.

Cronbach, L.J., 1963, Course improvement through evaluation, *Teachers College Record, 64*, 672-683.

Eisner, E.W., 1979, *The educational imagination: on the design and evaluation of school programs,* Macmillan, New York.

Flagg, B.N., 1990, *Formative evaluation for educational technologies,* Lawrence Erlbaum, Hillsdale, NJ.

Gray, D.E. and Black, T.R., 1994, Prototyping of computer-based training materials, *Computers and Education, 22* (3), 251-256.

Maslowski, R. and Visscher, A.J., 1997, *Methoden en technieken voor formatieve evaluatie in sociaal-wetenschappelijke ontwerpsituaties* [Methods and techniques for formative evaluation in social scientific design situations], University of Twente: Enschede.

Nathenson, M.B. and Henderson, E.S., 1980, *Using student feedback to improve learning materials,* Croom Helm, London.

Schön, D.A., 1983, *The reflective practitioner: how professional think in action,* Basic books, New York.

Schwab, J.J., 1970, *The practical: a language for curriculum,* National Education Association, Washington, DC.

Scriven, M., 1967, The methodology of evaluation, In Tyler, R.W., Gagné, R.M. and Scriven, M. (Eds.), *Perspectives on curriculum evaluation* (pp. 39-83), Rand McNally, Chicago.

Scriven, M., 1991a, *Evaluation Thesaurus,* Sage, London.

Scriven, M., 1991b, Beyond formative and summative evaluation, In McLaughlin, M.W. and Phillips, D.C. (Eds.), *Evaluation and education: At quarter century* (pp. 19-64), University of Chicago Press, Chicago, IL.

Stufflebeam, D.L. and Shinkfield, A.J., 1985, *Systematic evaluation: A self-instructional guide to theory and practice.* Kluwer-Nijhoff, Boston.

Tessmer, M., 1993, *Planning and conducting formative evaluations: improving the quality of education and training,* Kogan Page, London.

Thiagarajan, S., 1991, Formative evaluation in performance technology, *Performance Improvement Quarterly, 4* (2), 22-34.

Tripp, S.D. and Bichelmeyer, B., 1990, Rapid prototyping: An alternative instructional design strategy, *Educational Technology Research and Development, 38* (1), 31-44.

Tyler, R.W., 1949, *Basic principles of curriculum and instruction,* University of Chicago Press, Chicago.

Visscher, I. and Schulten, E., 1997, *Design approaches in training and education: Insights from practice,* Paper presented at the annual meeting of the American Educational Research Association, March 1997, Chicago.

Walker, D., 1990, *Fundamentals of curriculum,* Harcourt Brace Jovanovich, San Diego.

Chapter 12

Instructional Design: A Conceptual Parallel Processor for Navigating Learning Space

Robert Maribe Branch
The University of Georgia

Key words: Instructional design, Instructional development, Instructional systems, Learner-centered, Parallel processing, Curriculum development

Abstract

Instructional design is a process used to generate curriculum, courses, teaching units, and single episodes of guided learning. While instructional design is applicable for curriculum development, and in many types of teaching environments, this is a presentation about how instructional design can be conceived as a way to navigate in spaces dedicated to intentional [or guided] learning. Guided learning is purported to be complex because the student, the content, the media, the teacher, peers, and the context, all interacting within a discrete period of time; while moving toward a common goal. The instructional design concept suggested here is intended to be a way to respond to the complexities of guided learning. The contention is that the fundamental purpose of instructional design is to 1) respond to a performance discrepancy that is attributable to a lack of knowledge and skills, 2) generate learner-centered strategies, and 3) validate the products and procedures used for episodes of guided learning; and to accomplish theses processes parallel to each other.

1. INTRODUCTION

Instructional design is a process used to generate curriculum, courses, teaching units, and single episodes of guided learning. While instructional design is applicable for curriculum development, and in many types of teaching environments, this is a grounded discussion about how one can meet the challenge of the changing realities of education by focusing on ways to navigate in spaces dedicated to intentional [or guided] learning. Instructional design is perceived as a system composed of procedures dedicated to planning episodes of learning in shared spaces. Rowland and Adams (1999) in chapter 3 of this book suggest that a system should connote wholeness and interdependence. General systems theory (Banathy, 1987, 1991) allows an instructional design process to be interdependent, synergistic, dynamic, cybernetic, systematic, and systemic. Procedures can be defined as "a particular way of accomplishing something ... a series of steps ... a traditional or established way of doing things" (Merriam-Webster, 1991, p. 937). Because the procedures of instructional design are conceived in the mind, which are abstract or generic ideas generalized from particular instances (Merriam-Webster, 1991), the procedures are manifested through conceptual processes and a variety of products. "The [Instructional Systems Design] process is as important as the product, because confidence in the product is based on the process" (Seels & Glasgow, 1998). Spaces shared for guided learning should be conceived as a any place (physical or virtual) where multiple interactions occur between several entities that are guided by defined expectations. While learning is a covert state, unique to the individual, and extremely personal, teaching is an overt attempt to sequence events, and to order experiences in learning space.

Learner-centered instruction means that the student is the focal point of all learning activities. While learning is omnipotent, and active all the time, instructional design is concerned with those episodes of learning that are purposeful and guided. Guided learning is the quest for mutually agreed upon expectations for the student and the teacher. A mutual agreement within the context of instruction presupposes a learner shares the responsibility for his or her learning or the legal guardian acts as the learner's surrogate when the learner is unable to assume responsibility for his or her own learning. While an individual is in a perpetual state of learning, guided learning refers to the construction of knowledge that occurs in shared learning space, physical or otherwise. Guided learning space is complex because its essential entities: student, content, media, teacher, and context are within themselves complex. Further, the multiple interrelationships among the essential entities exponentially increases the complexity of guided learning (Branch, 1996). Therefore, an equally complex system is required to match the complexities of guided learning. Instructional design processes are capable of the necessary complexity because its "steps are interwoven in a *system* that can only be categorized as complex" (Braden, 1996, p. 20). Teacher planning routines (Earle, 1998), instructional planning (Reiser & Dick), 1996), and rapid prototyping (Tripp & Bichelmeyer, 1990) are several types of vehicles constructed to navigate learning space. Instructional design is also a type of vehicle that can navigate through the complexities of guided learning while remaining learner-centered because of the multiple activities are occurring in parallel.

Learning space can be situated at remote sites, accessed at convenient times, and personalized to match the capability of individual learners. While students may continue to 'meet' to study the same subject, the location, time, and pace are now dynamic for each episode. Learning space is characterized by several entities: the student, the content, the media, the teacher, peers, and the context, all interacting within a discrete period of time; while moving toward a common goal. Gagné, Briggs and Wager (1992) purport instructional design as centering on individual learning, having immediate and long-range phases, as being systematic, and using a systems approach to knowledge construction and human learning. *Figure 1* presents a paradigm for guided learning space. Learner-centered spaces, wherever they are located, represent an epistemological shift from treating students as receptacles into which instruction is poured, to valuing the individual's personal construction of knowledge as an asset during instruction.

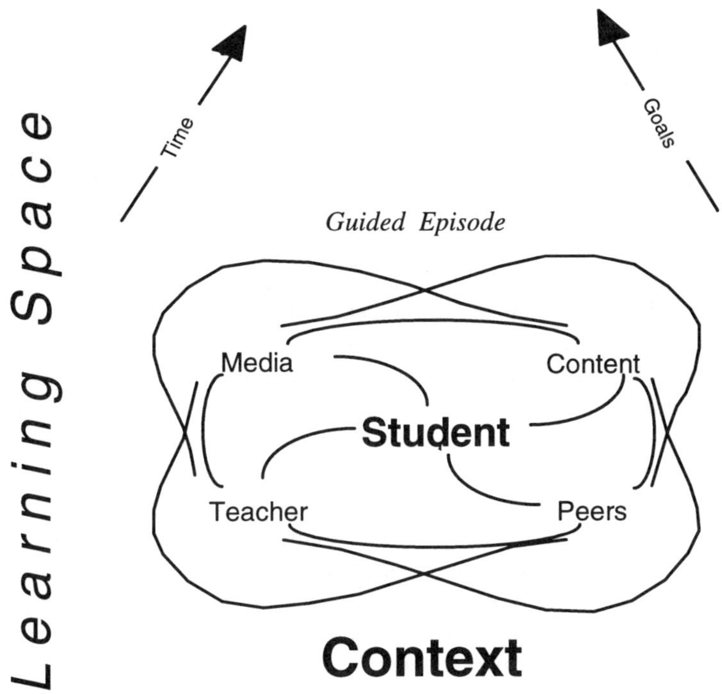

Figure 1. A paradigm for guided learning space

The purpose of this chapter is to introduce three concepts essential for the effective practice of instructional design. The contention is that people who apply instructional design should judge the quality of instructional design models by their potential to concurrently:

- respond to changes in the ecology of learning space;
- generate strategies for creating episodes of guided learning; and
- validate instructional products and the procedures used to create those products.

Models are used to manifest the conceptual processes of instructional design. Instructional designers commonly employ models to guide their practice (see chapter 2 of this book), albeit the explicit use of models is less obvious among experienced instructional designers. While instructional design processes are modeled in a variety of ways (Andrews & Goodson, 1980; Braden, 1996; Branch, 1997; Branch & Gustafson, 1998; Edmonds, Branch & Mukherjee, 1994; Gustafson, 1981, 1991; Gustafson & Branch, 1997a, 1997b; Taylor, 1986), evidence about their effectiveness is usually limited to anecdotal testimony rather than empirical support derived from research studies. This chapter is intended for education professionals who are considering the use of an instructional systems design approach to planning episodes of guided learning. However, rather than present another instructional design model here, this chapter focuses on three conceptual processes that any instructional design approach should effect during its implementation: responsiveness, generative and validating.

2. A RESPONSIVE PROCESS

Instructional design is responsive because it accepts whatever goals are established as its orientation. Learning needs are initially based on prima facie evidence that an outcome(s) is not being achieved, thus, requiring a needs assessment (Rossett, 1993) to determine whether there is a performance discrepancy and to identify its causes. Responsive is denoted as "something constituting a reply or a reaction" (Merriam-Webster, 1991, p. 1005). Instructional designers often initiate contact with stakeholders in order to anticipate learner needs or to respond to the concerns of other participants in the process. The major responsive characteristics of an instructional design process reside in its theoretical origin. Many theories contribute to instructional design, including communication theory, curriculum theory, theories related to cognitive psychology, social learning theory, and human development theories. Seels and Glasgow (1998) identify psychology, systems, and communications as the major theories that support instructional design. A theory that has been successfully applied to instructional design is general systems because of its characteristics of: interdependence, synergism, dynamic, cybernetic, systematic, and systemic.

An instructional design process that stresses *interdependence* can respond to the inextricably connected relationships among pairwise, and otherwise multi-bonded entities joined in learning space. A *synergistic* quality presents the sum of the parts as being greater than the whole, thereby, increasing the function of each entity beyond its individual value, thus, exponentially increasing the probability of achieving the desired goal. Instructional design is *dynamic* in order to respond to the changing variables within learning space. A good instructional design process governs, guides, automates, replicates, and prevents failure of the entire process; it is *cybernetic*. The *systematic* quality of instructional design establishes rules and procedures; as well as the protocol for establishing the rules and procedures, and helps to constitute responsible approaches to designing instruction. A *systemic* quality is one where all components of the process respond to any stimulus: or at least have the opportunity to respond to any stimulus. General systems theory makes it possible to respond to the multiple backgrounds of the learner audience, the multiple interactions that occur between the primary entitles revolving in learning space, and the multiple contexts in which guided learning may exist. While instructional design is concerned with building capacity for students to navigate learning spaces, the success of instructional design is also dependent upon being able to generate new ideas.

3. A GENERATIVE PROCESS

Instructional design is generative because it is a creative process. Instructional design utilizes the experiential knowledge of stakeholders participating in the process as well as incorporating documented successful teaching and learning strategies to describe best practices. Instructional design adopts an **Input → Process → Output** paradigm as a way to generate products and procedures. The *input* phase reacts to the variables identified in the learning context by accepting data, information, and knowledge which is organized into some meaningful way. The *process* phase seeks ways to stimulate creative and divergent thinking of learners by utilizing procedures to interpret, explain, configure, and display multiple approaches to events that are likely to occur in learning space. The *product* phase delivers the results of the process by explicitly presenting ways of knowing things that are translated into ways of doing things. Each step in the instructional design process should generate a deliverable that represents the collective thoughts of all the stakeholders. Each deliverable is then tested prior to becoming input for the next step in the instructional design process. Rowland and Adams (chapter 3 of this book) conducted a study among 19 professionals who had between 5 and 30 years of experience as instructional designers and reported a fairly consistent set of instructional system components categorized as: Problem Situation, Goal, Learners, Content, Instructional Setting, Instructional Method, and Assessment. *Figure 2* presents an application of the input → process → output concept for a set of instructional design components. While the terminology will vary from context to context, the components of instructional design are intended to generate products and procedures that facilitate the navigation of learning space. Once products and procedures are recommended, rigorous validation of instructional design products and procedures is required.

4. A VALIDATION PROCESS

An instructional design approach should seek to confirm that the products and procedures associated with the development of learning episodes are both authoritative, convincing, relevant and meaningful: valid. Validation is the guiding trait of instructional design and adds credibility through procedures that are analytical, evaluative, and philosophical. Valid paradigms of instructional design are goal oriented; however, different learning goals require different instructional strategies (Dick & Carey, 1996). Valid learning goals should reflect the reality in which students will be expected to perform, and by so doing, maintain a high degree of congruence between the learning space and the performance space where actual tasks are performed. A valid instructional design process also considers the emotional and instinctive feelings of the student as well as the primary participants in the process. The philosophical origin of any process is important as a way to validate the reasons and motivations of the participants engaged in a process. A valid process can trace reasons for participation (Balanchandran & Branch, 1997) and the influence of an individual's belief system on their professional practice (Moallem & Earle, 1998). Validation procedures should exhibit evidence of being situated, authentic, based on common sense, and account for prominent visceral feelings as well.

Instructional design seeks consensus and solidarity among the participants of the process. Systematically building consensus increases ownership, and meaningfulness among the participants involved in the process, as well as those absent from the process, but who are affected by its outcomes. The concept of building consensus is consistent with Kessels' relational approach to curriculum design (explained in detail in chapter 2 of this book), and Walker's Deliberative Approach to curriculum development (explained in detail in chapter 6 of this book). Consensus-building procedures that respect the thoughts, feelings, and actions of all participants involved in the final outcome reduce the risk of unexpected reactions during journeys in learning space. However, unanticipated incidents will occur, and should be expected.

Instructional Design Concept			
Component	***Input***	***Process***	***Output***
Conduct Situational Assessment	Evidence goals are not being achieved An existing performance discrepancy Request to develop instruction Request for training	Present prima facie evidence Collect assessment data Analyze capabilities of the intended learner audience Interview stakeholders Identify causes Draft a project management plan	Context Description Learner Profile Performance Assessment Purpose Statement Educational Goal Draft Instructional Goals
Validate Instructional Goals	Context Description Learner Profile Performance Assessment Purpose Statement Educational Goal Draft Instructional Goals		Validated Instructional Goals
Construct Content Analysis	Validated Instructional Goals	Conduct a goal analysis Conduct nominal group Identify primary learning tasks techniques	Content Analysis Diagram

Generate Objectives	Purpose Statement Content Analysis Diagram	Transform learning tasks into measurable actions State the performance task Describe the condition Identify the judgement criteria	Criterion-referenced Objectives
Generate Test Items	Criterion-referenced Objectives	Simulate the performance reality Generate authentic assessment	Criterion-referenced Test Items
Generate Instructional Strategies	Purpose statement Instructional goals Content analysis diagram Criterion-referenced Objectives Criterion-referenced Test Items	Identify a theoretical framework Organize strategies that will organize the events of instruction Strive for congruence throughout	Lesson Plans Study Guide Instructor Guide
Select and Develop Media and Materials	Criterion-referenced Objectives Criterion-referenced Test Items Lesson Plans	Establish objectives as primary reference for selection process Consult with media specialists Consult with materials developers Select or develop media for achieving objectives	Media & Materials that Support the Objectives
Conduct Formative Evaluation	Context Description Learner Profile Performance Assessment Purpose Statement Educational Goal Validated Instructional Goals Content Analysis Diagram Criterion-referenced Objectives Criterion-referenced Test Items Lesson Plans Media and Materials that Support the Objectives Study Guide Instructor Guide	Collect data Analyze data Summarize data Identify product strengths Identify product weaknesses Identify process strengths Identify process weaknesses Effect necessary changes Identify potential areas that require additional attention	Revision Analysis Summary
Propose Summative Evaluation	Same as above Revision Analysis Summary	Timely internal assessment Timely external assessment	Post-implementation Maintenance Plan

Figure 2. **Input → process → output** matrix for a common set of instructional design components

Determining the value or specific worth of an instructional design product or procedure requires a process that is inherently evaluative. Evaluation in instructional design is intended to strengthen a product or procedure, rather than merely ratify an existing way of knowing something or an existing way of doing something. While the diagnosis of an instructional product's deficits can include knowledge of its underlying causes, there may be no obvious explanation, and therefore, require systematic investigation (Maslowski & Visscher in chapter 11 of this book). Evaluation procedures should initiate, permeate, and conclude any instructional design process so as to afford ample opportunities to intervene during the process for the purpose of improving the product and the process. Revisions to planned instruction should be substantiated with empirical evidence acquired during formative evaluation, thus, increasing the validity of the process.

The contention here is that the essence of any validation process is its ability to identify its philosophical origin. Suggested here is a locus of motivation by using a hierarchy similar to the following:

Phenomena ⇨ Philosophy ⇨ Theory ⇨ Model ⇨ Paradigm.

Phenomenology allows instructional designers to apply procedures that identify abstract and observable knowledge, and events. An example of an instructional design phenomenon is the supposition that the essence of human existence depends on one being a life-long learner. Instructional design *philosophy* posits arguments that provide logical reasoning for organizing events of instruction. An example of a philosophical supposition is that learning can be guided through instruction. *Theory* is used to organize the principles, suppositions, empirical evidence, and philosophy that form processes dedicated to instructional design. Instructional design theories are often marked by a common thematic element that represents a range of thought contained within a theory; such as behaviorism, positivism, objectivism, constructionism, and constructivism. Instructional design *models* permit the manifestation of propositions dedicated to representing the many realities encountered in learning space and provide opportunities for testing instructional theories. *Paradigms* are a coherent set of concepts, principles, assumptions, and basic axioms that have come to be accepted by a sufficiently significant number of researchers in a field (Kuhn, 1962, 1970), and "are sociological phenomena, as well as psychological" (Dills & Romiszowski, 1997, p. 12). A paradigm, such as guided learning space, provides a macro-level frame of reference for instructional design. *Figure 3* presents a hierarchy of conscientiousness that can be used to identify the phenomenological development of ideas used to know the instructional design process.

5. CONCLUSION

Guided learning should be inspirational, and encourage personal empowerment. The processes of instructional design should be dedicated to facilitating negotiated meaning, constructing knowledge, acquiring skills, and managing diversity. Instructional design should respond to the learning context, generate creativity, and validate products and procedures. Future research on the effectiveness of an instructional design model, and the quality of an instructional design approach, should be based on its ability to 1) respond to a performance discrepancy that is attributable to a lack of knowledge and skills, 2) generate learner-centered strategies, and 3) validate the products and procedures used for episodes of guided learning; and to accomplish theses processes parallel to each other. Navigating in learning space requires the ability to process several complex procedures simultaneously. Instructional design should be measured by its potential to act as a conceptual parallel processor that provides people with the capability of navigating through learning space.

Construct	Definition	Attributes
Paradigm	A way of knowing: a conceptual pattern or example that verbally or diagrammatically describes recurring features of phenomena.	• Basis or referent for action. • Illustrates fundamental interrelationships. • Allows for variation in the way a concept is modeled. • Facilitates replication of the fundamental concept.
Model	A way of doing: an explicit representation of a reality. An example or pattern that prescribes relationships in a normative sense.	• Explains or predicts abstract, and observable phenomena. • Varies the levels of generality. • The greater the fidelity between application and supporting theory, the higher the relative level of generality. The more conditions required for application, the lower the relative level of generality. • Manifests a set of propositions.
Theory	An interpretation about a set of organized principles based on empirical evidence; or opinion, thought, observation, and supposition.	• Ability to generate hypotheses; and make predictions. • Provides conceptual explanations founded by philosophical argument. • Explains a phenomenon. • Yet to be disproved. • Provides the motivation for modeling.
Philosophy	Arguments posited in the search for truth through logical reasoning. Ways in which an individual [or community] rationalizes propositions.	• Contains numerous interconnecting theories. • Uses precise terminology, analytical statements and narrative examples to illustrate conditions of an argument. • Characterized as • *Ethical*: System of values governing conduct and expressions of moral approval. • *Epistemological*: Belief system conjured from multi-sensory interactions with the universe. • *Metaphysical*: System of relating to the reality beyond what is perceptible to the senses.
Phenomenology	Study of the development of human consciousness, and self-awareness.	• Based on human perceptions about the universe. • Units of life experiences. • Motivation for human action. • Based on perceived or observable facts and events

Figure 3. A hierarchy of conscientiousness with phenomenology as its origin.

REFERENCES

Andrews, D.H. and Goodson, L.A., 1980, A comparative analysis of models of instructional design, *Journal of Instructional Development, 3* (4), 2-16.

Balachandran, I. and Branch, R., 1997, Continuing professional education among Cytotechnologists: Reasons for participation, *The Journal of Continuing Education in the Health Professions, 17*, 81-96.

Banathy, B.H., 1987, Instructional systems design, In Gagné, R.M. (Ed.), *Instructional technology: Foundations* (pp. 85-112), Lawrence Earlbaum Associates, Publishers, Hillsdale, NJ.

Banathy, B.H., 1991, *Educational systems design: A journey to create the future,* Educational Technology Publications, Englewood Cliffs, New Jersey.

Braden, R.A., 1996, The case for linear instructional design and development: A commentary on models, challenges, and myths, *Educational Technology, 26*(3), 5-23.

Branch, R., 1996, Instructional design as a response to the complexities of instruction, In Venkataiah, N. (Ed.), *Educational Technology* (pp. 21-49), Ashish Publishing House, New Delhi.

Branch, R., 1997, Perceptions of instructional design process models, In Griffin, R.E., Beauchamp, D.G., Hunter, J.M. and Schiffman, C.B. (Eds.), *Selected Readings of the 28th Annual Convention of the International Visual Literacy Association,* Cheyenne, WY.

Branch, R. and Gustafson, K.L., 1998, *Re-visioning models of instructional development,* A paper presented at the annual meeting of the Association for Educational Communications and Technology, St. Louis, MI.

Dick, W. and Carey, L., 1996, *The systematic design of instruction* (4th ed.), Scott, Foresman and Company, Glenview, IL.

Dills, C.R. and Romiszowski, A.J., 1997, The instructional development paradigm: An introduction, In Dills, C.R. and Romiszowski, A.J. (Eds.), *Instructional development paradigms* (pp. 5-30), Educational Technology Publications, Englewood Cliffs, NJ.

Earle, R., 1998, Instructional design and teacher planning: Reflections and perspectives, In Branch, R. and Fitzgerald, M. (Ed.), *Educational technology and media yearbook* (pp. 29-41), Libraries Unlimited, Englewood, CO.

Edmonds, G., Branch, R. and Mukherjee, P., 1994, A conceptual framework for comparing instructional design models, *Educational Technology Research and Development, 42*(4), 55-62.

Gagné, R.M., Briggs, L.J. and Wager, W.W., 1992, *Principles of instructional design* (4th ed.), Holt, Rinehart and Winston, New York.

Gustafson, K.L., 1981, *Survey of instructional development models,* ERIC Clearinghouse on Information Resources, Syracuse University (IR - 81).

Gustafson, K.L., 1991, *Survey of instructional development models* (2nd ed.) [with an annotated ERIC bibliography by G.C. Powell], ERIC Clearinghouse on Information and Technology, Syracuse University: (IR - 91).

Gustafson, K.L. and Branch, R., 1997a, *Survey of instructional development models* (3rd ed.), ERIC Clearinghouse on Information and Technology, Syracuse University (IR - 103).

Gustafson, K.L. and Branch, R., 1997b, Revisioning models of instructional development, *Educational Technology Research and Development, 45*(3), 73-89.

Kuhn, T.S., 1962, 1970, *The structure of scientific resolutions,* University of Chicago Press, Chicago, IL.

Merriam-Webster, 1991, *Webster's ninth new collegiate dictionary,* Merriam-Webster, Incorporated, Springfield, MA.

Moallem, M. and Earle, R.S., 1998, Instructional design models and teacher thinking: Toward a new conceptual model for research and development, *Educational Technology, 38*(2), 5-22.

Reiser, R.A. and Dick, W., 1996, *Instructional planning: A guide for teachers* (2nd ed.), Allyn and Bacon, Boston, MA.

Rossett, A., 1993, Needs assessment, In Anglin, G.J. (Ed.), *Instructional technology: Past, present, and future* (2nd ed.) (pp. 156-169), Libraries Unlimited, Englewood, CO.

Seels, B. and Glasgow, Z., 1998, *Making instructional design decisions,* Merrill, Upper Saddle River, NJ.

Taylor, R.L.M., 1986, *An analysis of development and design models for microcomputer-based instruction,* (Doctoral dissertation, Syracuse University, 1986), University Microfilms International, Ann Arbor, Michigan.

Tripp, S. and Bichelmeyer, B., 1990, Rapid prototyping: An alternative instructional design strategy, *Educational Technology Research and Development, 38*(1), 31-44.

Chapter 13

Characteristics of Computer-based Tools for Education and Training Development: An Introduction

Nienke Nieveen
University of Twente

Kent Gustafson
The University of Georgia

Key words: Education development, Training development, Computer support tools, Electronic performance support systems

Abstract

Currently, Information and Communication Technology (ICT) permeates almost every profession, including those of designers and developers in education and training. This part of the book consists of ten examples of recently created design and development tools (see chapter 14 through 23). These tools can be classified in many ways. A conceptual framework can be of help when organizing the tools into different types. In this introductory chapter such a framework is presented with the following set of attributes: type of output; purpose and evidence of benefits; type of development process supported and any underlying theory; task support; and intended user group. The framework appeared to be useful in analyzing the tools presented in the following chapters. Results of the analysis are presented in an extended table. The chapter concludes with a discussion of the interpretations of the analysis and projected future trends.

1. INTRODUCTION

A growing number of designers and developers in education and training are now being supported during their daily work by computer-based tools. More efficient development, more effective learning programs, and increasingly competent designers are all potential benefits that make these computer-based support tools attractive to many organizations; especially in an era in which knowledge, as an economic factor, becomes more crucial in the competition among organizations. For individuals who work in these organizations the distinction between working and learning is becoming less clear. A positive learning climate in which they are encouraged to work together and are equipped with various kinds of support is important to them. Design and development tools fit into this latter characteristic of the ideal work environment of education and training developers.

The following chapters in this part of the book will describe some of these tools and present issues related to their use. In this introductory chapter we first present (in section 2) a conceptual framework and link it to the tools described in the following chapters. The chapter ends with an interpretation of the analysis, several conclusions and future trends (section 3).

2. CONCEPTUAL FRAMEWORK

Computer-based support tools for developers in the field of training and education can be classified in many ways. In this paragraph a framework is given with the following set of attributes:

A. type of output;
B. purpose and evidence of benefits;
C. type of development process supported and any underlying theory;
D. task support; and
E. intended user group.

Sections 2.1 to 2.5 elaborate the categories and subcategories of the conceptual framework. The framework was used to examine the tools that are described in the remaining chapters of this section of the book (see *Table 1*). The framework should be judged solely on its utility as a schema for examining and selecting from among tools and is not intended to be a scientifically valid taxonomy.

It should be noted that the analysis is based on limited information provided by the developers of the tools and is not based on personal experience with most of the tools. Moreover, for an individual who wants to select a tool based on the information in the framework, technical issues such as needed operating system; software; and hardware would be critical, as well. However, due to constraints on chapter length, the authors of the tools were not able to provide enough information to be able to look at them from these perspectives. The same also is true for other issues such as time needed to learn the tool; time needed until a user starts to be productive with the tool; the costs of a tool; and its general availability.

		GAIDA (Ch. 14)	**QIPP** (Ch. 15)	**PLATO** (Ch. 16)	**MediaPlant** (Ch. 17)	**SimQuest** (Ch. 18)	**CASCADE-SEA** (Ch. 19)
A	Curriculum level	Product	Product	Product	Product	Product	Course and lesson
	Characteristics of results	- Learner-based - Site-specific Paper-based/ Computer-based	- Learner-based - Site-specific - Paper-based	- Learner-based - Generic - Computer-based and Web-based	- Learner-based - Site-specific/ Generic - Computer-based	- Learner-based - Site-specific/ Generic - Computer-based	- Learner-based Teacher-based - Site-specific - Paper-based
B	Purpose of tool	- Better transfer - Performance improvement	- Better transfer - Performance improvement	- Performance improvement - Organizational learning	- Performance improvement	- Better transfer - Performance improvement	- Better transfer - Performance improvement - Organizational learning
	Evidence	Results available	Results available	Results available	IIA	IIA	IIA
C	Development paradigm	Instrumental	Pragmatic	Pragmatic	Pragmatic	Pragmatic	Pragmatic
	Elements of systematic approach	- Design	- Analysis - Design - Development - Implementation - Evaluation	- Analysis - Design - Development - Evaluation	- Design - Development	- Design - Development	- Analysis - Design - Development - Implementation - Evaluation
	Teaching/learning theory	Behavioristic	Cognitivistic	Cognitivistic	Constructivistic	Cognitivistic	Cognitivistic
D	Types of support	- Job aid: toolbox	- Job aid: toolbox, do-it-yourself kit	- Job aid: toolbox, do-it-yourself kit - Communication aids	- Job aids: do-it-yourself kit	- Job aids: do-it-yourself kit	- Job aid: toolbox, do-it-yourself kit
	Adaptability of support	Closed	Inside networked tool	Inside networked tool	Inside tool	Inside tool	Inside tool
E	Type of user group	- SME's	- Designers	- Team with range of roles	- Designers - Learners	- SME's - Teachers	- Teachers
	Scope	Various organizations	One organization	Various organizations	Various organizations	Various organizations	Various organizations
	Computer experience	Low	Medium	High	Medium	Medium	Low

Table 1 continues

		DST (Ch. 20)	**Mercator** (Ch. 21)	**IDXelerator** (Ch. 22)	**AGD** (Ch. 23)	**GTE** (Ch. 23)
A	Curriculum level	Course	Course	Product	Course	Course
	Characteristics of results	- Learner-based, Teacher-based - Site-specific - WWW-based	- Learner-based - Site-specific - Paper-/Computer-/ Web-based	- Learner-based - Site-specific - Computer-based	- Learner-based, Teacher-based - Site-specific - Computer-based	- Learner/based - Site-specific - Computer-based
B	Purpose of tool	- Performance improvement	- Performance improvement	- Performance improvement	- Better transfer - Performance improvement	- Performance improvement
	Evidence	Results available	Results available	IIA	IIA	N/A
C	Development paradigm	Pragmatic	Instrumental	Instrumental	Pragmatic	Instrumental
	Elements of systematic approach	- Design	- Design - Development	- Development	- Analysis - Design	- Analysis - Design - Development
	Teaching/learning theory	Cognitivistic	Cognitivistic	Behavioristic	Cognitive	Cognitivistic
D	Types of support	Job aid: automatic washing machine	Job aid: do-it-yourself kit	Job aids: automatic washing machine	Job aids: toolbox	Job aid: toolbox, automatic washing machine
	Adaptability of support	Closed	Inside tool	Closed	Closed	Closed
E	Type of user group	- Designers - Teachers	- Designers	- SME's	- SME's	- Teachers
	Scope	Specific organization	Various organizations	Various organizations	Various organizations	Various organizations
	Computer experience	Low	Medium	Medium	Medium	Medium

Note: IIA = Insufficient Information Available

Table 1. Conceptual framework

2.1 Type of output (A)

Category A	**Characteristics**
1. Curriculum level	- Few lessons - Product - Course - Collection of courses
2. Characteristics of results	- Target group (Learner-based, teacher-based) - Form (paper-based, computer-based, www-based) - Extensiveness (site specific, generic)

A1: Curriculum level

When designing the framework, a major category (Curriculum levels) was taken as a starting point. A basic definition of curriculum is a 'plan for learning' (cf. Taba, 1962). Such plans can be found at different levels in various educational settings (cf. Gustafson & Branch, 1997). A curriculum can refer to one or a few classroom *lessons* that would be developed and delivered by the same person or to instructional *products* that are developed for distribution and are often, but not always self-instructional. People can also refer to a curriculum when they mean a *course* or unit that would contain managerial elements in addition to the instructional components. And finally, the curriculum concept can refer to an interrelated *collection of courses*. Tools for designers can focus on one or more of these levels.

A2: Characteristics of results

The results of using a development tool can be characterized with respect to their target group, their extensiveness, and their form. The *target group* of the product that is constructed with the use of a support tool could be learner-based or teacher-based. Learner-based materials support students in learning (for example during the execution of assignments). They may also include technical information such as learning how to use certain instruments (e.g. an eyedropper or a computer). Teacher-based materials support teachers with instructional information (such as concrete support for lesson preparation and execution), information regarding the content, and technical information (for example, information on how to operate a video cassette recorder).

The *form* in which such materials appear may range from *paper-based* (e.g., textbooks, workbooks, quick reference cards), to *computer-based* (e.g., drill and practice or simulation), to *world wide web*-based (for instance live interactive video conversations).

The *extensiveness* of the output of a design process can be characterized as being 'site specific' or 'generic' (cf. Walker, 1990). In situations where the target group for a course is small and all persons involved (such as students or trainees, clients and developers) are in relatively close proximity to each other and know each other quite well one can speak of a site-specific course. Many corporate training courses correspond to these characteristics. In contrast, a generic course is developed for a large and diverse target group and the physical distance between the participants is often great. Lesson materials distributed by educational publishers and meant for widespread use in formal educational systems fall in the latter category.

2.2 Purpose and evidence of benefits (B)

Category B	Characteristics
1. Purpose of tool	- Transfer of knowledge and skills - Improved task performance - Organizational learning
2. Evidence of claimed benefits	- Validity - Practicality - Effectiveness

B1: Purpose of the tool

Computer support tools may be designed with various rationales in mind. In this section three major purposes are mentioned (transfer of knowledge and skills, improved task performance; organizational learning).

First, it is assumed that the provision of these tools to developers at the location where they normally perform their jobs will encourage the integration of working and learning in the job context. This is supposed to lead to *better transfer of knowledge and skills* to the actual task performance and may lead to cost reduction (Gery, 1991).

Just-in-time access to different kinds of support is another assumed advantage of computer-support systems. This type of access may lead to an extension of the employee long-term memory and a reduction of the working load memory (Law, 1994). Individuals who use a support system do not constantly need to remember all relevant aspects related to their work. By integrating all support in a single system, users can quickly find the required support, without being distracted. Assuming this takes place, the use of a support system could lead to *better task performance* with less mistakes and better quality outcomes (Gery, 1991).

The use of computer support systems could even lead to *organizational learning* (cf. Raybould, 1995). When individuals apply their knowledge during task performance, they can end up with new knowledge with respect to the task. The newly acquired knowledge can be turned into explicit information (for instance descriptions of concrete techniques or instruments). This new information then can be made available to the whole organization, so others can make use of it as well.

B2: Evidence of claimed benefits

Oftentimes the claims of possible advantages of computer support systems for curriculum development remain assumptions. So far, few data are available which demonstrate the practicality and effectiveness of these systems. While anecdotal evidence from the tools' creators is of some value, the needed next step in the emergence of this genre of computer-based tools will require more substantial and substantive data. Of course much of the evidence must be provided by the eventual users of the tools rather than their creators. Hopefully future evaluation of these and other tools will focus on their ability to: increase the effectiveness of learning and performance; enhance the appeal of the instruction to learners, provide learners with more options and ore control over their learning; increase the efficiency of the design and development process, and increase the relevance of learning. For a more complete discussion of why there is such limited evidence available for these and other performance support tools and systems see Gustafson (in press).

2.3 Type of development process and underlying theory (C)

Category C	Characteristics
1. Paradigm for engaging in education and training development	- Instrumental - Communicative - Pragmatic - Artistic
2. Underlying elements of the systematic approach to development of education and training	- Analysis - Design - Development - Implementation - Evaluation
3. Underlying teaching/learning theory	- Behaviorism - Cognitivism - Constructivism

C1: Paradigm for engaging in education and training development

As mentioned earlier, exactly how individual instructional designers go about creating instruction can vary. These variations may be due to project requirements, client preferences, or the designers training and expertise. Visscher and Schulten (1997) believe that there are at least four different perspectives on the development process, but that all draw on the same fundamentals. Their four paradigms (instrumental, communicative, pragmatic, and artistic) are described in chapter 2 of this book and thus are only briefly presented here. Although the four perspectives are presented independently and may be applied that way, often projects will use elements of more than one to achieve their goals. However, tools tend to be based on one of these paradigms.

The *instrumental paradigm* is typified by a rather linear completion of each of the five basic elements (analysis, design, development, implementation, and evaluation). Great emphasis is placed on careful and complete analysis of the setting to determine specific learning objectives and other system requirements before any decisions are made as to how the instruction will be designed. A similar amount of effort is put into designing and developing the instruction which is then subjected to formative evaluation with the expectation that only minor revisions will be required before it can be implemented and summatively evaluated (e.g., Dick & Carey, 1996). The instrumental paradigm is particularly attractive to those involved in training for technical or time-critical skills.

Instructional designers using the *communicative paradigm* place greater emphasis on determining group consensus regarding what is to be developed as well as determining the specific instructional strategies and media to be employed. This paradigm assumes that if the client is not satisfied with the instruction that is created it will not be used, or perhaps misused, and thereby not achieves its purposes. Great reliance is placed on the group's ability to accurately identify its needs and to determine appropriate instructional strategies. The communicative paradigm is attractive to those who have great content knowledge and who also have control over the setting in which learning will take place. A team of university faculty developing a course for their common use would be a typical group of users of the communicative paradigm.

The *pragmatic paradigm* is characterized by early and frequent use of tryout and revision processes to shape both content and instructional strategies. The pragmatic approach, sometimes called rapid prototyping (Tripp & Bichelmeyer, 1990), engages learners in the process almost from the beginning and draws heavily on their advice and suggestions as well as their performance data from its actual use. Many (perhaps even competing) prototypes might be created and tested with no commitment to their later use in the final product. The pragmatic approach has great appeal to designers of computer-based instruction; especially for designing the user interface, user feedback, and management components of the system.

The *artistic paradigm* is characterized by deliberately creating an environment that promotes "outside the box" thinking about how to create the instruction. Brainstorming, seeking input from many and varied sources, and ultimately relying on one's individual, and often intuitive, judgement also characterize this approach. Evaluation tends to be conducted more from the perspective of connoisseurship than objective and external criteria. In recent years a number of highly imaginative and creative multimedia instructional products have be developed using this artistic approach (e.g., Hedberg, Harper, Brown & Corderoy, 1994; Henderson, 1998).

C2: Underlying elements of the systematic approach to development of education and training

Although there are numerous variations on specific details of the systematic design and development process, most authors include five major elements: analysis, design, development, implementation, and evaluation (ADDIE) (Gustafson & Branch, 1997). Procedural models abound that subdivide these five elements in various ways and provide varying amount of detail as to how the resulting tasks should be accomplished (e.g., Dick & Carey, 1996; Smith & Ragan, 1999). There is some disagreement on the degree to which the process of completing these tasks is linear or simultaneous. Some constructivist oriented critics of the systematic design process (e.g., von Glasersfeld, 1995) go so far as to state that it is not valid and the resulting education or training is inappropriate at best and perhaps even destructive to learners achieving worthwhile goals. However, other constructivists (e.g., Wilson, 1997) have provided thoughtful comments on how constructivism and systematic design can complement each other. Although some models focus primarily on only one type of instruction such as interactive video or computer-based instruction (e.g., Bergman & Moore, 1990), the more general models support creating a variety of forms of instruction that can be replicated with some degree of predictability and consistency. Tools for designers may focus on one or all elements of the systematic process.

The five major elements of the systematic approach (analysis, design, development, implementation, and evaluation) are now briefly reviewed. For a more extensive discussion of each see Gustafson and Branch (1997) and Seels and Richey (1994).

Analysis usually is the first step in the process of creating education or training and is completed before other steps are taken. However, analysis may be an ongoing activity occurring simultaneously with other tasks. In any event the general outcomes of analysis are to identify in relatively precise terms: what the intended outcomes are, who is the audience for the instruction, the expected conditions under which the learning is to occur, and how the learning is to be applied (transferred) to "real" settings.

Design may be conducted as an independent or concurrent task with its major outcomes being detailed specifications for the types of instructional strategies, media, management structures, and student support components being created. *Development* (sometimes called production) includes preparation of the actual instruction in its various forms. These might include printed materials, visuals to support an instructor, audio and video materials, simulations and games, web-based materials, or computer-based courseware.

Implementation typically follows development, but when rapid prototyping is being used, implementation may occur simultaneously with development and formative evaluation. Implementation usually involves distributing the instruction across the entire range of intended users in the belief it is at the point where it will be effective for most (if not all) learners. For example, a large corporation may create a one-week sales training program about a new product it is releasing and distribute the training to all its sales offices for simultaneous use. Or, a university might implement a web-based course over a large geographic area once it has been tested and revised with smaller groups in more controlled settings.

The *evaluation* element of the process typically is thought of as having two sub-components; formative and summative evaluation. Formative evaluation serves the purpose of identifying needed changes to the instruction so that it may better accomplish its intended outcomes. Summative evaluation is used to determine that the instruction works as intended outside the controlled environment typically used during formative evaluation.

C3: Underlying teaching/learning theory

Any attempt to define complex concepts such as different learning theories is bound to be less than complete and this brief description is no exception. Further, even those who claim adherence to any theoretical perspective differ (often substantially) in their definitions. Nonetheless, we attempt here to provide short descriptions of what we believe are the main tenets of each of three theories (behaviorism, cognitivism, and constructivism) based on noted authors in the field. Readers interested in more detail on any of the theories should refer to the original citations since each summarizes literature on the respective theories. Another informative source is Ertmer and Newby's article that compares the three theories from the perspective of instructional design (1993). Our sole purpose here is to present the perspectives on these theories that we have applied in reviewing the computer based tools described in this section of the book.

Behaviorism is often linked exclusively to Skinner, however, its origins more likely should be connected to Thorndike and other earlier psychologists (Burton, Moore & Magliaro, 1996). Philosophically, behaviorists believe that the direct study of observable behavior provides a means of understanding how learning takes place, and from that, how to design instruction can be inferred and then empirically tested. Learning is generally defined as establishing a link between a stimulus and a response. Thus learning represents a measurable change in behavior. This leads to the belief that conditions can be established that promote, but not necessarily guarantee, learning will take place. Thus, contingencies can be created for different learners who are expected to be active participants in the learning process. (This latter point is often misunderstood by critics of behaviorism who assume it expects the learner to be a passive vessel into which the teacher pours knowledge.) Behaviorists believe that complex content can be analyzed into its subcomponents that can then be sequenced to facilitate learning and that learning of "prerequisite" content makes for efficient instruction. Since behaviorist instructional designers believe behavior is generally predicable, they rely heavily on task analysis, instructional strategy selection based on available research, and tryout and revision to create instruction that will produce predicable results. Thus, analysis and testing are central activities within the behaviorist paradigm. Virtually all classic instructional design/development models have their roots in behaviorism.

Definitions of *cognitive* psychology vary widely and we have chosen to use one described by Winn and Snyder (1996). They comment that cognitivism is in part a reaction to behaviorism, but at the same time the two theories share many common characteristics. Like behaviorists, cognitivists believe that behavior can be studied, but that there are internal mental processes and mental representations that greatly influence how learning takes place. However, since these processes and representation occur within the individual they can only be studied indirectly. Most study of mental processes by cognitivists has been via the concept of information processing with a particular focus on short and long-term memory. Mental representations have been studied using various concepts including: schema, information mapping, and mental models.

Cognitivists differ from behaviorists in believing that most behavior is not predictable and thus that there will be varying effects of the same stimulus for different individuals. They also believe that learners can mediate their own behavior via metacognitive strategies and can make intuitive leaps without guidance. The ability to make reasonable estimates and decisions on incomplete information are other evidence of "cognition." Cognitivist instructional designers may follow the same general development models as behaviorists, but use different procedures for each step. For example, cognitive task analysis attempts to identify the underlying knowledge related to a step in a process rather than just identifying the step as would a behaviorist. Providing learners with a schema (or encouraging them to relate new learning to the learner's existing one), promoting recall of prior learning, and assessing learners' mental models are other examples of cognitivist instructional design activities. With various degrees of modification, classic instructional design/development models and accompanying procedures are employed by cognitivist instructional designers.

Constructivism, like behaviorism and cognitivism has many adherents presenting somewhat different perspectives. We base this definition on the writings of Duffy and Cunningham (1996). According to those authors, constructivists believe that learning requires active construction rather than acquisition of knowledge by the learner and that instruction involves supporting that construction rather than communicating knowledge. They state that there are two main branches of constructivism: cognitive and socialcultural. Cognitive constructivists consider learning from the perspective of each unique individual whereas socialcultural contructivists consider learning from the perspective of the individual-within-the group. Consequently, cognitive constructivists consider learning to be a process of "active cognitive reorganization" from existing knowledge whereas socialcultural constructivists consider it as "acculturation into an established community of practice" (Duffy & Cunningham, 175). Thus, the role of community in determining both the curriculum and "instructional" methods varies between the two perspectives. However, both believe that knowledge is individually constructed and that there is no external "truth." Formal learning environments can only be expected to achieve consensus, but not congruence among learners and with the instructor.

Among the commonly promoted "instructional" methods are: collaborative learning, problem based learning, discovery learning, cognitive apprenticeships and coaching (Duffy & Cunningham). Classic instructional design models and design/development procedures are not seen as relevant when creating constructivist learning environments. Rather than identifying specific objectives and outcomes measures and then designing and empirically testing strategies to achieve them, constructivist seek to create "supportive structures embedded in the problem tasks themselves, tools that may support and transform participation, and outcomes" (Duffy & Cunningham, p. 195).

2.4 Task support (D)

Category D	Characteristics
1. Types of support	- Communication aids - Job aids - Training aids
2. Adaptability of support	- Outside the tool - Inside the tool - Inside networked tool - Closed

D1: Types of support

Becoming competent in a certain domain is a complex process often requiring external support. Individuals interact with other people, attend training or other learning situations, and make use of handbooks with facts and procedures to become more competent. Meanwhile they gain experience by carrying out different tasks in practice, looking for feedback on their performance, and sharing ideas. Interaction with other people (such as informal mentoring or expert advice), learning facilities, and job aids (such as handbooks, manuals and checklists) are three main types of external performance support from which an individual could benefit when carrying out a task. Computer-based technologies have impacted all three types of external support.

The *communication* among people over long distances (varying from people in the same hallway to people on the other side of the world) has been growing enormously over the last ten years. Making a phone call used to be the major way of communicating with other people over a distance without delay (synchronously). Today's computer technologies make text-conferencing, audio-conferencing and even video-conferencing possible. With respect to asynchronous communication people used to write letters or send faxes. Again, due to new technologies people have additional means (such as electronic mail and news groups) for asynchronous communication.

As a consequence of the new technologies, innovative *learning facilities* have been developed. The applications go by many names, such as computer-based training or computer-assisted instruction. Gery (1987, p. 214) defines computer-based training (CBT) as "an interactive learning experience between a learner and a computer in which the computer provides the majority of the stimulus, the learner must respond, and the computer analyzes the response and provides feedback to the learner."

Many different types of CBT applications exist: drill and practice exercises, tutorials, and simulations, to name just a few. More recent applications may incorporate audio and video. To supplement relatively passive (often paper-based) job aids, people have started developing more *active job aids*, called on-line help systems or reference systems (Carr, 1989; Gery, 1991; Rossett, 1996; Stevens & Stevens, 1995). Help systems are sometimes embedded in computer software. For instance, in case users need support in using a word processor, they can access the help system to find the information related to the problem. Although this kind of support is still rather passive, it may respond to the needs of its users (Carr, 1989). Through the years more intelligent job aids, such as expert systems, have been developed. An expert system serves as an intellectual adviser (Lippert, 1989; Rossett, 1996). For instance, when a medical student does not know what the best treatment will be in order to cure a patient, an expert system may support this student. It will ask a series of questions in order to diagnose the illness. Based on the answers and rules stored in the expert system, it will then recommend a treatment.

Job aids offered by a computer system may have different characteristics (Gayeski, 1991; Halff, 1993; Jonassen & Wilson, 1990). The following metaphors can be used to distinguish between support systems with respect to the amount of user control or computer control. The computer support can be characterized as a *toolbox* containing a collection of instruments. Users are free to apply those instruments that they need during the development process and the results of using them depend on their expertise and experience. This leads to the assumption that these kinds of systems may best fit the work of professional curriculum developers who wish a large degree of freedom and know how to select and use the most appropriate tools. A second metaphor for computer support is a *do-it-yourself kit,* containing a collection of pre-formatted parts of a product that can be put together; accompanied by advice on using the components. In this case user freedom is somewhat restricted by the available components. Moreover, these kinds of support systems are based on the idea that the content and appearance of parts of the final product can be known and provided in a standardized way. This idea may not mesh with the practice of complex curriculum development endeavors. A third metaphor compares the support to a *cookbook*, containing a collection of guidelines prescribing how to reach a well-defined result. This metaphor fits the advisory approach (Halff, 1993). In this approach the computer advises the user, while the implementation of the advice is still the responsibility of the user. Although there are hardly any algorithms in the field of curriculum development, a 'cookbook' based on heuristic guidelines may be suitable in some instances. The latter point also applies to the final metaphor, concerning an *automatic washing machine*. With these systems, the design task is fully performed and controlled by the support system. The user only sets some parameters and the machine "guarantees" a successful result. The user does not need to have prior knowledge of the process, since the machine takes over almost all tasks. This situation relates to a generative approach (Halff, 1993). That is, the developer relies on the computer to generate the curriculum. In theory, this approach could be extended to an even more ambitious support system, which adapts to the user in making a distinction between novice and expert designers, and provides a variety of content-specific design aids (Tennyson, 1993).

Today's computer and networking facilities can provide the three forms of performance support (job aids, communication, and learning facilities), and can even integrate them. Instead of separately providing different ways of support to individuals, an electronic performance support system (EPSS) provides integrated information, advice and learning opportunities to improve user performance (Gery, 1991; Raybould, 1995).

D2: Adaptability of support

Oftentimes support systems place responsibility on its users for adapting the support (such as: instruments, procedural guidelines, etc.) to their situations. The following four types of adaptability can be distinguished: closed tools; adapting support outside the tool; adapting support within one tool; adapting support of a networked tool. In a *closed* support system it is not possible for the user to adapt the support to users' wishes and enrich it with their experiences. These systems are only modifiable by their creators or managers.

Users might be able to adapt the support *outside* the tool. In order to stimulate this the support may be provided in the form of text files. For example, when users indicate that they need certain support, the system may start a word processor and open the specific support file. In this way, all functionality of the word processor is available. The users can edit the support file to their own needs and wishes.

Support systems that allow adaptations *within* the support system go a step further. These systems are designed as expandable libraries and provide users with the opportunity to add their own instruments, guidelines, etc. to the existing library. This may encourage the re-use of these files later.

If a *networked* version of a tool is available, new support files or adapted files become available to other colleagues, as well. This may stimulate learning from the experiences of others in the same organization and promote joint capturing and sharing of data and other design and development activities.

2.5 Intended user group (E)

Category E	Characteristics
1. Expertise of user group	- Professional designer (ISD) - Subject matter expert - Teacher - Learner
2. Scope of intended user group	- Various organizations - Specific organization
3. Computer experience	- Low - High

E1: Expertise of user group

Design and development activities in the field of education and training require "the integration of practical judgment with theoretical insight" (Walker, 1990, p. 499). The developers' experiences in practical matters (what will and will not work in the classroom), and their subject matter knowledge, pedagogical insights, design knowledge and creativity influence the choices and decisions made in curriculum development. Since only few people have expertise in all of these domains, oftentimes a team of designers with complementary expertise is needed. For instance, *subject matter experts* have much knowledge about a certain subject, but oftentimes lack sufficient design expertise. *Professional designers* may lack sufficient subject matter expertise and pedagogical insights. *Teachers* on the other hand usually have many practical instructional experiences with the subject but may lack design knowledge. If computer-based materials are to be designed, a programmer might have a specific role in the design team, as well. In a constructivistic setting, *learners* may be invited to design their own learning situation (cf. Jonassen & Reeves, 1996), again with the help of others. To design high quality education, people with different expertise are needed. To characterize a support tool, it is important to analyze what skills are expected in the intended user group.

E2: Scope of intended user group

Computer-based design and development tools may be intended for developers of only one organization or various organizations. When the support system is developed for a *specific organization*, the support system might use organization-specific terminology, contain customized support (such as organization-specific procedures and instruments) and specify output requirements. Especially in organizations with standardized approaches regarding instructional design and development, the practicality of the support system may increase when customized support is included in the system. In contrast, support systems that are meant for members of *various organizations* need to be more general.

E3: Computer experience

In the case of a computer-based tool it is also important to look at the amount of computer experience needed to be able to be productive with the tool. People who lack computer experience can have problems using certain computer support systems. Computer experience can become a major factor when selecting tools to use for any given project. Our framework characterizes the amount of computer experience required to use the support tools as either low or high.

3. CONCLUSIONS AND TRENDS

The framework presented in this chapter was used to examine the tools that are described in the remaining chapters of this part of the book (see *Table 1*). Here we provide a brief interpretation of the analysis together with some projected future trends.

A. Type of output

In general, the use of most of the tools results in instructional products or courses that are computer-based and have learners as their main target group. Although it might be argued that many, if not all, of the tools could be used for creating many different forms of instruction, they do appear to lend themselves best to one or only a few forms of output. None of the systems seem to focus primarily on the development of interrelated collections of courses. Only a few of them may lead to paper-based materials (GAIDA, CASCADE-SEA and Mercator), and/or web-based materials (PLATO, DST, and Mercator) and only three tools support the development of teacher-based materials (CASCADE-SEA, DST, and AGD).

B. Purpose of the tool

Generally speaking, all tools are designed with the expectation of improving the performance of developers of training and education. In describing the tools, some authors anticipate that their tool will lead to a better transfer of knowledge and skills to the actual task performance because it makes the rationale of the tool (and thus that of the design process) explicit (cf. QIPP, SimQuest, CASCADE-SEA, and AGD). Others expect that the use of their tool will lead to organizational learning, since the tool invites users to make newly acquired information available to their whole organization (PLATO and CASCADE-SEA). However, it should be noted that most of these claims remain assumptions, since few data are available that demonstrates the actual benefits of these tools.

C. Type of process supported and underlying theory

From an overall view, the tools tend to be based on only two of the four paradigms mentioned before (instrumental, communicative, pragmatic, and artistic) for engaging in education and training development. Many tools that were analyzed make extensive use of a prototyping approach, which refers to a pragmatic paradigm. This is with the exception of GAIDA, Mercator, IDXelerator and GTE: these tools seem to be based on the instrumental paradigm that follows a more linear completion of the instructional design process.

When looking closer at the underlying elements of the systematic approach to development of education and training it appears that two tools (QIPP and CASCADE-SEA) intend to support the designer during the entire process (from analysis through evaluation). All other tools support specific elements of the process, of which design and development get the most attention.

When reviewing the tools with respect to the underlying teaching/learning theory, it appears that most tools are based (to various degrees) on a cognitivistic theory. Two tools seem to be based on a more behavioristic theory (GAIDA and IDXelerator) and one starts from constructivism (MediaPlant).

D. Task support

All tools contain job aids to support users in their development activities. The metaphor of a toolbox and a do-it-yourself kit fits most tools. Two systems fully perform and control a part of the users' tasks and thus could be compared with an automatic washing machine (DST and IDXelerator). It should be noted that GAIDA and DST support the design process only and IDXelerator only focuses on supporting the development process. None of the systems that were analyzed have the ability to automate the entire instructional design process. In all cases, considerable human skills are needed to make effective instructional products and courses.

Concerning the two other potential types of support (communication and training aids), only one tool (PLATO) supports the users with communication aids. This may be related to the fact that PLATO is also the only tool that supports an entire development team (see characteristic E1). It is noteworthy that none of the tools seem to include explicit learning facilities for designers who express a need for learning a specific design task. For novice designers, with the possible exception of GAIDA, the tools seem to count on an informal learning process of learning-by-doing or some form of external assistance.

Looking more closely at the possibility of adapting the support provided by the tools leads to a varied picture. It appears that GAIDA, DST, IDXelerator, AGD and GTE are closed tools: it is not possible for the user to adapt the support to their wishes and enrich it with their experiences. MediaPlant, SimQuest, CASCADE-SEA and Mercator seem to allow adaptations within the support system. QIPP and PLATO seem to be the only tools that make adapted or new files available to colleague users.

E. Intended user group

Generally speaking, it appears the designers of all tools started creating the tools with a specific organization in mind. However, for most tools the scope was extended to various organizations. In an overall view, the tools are intended for one or two of the following user groups: professional designers, subject matter experts, teachers and/or learners. This means that the support tools need to contain (parts of) team members' expertise that would have been needed in times when the tool was not available. For instance, if the support system is meant for teachers who want to design their own instructional materials without help of others (cf. CASCADE-SEA and GTE), it means that the tool should include design support that the professional designer would bring into the design team. Tools that support subject matter experts in designing computer-based products and courses without the help of professional designers and programmers (cf. SimQuest, IDXelerator, and AGD) should include design support and an easy-to-use authoring system. In contrast, tools that are intended for an entire design team (e.g. PLATO) help each specific user group of the team in improving the performance of their specific tasks: the support components included in these tools do not intend to take over other team members' roles.

Future trends

In looking ahead it seems clear that there are several already established trends that will dramatically impact the form and substance of future computer-based tools for developing education and training. Among these trends are:

- increasing interest in these tools from many different individuals and organizations and for many different reasons;
- creating of a wider variety of tools ranging from the very narrow and specialized to the very comprehensive and generic;
- increasing commercial distribution and support of tools;
- increasing number of tools to support team development activities;
- increasing number of tools that support different forms of the development process;
- increasing number of tools that support different perspectives on learning;
- increasing number of tools that promote learning by their users; and
- increasing number of tools with "intelligence".

Each of these trends is briefly discussed below.

Computer-based development tools have a relatively brief history. The term Electronic Performance Support Systems (EPSS) was popularized by Gery in 1991. In those days there were few tools available and most were capable of supporting only one or a few of the tasks associated with designing instruction. Users were expected to know both how to use the tools and be expert in the underlying process they supported. From those modest beginnings have blossomed a wide variety of tools in a relatively short period of time and there is every indication there will be many more in the near future. Everyone from novice subject matter experts with no knowledge of instructional design to experts in instructional design has become interested in how these tools can make them more effective and productive. Although the first interest was largely centered in business, industry, and military training development, many of the new tools support educators at all levels from kindergarten to college.

As mentioned, the first tools were almost all created to support only one or a few tasks related to the instructional design process. Although there are now some tools that support many different tasks, none are completely adequate for all tasks on different types of projects despite the sometimes excessive claims of some of their creators. What we now see is expansion in several directions. The sheer number of tools is increasing, but so is the number that attempt to integrate multiple tasks. However, at the same time the number of single purpose tools for highly specialized situations is also increasing. So the array of tools and their degrees of sophistication continues to expand with no end in sight.

As the interest of users in tools has increased, so has the number of organizations producing them on a commercial basis. While many of the early tools where not marketed, indeed some where not even available outside the originating organization, today there is an increasing supply of tools available for purchase. While some might lament the early days when many tools were available free, we feel that commercialization will have major benefits by increasing both tool quality and user support systems. And if the history of other forms of software is any indication, competition will result in users having many more options and the price of sophisticated tools will fall as their creators recover their development costs through volume sales.

Although some tools already support team efforts, more will become network compatible and facilitate work by team members who are widely separated in time and space. Networking and team support are critical to large scale, complex projects, especially those that will result in technology-based instruction such as multi-media or web-based courses. Reducing travel time, speeding tryout and revision, and facilitating distribution will all reduce development time and costs and enhance the attractiveness of team-based tools.

As was briefly described earlier in this chapter, and is discussed in detail in the chapter 2, there are at least four distinct paradigms (instrumental, communicative, pragmatic, and artistic) for how the development process might be performed. Although most early tools best supported the instrumental approach, many of the newer ones can be used to support other paradigms. In particular, tools to support activities within the pragmatic approach (e.g., design of quick and inexpensive prototypes and collecting tryout data) are becoming more available. Those using the communicative approach can also benefit from tools that facilitate interaction, consensus building, and experimentation. Due to the unique nature of many artistic development projects, this may be the last approach for which a large number of tools become available. However these artists may find ways to use tools created for others to support their activities.

While many of the early tools best supported instruction consistent with behaviorist theories of learning, many recent tools are specifically designed from other perspectives. Several of the tools reviewed here seem to best fit the cognitivist perspective and Wright, Hedberg, and Harper go to some lengths to describe how their tool was conceived from the constructivist perspective. The future will see many more tools created to support each of these theoretical perspectives since all have a place in some form of learning or another.

While many early tools where "black box" in nature, many of the current tools are being designed as "glass boxes." By black box we mean that creators of the tool have no intent facilitating learning by the tool's user. Black boxes are designed solely to "get the job done" as effectively and efficiently as possible with no concern for whether the user becomes more competent or not (other than in operating the tool). Thus, the user continues to be completely dependent on the tool to complete the task. Of course users might become more competent, but that is not a stated benefit of its use. Glass box tool design on the other hand has as one of goals that the user will become more competent and perhaps either stop using the tool or become able to use some of more advanced functions not readily accessible to novices. Organizations desiring to invest in their human capital will be more interested in these glass box tools, even though they may be more expensive and less efficient in the short run than their black box counterparts.

One of the most challenging areas of tool development will be for those that contain intelligence. Although the early promises, and sometimes wild claims, made for the power of artificial intelligence remain largely unfulfilled, slow progress is being made on a number of fronts. The work of Merrill on tools to support subject matter expert creation of computer-based training is of particular note. While currently intelligent tools have limited ability to assist or perform complex or integrated tasks, research and development continue and slow, but steady progress is anticipated.

For all of these reasons, we believe the future of computer-based support tools is very bright. The emergence and expansion of tool creation and use that we have witnessed over the last ten years will pale in comparison to what will happen during the next decade. We have no doubt that future tools will be as different from current ones as current desktop computers are from their predecessors of ten years ago. Continuing advances in computers, digital processing, and communication technology will both add to the demand for a complementary set of development tools and support new features we can only dream about today.

REFERENCES

Bergman, R. and Moore, T., 1990, *Managing interactive video/multimedia projects,* Educational Technology Publications, Englewood Cliffs, NJ.

Burton, J., Moore, D. and Magliaro, S., 1996, Behaviorism and instructional technology, In Jonassen, D. (Ed.), *Handbook of research for educational communications and technology,* pp. 46-73, Macmillan, New York.

Carr, C., 1989, Expert systems: What do they mean for training?, *Training, 26*(12), 41-48.

Dick, W. and Carey, L., 1996, *The systematic design of instruction* (4th ed.), , Harper Collins College Publishers, New York.

Duffy, T. and Cunningham, D., 1996, Constructivism: Implications for the design and delivery of instruction, pp. 170-198, In Jonassen, D. (Ed.), *Handbook of research for educational communications and technology,* Macmillan, New York.

Ertmer, P. and Newby, T., 1993, Behaviorism, cognitivism, constructivism: Comparing critical features from an instructional design perspective, *Performance Improvement Quarterly,* 6(4), 50-72.

Gayeski, D.M., 1991, Software tools for empowering instructional developers, *Performance Improvement Quarterly, 4*(4), 21-35.

Gery, G., 1987, *Making CBT happen: Prescriptions for successful implementation of computer based training in your organization,* Weingarten, Boston, MA.

Gery, G.J., 1991, *Electronic performance support systems: How and why to remake the workplace through strategic application of technology,* Boston, MA, Weingarten.

Glasersfeld, E. von, 1995, *Radical constructivism: a way of knowing and learning,* Falmer Press, London.

Gustafson, K., in press, Designing technology-based performance support, *Educational Technology.*

Gustafson, K. and Branch, R., 1997, *Survey of instructional development models* (3rd ed.), Syracuse, NY, ERIC Clearinghouse on Information Resources, Syracuse University.

Halff, H.M., 1993, Prospects for automating instructional design, In Spector, J.M., Polson, M.C. and Muraida, D.J. (Eds.), *Automating instructional design: Concepts and issues,* pp. 67-131, Englewood Cliffs, NJ: Educational Technology.

Hedberg, J., Harper, B., Brown, C. and Corderoy, R., 1994, Exploring user interfaces to improve outcomes, In Beatie, K., McNaught, C. and Wills, S. (Eds.), *Interactive multimedia in university education: Designing for change in teaching and learning,* pp. 15-29, Elsevier, Amsterdam.

Henderson, J.V., 1998, Comprehensive, technology-based clinical education: The virtual practicum, *International Journal in Psychiatry in Medicine, 28*(1), 41-79.

Jonassen, D.H. and Reeves, T.C., 1996, Learning with technology: Using computers as cognitive tools, In Jonassen, D. (Ed.), *Handbook of research for educational communications and technology,* pp. 695-719, Macmillan, New York.

Jonassen, D.H. and Wilson, B.G., 1990, Automated instructional systems design: A review of prototype systems, *Journal of Artificial Intelligence in Education,* 2(2), 17-30.

Law, M.P., 1994, *Electronic performance support systems: Cognitive "training wheels" for the acquisition of skilled performance,* Paper presented at the Association for Educational Communications and Technology conference, February 16-20, Nashville.

Lippert, R.C., 1989, Expert systems: Tutors, tools, and tutees, *Journal of Computer-based Instruction, 16*(1), 11-19.

Raybould, B., 1995, Performance support engineering: An emerging development methodology for enabling organizational learning, *Performance Improvement Quarterly, 8*(1), 7-22.

Rossett, A., 1996, Job aids, In Plomp, Tj. and Ely, D.P. (Eds.), *International encyclopedia of educational technology,* pp. 132-136, Pergamon, Oxford.

Seels, B.B. and Richey, R.C., 1994, *Instructional technology: The definition and domains of the field,* Association for Educational Communications and Technology, Washington, DC.

Smith, P.L. and Ragan, T.J., 1999, *Instructional Design* (2nd ed.), Merrill, Upper Saddle River, NJ.

Stevens, G.H. and Stevens, E.F., 1995, *Designing electronic performance support tools: Improving workplace performance with hypertext, hypermedia and multimedia,* Educational Technology, Publications.Englewood Cliffs, NJ.

Taba, H., 1962, *Curriculum development: Theory and practice,* Harcourt, Brace & World. New York.

Tennyson, R.D., 1993, A framework for automating instructional design, In Spector, J.M., Polson, M.C. and Muraida, D.J. (Eds.), *Automating instructional design: Concepts and issues,* pp. 191-212, , Educational Technology, Englewood Cliffs, NJ.

Tripp, S.D. and Bichelmeyer, B., 1990, Rapid prototyping: An alternative instructional design strategy, *Educational Technology and Development, 38*(1), 31-44.

Visscher-Voerman, J.I.A. and Schulten, E., 1997, *Design approaches in training and education: Insights from practice,* Paper presented at the annual meeting of the American Educational Research Association, March 25-29, Chicago.

Walker, D.F., 1990, *Fundamentals of curriculum,* Harcourt Brace Jovanovich, San Diego, CA.

Wilson, B., 1997, Reflections on constructivism and instructional design, In Dills, C. and Romiszowski, A. (Eds.), *Instructional development paradigms,* 63-80, Educational Technology Publications, Englewood Cliffs, NJ.

Winn, W. and Snyder, D., 1996, Cognitive perspectives in psychology, In Jonassen, D. (Ed.*), Handbook of research for educational communications and technology,* pp. 112-142, Macmillan, New York.

Chapter 14

The Guided Approach to Instructional Design Advising (GAIDA): A Case-based Approach to Developing Instructional Design Expertise

Dennis Gettman
United States Air Force

Theresa McNelly
Development Dimensions International

Daniel Muraida
United States Air Force Research Laboratory

Key words: Instructional design, Instructional development, Automated instructional development, Electronic performance support systems

Abstract

The problem of developing instructional design expertise in Subject Matter Experts (SMEs) has persistently plagued training organizations. This chapter presents a description of the development of an electronic performance support system as one potential remedy. The Guided Approach to Instructional Design Advising (GAIDA) was developed at the Air Force Research Laboratory as part of the Advanced Instructional Design Advisor (AIDA) project. GAIDA offers an on-line, case-based approach to developing instructional design expertise in novice instructional developers. The history, theory, and design of GAIDA are described. Past and on-going evaluation efforts are presented along with recommendations and future direction.

1. INTRODUCTION

Developing instructional design expertise in technically competent individuals has been a persistent problem for training organizations. Typical practice is to select highly skilled technicians to act as trainers, training specialists, and instructional developers. The problem with this approach is that designing good instruction requires a degree of acumen in instructional design that is often lacking in Subject Matter Experts (SMEs). An erroneous assumption often made is that technical expertise is *the* vital component in the instructional development process. Individuals assigned to training functions are typically given an authoring system to learn and directed to develop instruction. The question of knowledge about designing effective instruction is often only a secondary consideration. In our view, there are at least two reasons why training organizations face the problem of trying to make instructional design experts out of their technical personnel. First, instructional design expertise is a cognitively complex skill that takes many years to master. Second, the cognitive complexity is frequently not understood or recognized by decision-makers in organizations.

One solution to the problem of more efficiently developing instructional design expertise is to imbed the instructional design expertise in computer-based instructional authoring systems. This approach has been taken in the Advanced Instructional Design Advisor (AIDA) project (e.g., Hickey, Spector & Muraida, 1992) sponsored by the Air Force Research Laboratory. The AIDA project is engaged with the development of computer-based shells that represent a variety of instructional strategies pertaining to different learning goals (Gagné, 1992). The ultimate goal of the AIDA project is to make the instructional design phase of the development process transparent to the user. A prototype system, the Experimental Advanced Instructional Design Advisor (XAIDA) has evolved into an instructional development and delivery tool that automatically generates interactive courseware. XAIDA was developed for SMEs with no experience in instructional design or interactive courseware development. XAIDA offers the SME a complete set of courseware development features for entering and editing instructional content, associating multimedia with the content, and modifying practice exercises. XAIDA is in the final phase of test and evaluation and has been transitioned to operational use in a variety of public and private organizations (Wenzel, Dirnberger & Ellis, 1998; Wenzel, Keeney, Licklider, Roberts & Reyna, H., 1997).

A complementary, second approach to the dearth of instructional design knowledge is to provide SMEs with a ready means of acquiring instructional design expertise as they are tasked with developing courseware. A spin-off technology from the AIDA project, The Guided Approach to Instructional Design Advising (GAIDA) was developed to provide on-line instructional design guidance for novice developers. GAIDA is not a courseware development system, but rather it is intended to complement development tools by providing on-line instructional design guidance.

2. THE GUIDED APPROACH TO INSTRUCTIONAL DESIGN ADVISING

The GAIDA system presents elaborated guidance for the application of Gagné's nine events of instruction (e.g., Gagné, Briggs & Wager, 1992) to the design of interactive courseware and other instructional media including paperware and classroom lesson plans for traditional stand-up instruction. Guidance is provided in the context of specific instructional objectives such as identification, classification, checklist procedures, memory-based procedures, etc. Gagné and Perkins-Driskoll (1988) theorized that instruction consists of a set of events external to the learner that occur in a generally sequential manner. The set of events are: (1) gain the student's attention, (2) present the learning objective or describe the goal of the lesson, (3) reference aspects of prior knowledge by stimulating recall of relevant prior learning, (4) present the content of the lesson, (5) provide guidance for learning (6) elicit performance through demonstration of action sequences, (7) provide informative and corrective feedback, (8) assess performance, and (9) enhance retention and transfer. A skilled instructional designer must be able to identify the content of each event.

GAIDA has two modes of operation: Guidance and Lesson. The Guidance mode presents general instructional design advice based on the nine events (See *Figure 1*). The user has the options of studying the events in sequence or browsing the events in any order. Further guidance and elaboration of each event can be accessed through the use of "hot" words. The Guidance mode also provides advice specific to the use of multimedia in courseware.

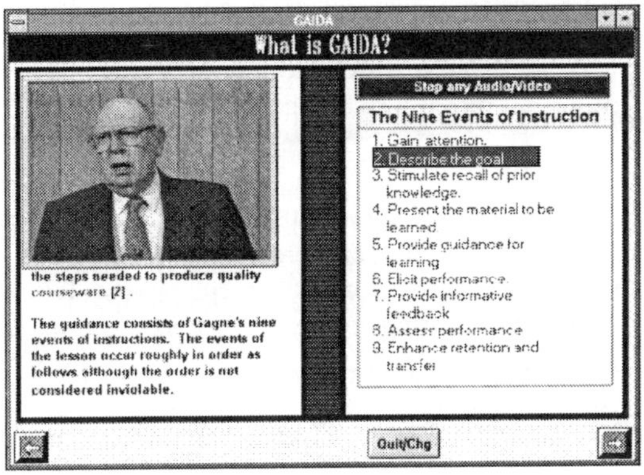

Figure 1. Guidance Mode

The Lesson mode presents examples of interactive courseware developed in the framework of the nine events (See *Figure 2*). These examples are referred to as cases. A case library drives the Lesson mode. The user can select the cases to be presented and view any number of cases from the library. The cases provide examples of how to apply the advice given in the Guidance mode. Early in the process of using GAIDA the user is asked to select a case that is most pertinent to the particular design problem at hand. The user can then move freely between the Guidance and Lesson modes. *Figure 1* depicts the introduction of the nine events by Gagné. The user can skip the general introduction and go directly to an instructional event of interest by clicking the mouse on the desired event on the right half of the screen depicted in *Figure 1*. Once the user has moved to the desired event, for example event 2, "Describe the goal," a click of the mouse on the "Go to Sample Lesson" button will send the user to a case in the Lesson mode where the event is demonstrated (*Figure 2*).

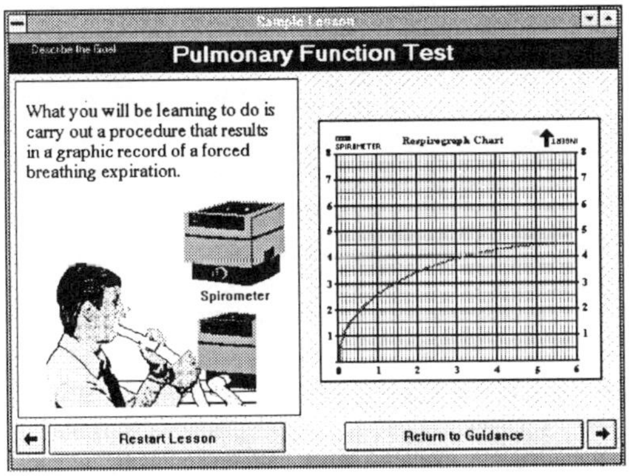

Figure 2. Lesson Mode - "Describe the Goal

The goal of GAIDA is to enable novice developers to become grounded in the basic principles and applications of instructional design *before* they begin implementing an instructional plan within a particular medium, (e.g., using an authoring system). Novice instructional developers are often in the difficult position of being required to learn how to design effective instruction while learning how to use an authoring system. Thus, novice developers tend to overlook principles of instructional design as critical to developing effective courseware. It is important to note that although GAIDA was designed with computer-based instruction in mind, it is appropriate for designing instruction for delivery with any instructional medium. GAIDA is a performance support tool for instructional planning and design.

Distribution of GAIDA

GAIDA has been demonstrated and distributed to a wide variety of organizations nationally and internationally. The reactions to GAIDA have been overwhelmingly positive. It was delivered to the Air Force's Education and Training Command in 1993 for testing at their technical training centers. As a result of this testing it has been adopted for use in the Computer-based Instructional Developers Course at Sheppard Air Force Base, and is being distributed with the updated Air Force policy manuals on instructional systems development. GAIDA lessons were also developed and transitioned to the Air Force's Academic Instructor School in 1995. This project was unique because the lessons developed were audio/visual cases of expert instructors presenting staged stand-up instruction lessons organized around the nine events of instruction. This project illustrates the flexibility of a tool like GAIDA. The nine events of instruction and the relevant examples are applicable for any instructional delivery medium including classroom lessons. GAIDA received a federally sponsored award in 1996 for its potential as a promising transition technology to private industry.

3. EVALUATION: DOES GAIDA WORK?

Does the system actually enable novice designers to learn and apply instructional design knowledge effectively? Several formative evaluations of GAIDA have been conducted (e.g., Gagné, 1992; Tennyson, 1995; Tennyson & Gettman, 1995), and initial results have been positive. To date, however, these evaluations have been qualitative in nature, have used relatively small numbers of participants, and the results have been somewhat mixed. The following section will briefly describe past evaluations of GAIDA and ongoing evaluative research.

GAIDA represents one type of training program for novice instructional developers. The effectiveness of any training program can be evaluated in terms of four criteria: reaction, learning, behavior, and results (Kirkpatrick, 1987). Reaction criteria measure how well the participants or trainees liked the program, including, but not limited to, its content, the methods used, and the training environment. Learning criteria measure knowledge gained by the trainees. Knowledge acquisition is typically measured by paper-and-pencil tests (Wexley & Latham, 1991). Behavior criteria measure changes in trainees' overt behaviors upon returning to the job. Results criteria measure the actual benefit, at an organizational level, of the training program. Previous evaluations of GAIDA have mainly focused on reaction and learning criteria.

Gagné conducted the first formative evaluation of GAIDA in December of 1991 at Lowry Air Force Base. The evaluation, conducted with Air Force SMEs, focused on three areas: clarity, impact, and feasibility. The evaluation results showed that: (a) SMEs experienced no difficulty in understanding or using the instruction of the GAIDA lesson; (b) the lessons SMEs designed closely followed the model task provided in the cases (i.e., the instruction developed was both feasible and understandable); and (c) 5 of 6 (83%) of the SMEs judged the lessons they designed to be useful and effective. Only one SME took exception in judging the applicability of his lesson.

A more extensive evaluation of GAIDA was conducted from 1994-1995 (Tennyson, 1995; Tennyson & Gettman, 1995). The evaluation conducted with instructional science graduate students (N=15), focused on learning, attitudes and perceptions, and quality of the GAIDA program. The evaluation results revealed that, in general, learning the nine events is improved with the use of GAIDA over the use of strictly text. Student attitudes toward GAIDA shifted from "not much interest" to "a good deal of interest" by the end of the course. The shift in attitude was most apparent in novice designers as compared to competent designers. Results from an end-of-course questionnaire also indicated that students gave high ratings to the quality of the GAIDA program.

Tennyson also examined how graduate students perceived GAIDA within the broad instructional context of their educational experience. Results showed that, in the winter quarter when GAIDA's content was important to course content, GAIDA was perceived as being "very good." However, in the spring quarter when GAIDA was not a major component of course content, perceptions were "marginal."

The most complete evaluation of GAIDA to date is currently underway. This evaluation focuses on reaction, learning, and behavior criteria. Assessments will be made before and after a six-week exposure to GAIDA to assess reactions. Learning the nine events of instruction will be assessed as differences in pre- and post evaluation written test scores. Further, an assessment will be made of the manner in which participants organize their knowledge of the nine events of instruction. There has been a growing interest in measuring how knowledge is organized in memory (Dorsey & Foster, 1996). This interest suggests that the *organization* of knowledge stored in memory is perhaps of equal or greater importance than the *amount* of knowledge stored in memory (Johnson-Laird, 1983; Kraiger, Ford & Salas, 1992; Rouse & Morris, 1986). Knowledge organization, referred to as a mental model has been defined as "rich and elaborate structures, reflecting the user's understanding of what the system contains, how it works, and why it works that way. It can be conceived as knowledge about the system sufficient to permit the user to try out actions mentally before choosing one to execute" (Carroll & Olson, 1988, p. 51).

Behavior criteria will be assessed as an individual's ability to transfer what was learned in training to the operational setting. For this upcoming evaluation, behavior will be assessed as a difference score between ratings of the degree to which Gagné's nine events are incorporated into instruction developed before and after implementation of GAIDA. In addition to the reaction, learning, and behavioral measures, several measures of individual differences will be obtained. In particular, personality (e.g., conscientiousness) and cognitive ability will be assessed as a pre-measure.

4. FUTURE DIRECTIONS FOR TOOLS LIKE GAIDA

Several valuable lessons have been learned from demonstrating and distributing GAIDA. First, the strong intuitive appeal or face validity of GAIDA was apparent throughout the history of the project. This appeal had both positive and negative consequences. Representatives from training organizations that saw GAIDA demonstrated almost always asked to take the system back to work with them. We established a policy early of distributing the system on demand. We required signed letters of agreement stating that the user would report back to us on certain criteria. However, we do not recommend this approach. Agreements were rarely complied with and we lacked sufficient resources to be involved with and adequately track implementation of GAIDA. Our experience suggests limiting the distribution of an experimental tool such as GAIDA until sufficient implementation and evaluation resources are made available.

The computer-based technology used to program and develop GAIDA changed rapidly over the course of development. It was difficult to maintain a consistent version long enough to evaluate it. We discovered that once an experimental tool is developed its control is important. Programmers and other stakeholders may not recognize the importance of scientific controls, especially when the tool has high face validity. In the future we would more vigorously monitor and control software development on versions for evaluation.

GAIDA represents only a part of a complete training system for novice instructional developers. GAIDA is not intended to be the "solution" for training organizations. It is important to adhere to a systematic approach of conducting needs analyses of prospective users (i.e., person, tasks, and organization), (e.g., Goldstein, 1993; Wexley & Latham, 1991). Certain performance support tools that fit well in one organization are often not needed in other organizations. Moreover, it is often unclear as how to best introduce a new support technology into an organization. Our experience has been that organizations would request and receive GAIDA but would never use it operationally. It was not that the need no longer existed, or the organization necessarily found a better way. It was typically that organizations did not know how to effectively adopt the new technology in the context of their on-going work. Longitudinal research is needed to better understand technology transfer and the resulting changes that occur within an organization. GAIDA represents a first attempt at developing an on-line, case-based instructional design advisor. Future work with such tools should seek to integrate instructional design advice with automated approaches to instructional development such as XAIDA. Future work with case-based approaches should include investigations of techniques for developing and managing a case library. Due to extreme budget cuts, it is doubtful that the Air Force Research Laboratory will further develop GAIDA as a stand-alone performance support tool. This is unfortunate because we believe that, conceptually, GAIDA offers a unique and potentially valuable approach to speeding up the acquisition of instructional design skills.

REFERENCES

Carroll, J.M. and Olson, J.R., 1988, Mental models in human-computer interaction, In Helander, M. (Ed.), *Handbook of human-computer interaction*, Elsevier , Amsterdam.

Dorsey, D.W. and Foster, L.L., 1996, An application of knowledge structure assessment in the computer knowledge domain: Relations with experience and implications for training, Symposium paper presented in R. Klimoski's, *Knowledge Structures and Training: Measurement and Validity Issues*, 11th Annual Conference of the Society for Industrial and Organizational Psychology, San Diego, CA.

Gagné, R.M., 1992, *Tryout of an organizing strategy for lesson design: maintenance procedure with checklist*, Brooks AFB, TX, Air Force Research Laboratory (AL-TP-1992-0016).

Gagné, R.M., Briggs, L.J. and Wager, W.W., 1992, *Principles of instructional design* (4th ed.), Holt, Rinehart, and Winston, New York.

Gagné, R.M. and Perkins-Driskoll, M., 1988, *Essentials of learning for instruction* (2nd ed.), Prentice Hall, Englewood Cliffs, NJ.

Goldstein, I.L., 1993, *Training in organizations*, Brooks/Cole Publishing Company, Pacific Grove, CA.

Hickey, A.E., Spector, J.M. and Muraida, D.J., 1992, *Design specifications for the Advanced Instructional Design Advisor (AIDA),* Air Force Research Laboratory (AL-TR-1991-0085-Vol. 2), Brooks AFB, TX.

Johnson-Laird, P., 1983, *Mental models*, Cambridge, MA, Harvard University Press.

Kirkpatrick, D.L., 1987, Evaluation of training, In Craig, R.L. (Ed.), *Training and development handbook: A guide to human resource development* (3rd ed.), McGraw Hill, New York.

Kraiger, K., Ford, J.K. and Salas, E., 1992, Application of cognitive, skill-based, and affective theories of learning outcomes to new methods of training evaluation, *Journal of Applied Psychology,* 78, 311-328.

Rouse, W.B. and Morris, N.M., 1986, On looking into the black box: Prospects and limits in the search for mental models, *Psychological Bulletin,* 100, 349-363.

Tennyson, R.D. and Gettman, D.J., in press, *GAIDA evaluation report: Student learning, quality, attitudes, and perceptions,* Air Force Research Laboratory, Brooks AFB, TX.

Wenzel, B.M., Dirnberger, M.T. and Ellis, R., 1998, April, Evaluation of the instructional effectiveness of distributed interactive multimedia courseware for emergency medical technician training, Paper presented at the Sixteenth Biennial Applied Behavioral Sciences Symposium, USAF Academy Department of Behavioral Sciences and Leadership, Colorado Springs, CO.

Wenzel, B.M., Keeney, C., Licklider, J., Roberts, J.D. and Reyna, H., 1997, August, *Evaluation of the courseware developed and delivered using the Experimental Advanced Instructional Design Advisor,* Presentation at the annual meeting of the Texas Association of College Technical Educators, Austin, Texas.

Wexley, K.N. and Latham, G.P., 1991, *Developing and training human resources in organizations,* Harper Collins, New York.

Chapter 15

An EPSS for Instructional Design: NCR's Quality Information Products Process

Thomas Jury
WPI

Thomas Reeves
University of Georgia

Keywords: Instructional design, Online documentation, Performance support

Abstract

In response to the need to enhance the development of information products at NCR, a new development methodology called QIPP (Quality Information Products Process) was developed for use by information product developers, including both technical writers and instructional designers. QIPP, a systems approach that combines instructional systems design with technical documentation development, can be used to develop any type of information product ranging from printed or online technical manuals to interactive multimedia training CD-ROMs and web-sites. To support the integration of QIPP throughout NCR, a QIPP EPSS was developed by a collaboration of NCR personnel, external contractors, and university consultants. Evaluation results indicated that the EPSS did support the adoption and implementation of QIPP, especially at those sites where locally produced resources were incorporated into the EPSS.

1. BACKGROUND OF THE PROJECT

NCR Corporation designs, develops, markets, and services information technology products, services, systems, and solutions worldwide. With its primary markets in retail and financial businesses, NCR is a leader in manufacturing and selling automatic teller machines, bar code scanners, and point of sale terminals. NCR also develops and sells large scale computer systems that connect numerous financial terminals and devices and manage the vast array of consumer data collected during transactions.

Launched in 1884 as "The National Cash Register Company" in Dayton, Ohio, NCR today is a six and a half billion dollar global company that has seen more than it's share of corporate changes in the past decade (http://www.ncr.com). In 1992, it was the target of a hostile take-over by AT&T which eventually turned friendly after NCR stock price nearly tripled due to AT&T overtures. Unfortunately, during the brief but rocky marriage, NCR's losses totaled nearly $4 billion. Late in 1995, AT&T announced the 'spin off' of two start-up companies, NCR and Lucent (formerly Bell Labs). NCR steadily regained its autonomy, and as of 1997, once again became an independent company.

In the early 90's, NCR made major shifts in its product line. Prior to that time, the company's computer systems used proprietary operating systems and the retail, financial, large-sized, mid-sized, and workstation product lines all had different architectures. NCR redesigned its product lines, incorporating Intel chips and running open, non-proprietary operating systems, mainly UNIX and MS-DOS. All of NCR embraced "Open Cooperative Computing." (Ironically, this new strategy was probably what attracted other corporate suitors, including AT&T, to the prospects of a takeover.)

NCR had a difficult time incorporating all the changes the new product architectures implied and the company's training departments bore the brunt of the blame. The various education and training departments were expected to re-skill the sales force and the support personnel so that NCR could effectively sell and service their new product lines. As sales and support lagged behind expectations, "heads rolled" and an education task force was assembled to "evaluate NCR's approach to sales, support, and customer education, identify major deficiencies and underlying causes, and recommend changes." Among the recommended changes were:

- combining the function of product training and technical documentation into a single function, called "Information Products";
- creating an integrated Information Products Process which would be linked to the standard product development process;
- applying technology to the development and delivery of information products.

2. QUALITY INFORMATION PRODUCTS PROCESS (QIPP)

In a largely top-down manner, NCR's management created the Information Products Division, making it responsible for implementing and managing the new information products strategy. Information products (as contrasted with tangible products such as cash registers) included many different things, e.g., sales brochures, printed or online technical manuals, and even interactive multimedia training CD-ROMs and web-sites. In the past, some of these products were developed by technical writers whereas others were developed by instructional designers. Now everyone associated with information products would be known as an "information products developer." One of the first challenges was creating a process which would define how to develop and deliver Information Products defined as both training and documentation. The result was called the Quality Information Products Process (QIPP), a systems approach that combined instructional systems design with technical documentation development. This integrated development model emphasized user performance, quality, and customer requirements. QIPP contains eight phases as illustrated in *Figure 1*.

Phase	**Definition**	**Deliverables**
1 - Analysis	Analyzing audiences and tasks, planning information product deliverables, and testing for strategic errors	Audience descriptions, task descriptions, audience/task matrices, Information Products Plan
2 - Design	Creating successively finer models of the information products, testing models for structural errors	Substantive outlines, structured outlines, performance objectives, Unit Plans
3 - Development	Developing the information product prototypes, testing the prototypes for tactical errors and for usability	Prototypes of information products
4 – Validation	Conducting integration and usability testing with representative samples of the intended audiences	Validation report (usability, integration, Kirkpatrick's Level 1 and Level 2 evaluation results), revised prototypes
5 – Production	Creating reproducible masters	Reproducible masters, manufacturing forms
6 – Manufacturing	Training instructors and generating the required inventory stock	Orderable information products, trained instructors
7 – Delivery	Delivering the information products to the intended audiences in a timely and cost-effective manner	Ready-to-use information products, trained users
8 - Customer Satisfaction Assurance	Actively collecting and acting upon customer feedback	Customer satisfaction report (surveys, sites visits, focus groups, Kirkpatrick's Level 2 and Level 3 evaluation results)

Figure 1. QIPP Phases and Deliverables

3. SUPPORT STRATEGIES FOR QIPP

NCR implemented several programs to support the Quality Information Products Process and to encourage information product developers to use the process. First, a two-day introductory workshop was created and presented to several information product departments. In addition, a semi-annual company wide information products conference was instituted. *Information Products*, an in-house professional journal, was published quarterly. As illustrated in *Figure 2*, NCR tried to attend to all aspects of the information products infrastructure to enhance the use and acceptance of the new process.

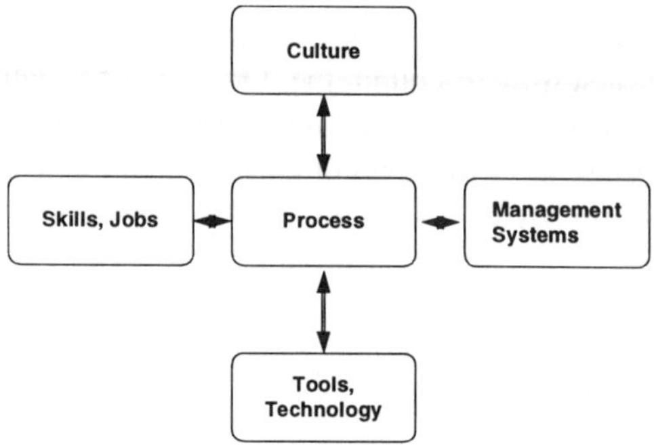

Figure 2. Process Model

Supporting programs were designed to address one or more of the infrastructure issues:

- Skills and Jobs: developers had to have the necessary skills and their assigned jobs had to require those skills.
- Tools and Technology: developers had to have the necessary information technologies and other tools in order to execute the QIPP process.
- Culture: developers had to be encouraged to use the QIPP process and publicly acknowledged for doing so.
- Management Systems: rewards (financial and other) had to be consistent with using the QIPP process.

4. THE EPSS

With hundreds of people involved in this change process, representing different "cultures" such as technical writing and instructional systems design, the need for a tool to help everyone work together in following the QIPP process was recognized. The proposed solution was an electronic performance support system (EPSS[1]) (Gery, 1991). EPSS's are powerful electronic systems that integrate the information and tools required for jobs or tasks such as marketing, sales, or hardware maintenance and repair, as well as for professional practices in fields as diverse as medicine, law, and business. EPSS's may include information, tools, tutorials, simulations, and expert systems as well as other types of productivity software needed to support the highest levels of performance.

5. PROJECT TEAM

To create the EPSS, NCR gathered a cross-functional design and development team made up of experts in the areas of instructional systems design, technical

---

[1] Additional information about EPSS can be found on the World Wide Web at: (http://itech1.coe.uga.edu/EPSS/EPSS.html).

communication, electronic performance support systems, and information technology. Analysts from NCR Corporation (including the first author of this chapter), faculty from the Department of Instructional Technology (http://itech1.coe.uga.edu/) at The University of Georgia (including the second author of this chapter), and consultants from Comware, Inc. (http://www.comware.com/) worked with representatives of various information product development groups at NCR during the analysis, design, development, and pilot testing of the QIPP EPSS to ensure that the completed system had 'buy in' from the users and met customer expectations. The consultants from The University of Georgia had previous experience with developing an EPSS for instructional designers (Gustafson & Reeves, 1990), and Comware, Inc. possessed unique expertise in developing EPSS's for other companies.

6. REQUIREMENTS OF THE EPSS

The EPSS project team began it work by interviewing several members of the constituent group, and then developing a "rapid prototype" of what the EPSS might look like (Gustafson & Branch, 1997; Witt & Wager, 1994). The team demonstrated the prototype to constituents, and discussed requirements, environments, and user skills with them. A number of issues were identified during this analysis, but three major items surfaced.

6.1 The need for the integration of applications into the EPSS

Information product developers use a wide variety of commercial software applications, including word processors, desktop publishing systems, spreadsheets, databases, graphics, and authoring systems. Those applications would have to be integrated into the EPSS and allow the developer to access them easily. It would have been easier to define a limited set of applications and support those with the EPSS, but that would not have met the needs of NCR's diverse documentation and training communities.

6.2 The need for local modifiability

The EPSS had to be functional in a wide variety of development sites around the globe. Local teams and individuals would have to be able to modify the EPSS to match their needs and capabilities. Several kinds of modifications were anticipated:

- substituting or adding additional applications to the EPSS (e.g., substituting WordPerfect for Word for Windows);
- deleting applications and the supporting information for components of the QIPP ISD process that might not be applicable at a site (e.g., eliminating CBT authoring systems at a site that does not develop CBT);
- adding examples that reflect local conditions, markets, and needs (e.g., replacing standard generic audience descriptions with industry specific ones); and
- adding new tools and variations developed locally (e.g., adding a printing costs estimation spreadsheet developed to reflect the prices charged by local vendors).

6.3 The need to include ISO 9000 implications

Most NCR business units have either passed ISO 9000 certification or are planning for it (Sanders, Sanders, & Johnson, 1993). Basically, to be ISO certified, the business unit must have documented processes and evidence that those processes are followed. The QIPP EPSS needed to support the ISO 9000 requirements. This means that policies, procedures, and work instructions must be:

- accurate and reflect current practices;
- accessible by those who use them; and
- modifiable by the responsible parties.

In addition, ISO 9000 requires that records of the results of all work activities must be stored and maintained for future reference.

7. A SCENARIO: SAMPLE USE OF THE EPSS

The QIPP EPSS prototype was refined and redesigned based upon constituent feedback and subsequently implemented as several beta sites around the USA and Europe. The following is a scenario showing how an information products developer could use the EPSS to help in a development project. Assume that the information product developer is working on a user manual to accompany a new cash register being produced at NCR that is code-named as the X101.

7.1 Open the project in the QIPP EPSS and select a phase, work activity, and job aid.

As illustrated in *Figure 3*, the Project Folder Screen identifies the specific phases, work activities, and job aids of the ISD process. The developer selects the phase and work activity to be performed and then picks one of the available job aids. In the following example, the developer has chosen the "Analysis" phase and the work activity of "Preparing Audience Definitions." Two job aids are listed, and the developer chooses the "Info Gathering Aid."

The Info Gathering Aid helps a developer decide when and how to gather more information about an audience. It contains a sample questionnaire, a sample interview protocol, and a sample focus group protocol. When the developer selects the Info Gathering Aid option in the "Job Aids" menu, the tool opens a Job Aid Access Screen for the Info Gathering Aid (see *Figure 4*).

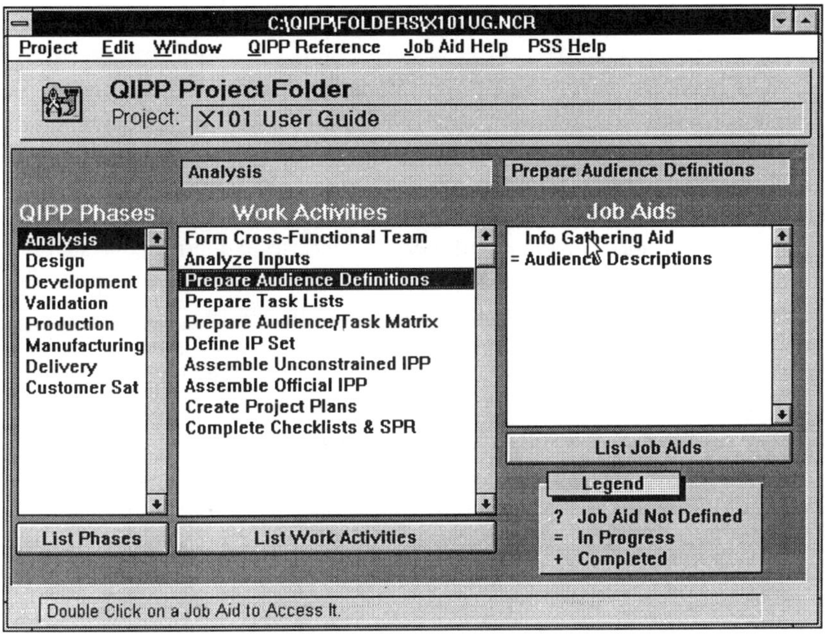

Figure 3. Project Folder Screen--Choosing a Phase, Work Activity, and Job Aid

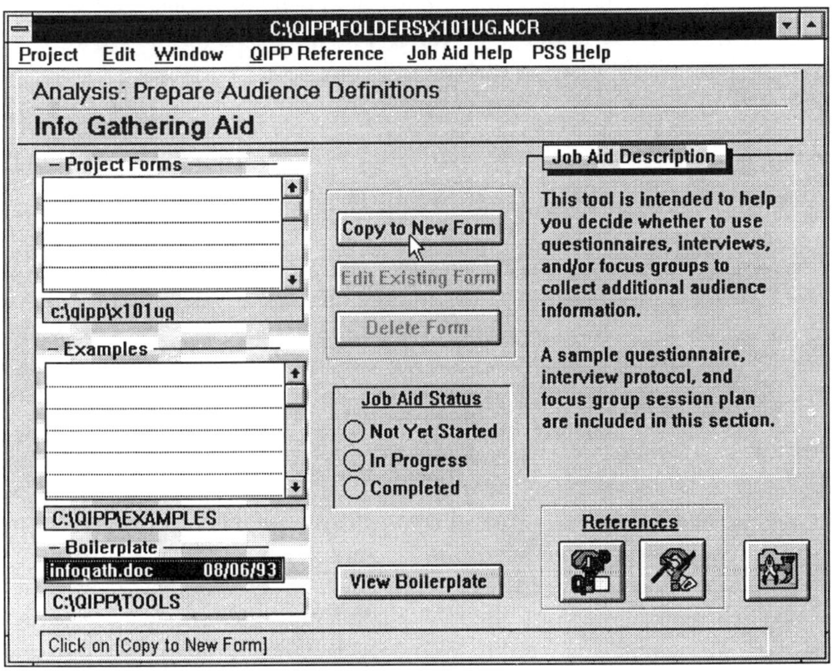

Figure 4. Job Aid Access Screen for Info Gathering Aid

7.2 Select help about the job aid

Each job aid has an associated help file that is organized into six sections: 1) What It Is, 2) Why It's Included, 3) Where in QIPP, 4) How to Use It, 5) References, and 6) Credits. As illustrated in *Figure 5*, the developer selects "How to Use It" in order to see how the Information Gathering Job Aid might be used to help in preparing audience definitions for the X101 cash register.

As illustrated in *Figure 5*, the developer can choose to learn about survey types, access directions for conducting surveys, or use a decision aid to help decide what survey type to use. The decision aid is interactive in that it asks the developer a series of questions (e.g., how much time do you have before you need the survey results) and then makes a recommendation about which type of survey to conduct.

7.3 Refer to an online ISD reference guide

In addition to help about the job aid, the developer can refer to an on-line manual explaining the ISD process. The access to this on-line manual is context-sensitive. For example, when the developer refers to the process manual, the EPSS jumps to a section that describes the work activity and phase the developer is trying to perform. (See *Figure 6*.)

In this case, the QIPP Reference Guide opens to a section that lists several types of background research that can be performed to support this QIPP activity. The developer can use any of the standard Windows help features (Contents, Search, Back, etc.) to access information in the QIPP Reference Guide. Both the QIPP Reference Guide and the Job Aid Help are available to the information product developer at any time while using this job aid, but they are passive help systems in that they must be selected by the developer.

7.4 Create a new project form using the job aid

After referring to the appropriate help (for the job aid and the QIPP Reference Guide) the developer is going to use the Info Gather Aid. This job aid is actually a Word for Windows file, or "boilerplate." To use the job aid, the developer selects the boilerplate ("infogath.doc") on the Job Aid Access Screen (see *Figure 4*) and clicks on the button "Copy to New Form." This copies the boilerplate to a new file, (referred to as a Project Form) and launches Word for Windows with the new file loaded (see *Figure 7*). The system prompts the developer to give the new file a name and the developer calls it "INFO1.DOC."

As illustrated in *Figure 7*, this particular word file contains links within it, and the developer can choose to explore the use of a questionnaire, interview, or focus group to gather audience information. Once a choice is made, the developer can create a list of questions for the specific instrument from the samples included in the job aid. Since this new file is simply a Word for Windows file, the developer can edit, modify, add, or delete any information in the form so that it meets the needs of his/her specific project.

Saving the file and exiting from Word, the developer goes back to the EPSS, and the new form INFO1.DOC is listed among the project forms for this work activity. (See *Figure 8*.) If the developer wants to revise this form later in the project cycle, clicking on the Project Form called INFO1.DOC will again launch Word with this file open and ready for additional editing.

Having completed this task, the EPSS user is now ready to move onto the next one by accessing another phase, activity, or job aid and its accompanying help files. Alternatively, the user may be working on several projects, and decide to move to a completely different project, e.g., the development of a multimedia training web-site for a different NCR product.

The QIPP EPSS allows the developer to maintain multiple projects, and within each project, multiple documents. There are more than fifty tools in the basic installation of the EPSS, and developers can add their own tools without difficulty. Most tools, such as the Info Gather Aid illustrated above, are templates. Others are productivity tools such as spreadsheets that allow the calculation of budgets for various activities, and a few tools provide assistance in making decisions.

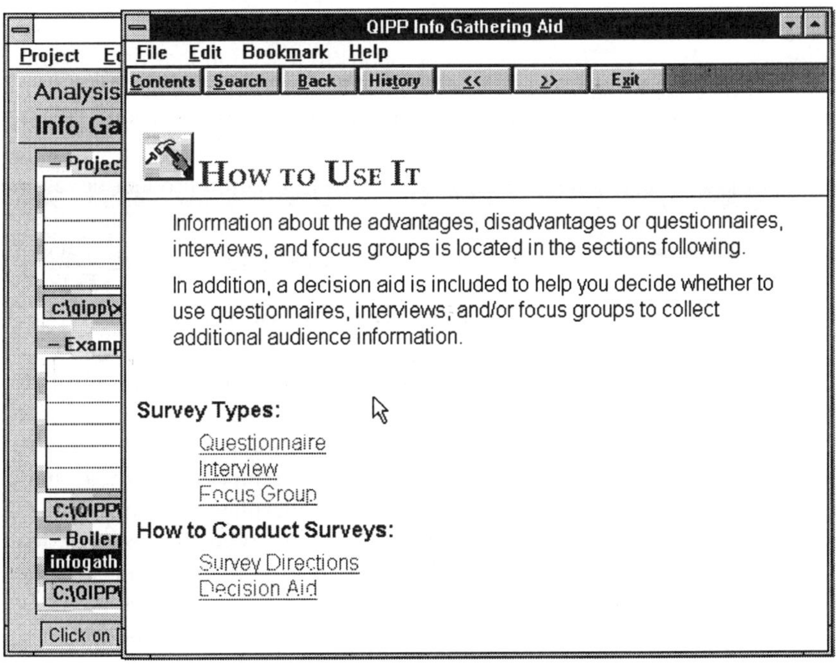

Figure 5. "How To Use It" Help for the Info Gathering Aid

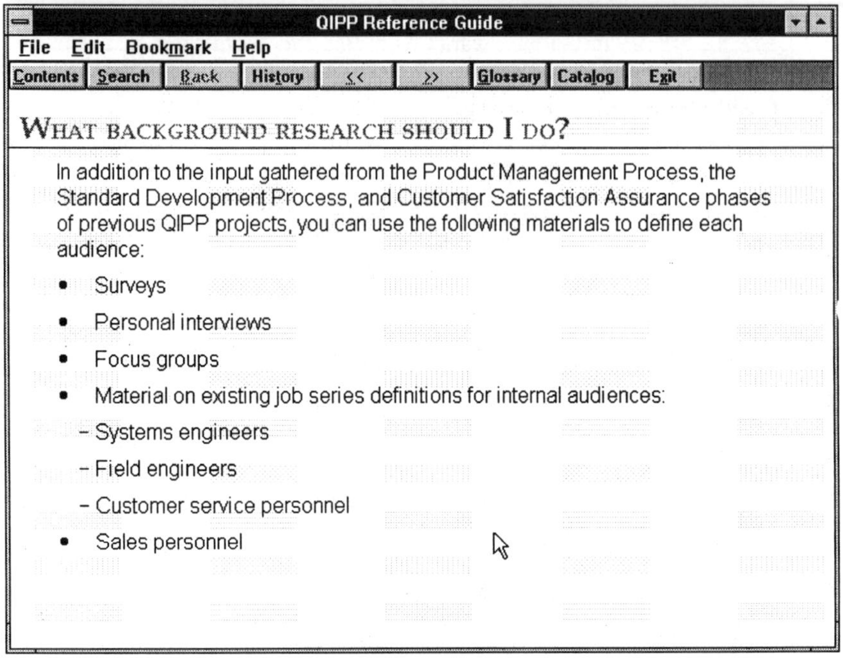

Figure 6. Sample screen from the QIPP On-line Reference Guide

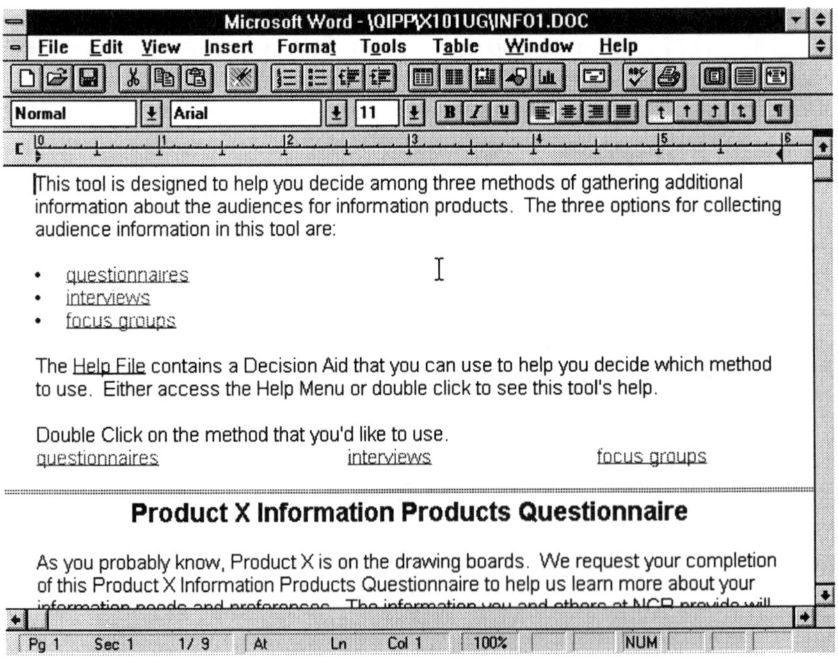

Figure 7. New File (INFO1.DOC) generated from the Info Gathering Job Aid

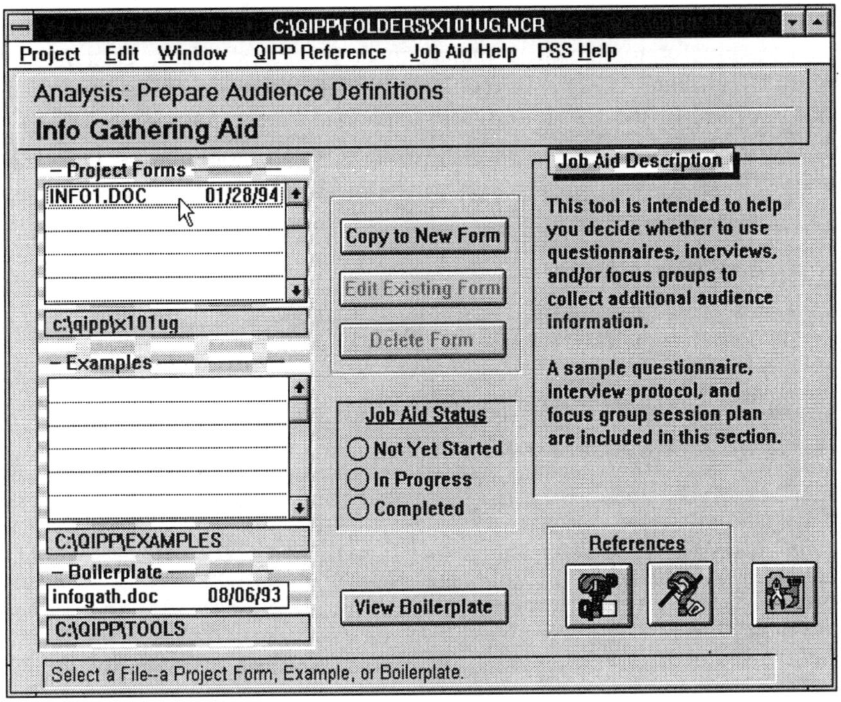

Figure 8. Job Aid Access with New Project Form

8. ORGANIZATIONAL RESULTS

At first, deployment of this EPSS was limited to only a selected few information product development organizations within NCR. The early adopters among NCR's information product developers were enthusiastic about the QIPP, and an external formative evaluation of the EPSS was conducted with positive results (Gustafson, Reeves, & Smith, 1995). No major problems were identified. The interface was easy to learn and it allowed efficient use of the tools. Minor weaknesses included a lack of sufficient explanation for certain processes in QIPP. Participants in the evaluation also expressed a desire for more samples to be included with each tool.

Although there was no hard empirical evidence, anecdotal data clearly indicated that there was a positive relationship between the degree to which QIPP EPSS was personalized at a site and the adoption of the QIPP process. In other words, at those sites where previously existing electronic tools (such as a printing costs calculation spreadsheet) were integrated into the QIPP EPPS, the local information product developers viewed the EPSS as their own and were more inclined to use other features of the EPSS. At those sites were little or no local adaptations were made, the EPSS and QIPP itself seemed to be viewed as top-down corporate solutions, and implementation of QIPP was minimized. This and similar findings from our evaluation indicate the importance of involving eventual users in the development of anything as complex as an EPSS.

Subsequently, the EPSS was made available throughout NCR. A U.S. Patent (5,655.086) was issued for the EPSS design as a "configurable electronic performance support system for total quality management processes" (Jury, Gustafson, Reeves, Cole, and Huffman, 1997). Interestingly, other groups at NCR later adopted the overall structure of the EPSS design to support different complex processes such as marketing and engineering. Not surprisingly, EPSS continues to be a significant technology at NCR.

REFERENCES

Gery, G., 1991, *Electronic performance support systems,* Weingarten Publications, Boston, MA.

Gustafson, K.L. and Branch, R.M., 1997, Revisioning models of instructional development, *Educational Technology Research and Development, 45*(3), 73-89.

Gustafson, K.L. and Reeves, T.C., 1990, IDioM: A platform for a course development expert system, *Educational Technology*, 30(3), 19-25.

Gustafson, K.L., Reeves, T.C. and Smith, M.L., 1995, Evaluation of an EPSS for instructional design in corporate training centers, In Mulder, M. (Ed.), *Training in business and industry* (pp. 53-67). University of Twente Press, Enschede.

Jury, T.W., Gustafson, K.L., Reeves, T.C., Cole, K.C. and Huffman, G., 1997, *Configurable electronic performance support system for total quality management processes* (U.S. Patent No. 5,655.086).

Sanders, D.A., Sanders, J.A. and Johnson, R.H., 1993, *How to qualify for ISO 9000,* American Management Association, New York.

Witt, C.L. and Wager, W., 1994, A comparison of instructional systems design and electronic performance systems design, *Educational Technology, 34*(6), 20-24.

Chapter 16

The PLATO® Courseware Development Environment

Frank Preese
Wellesley Foshay
TRO Learning

Key words: Computer-based Training, Authoring Systems, CBT Authoring Automation

Abstract

The PLATO® computer-based learning system is an example of large-scale, high-quality CBT courseware development using a wide variety of models, including tutorial, simulation, and constructivist learning environments.

To overcome the limitations of conventional CBT authoring technologies and linear ISD methodologies, TRO Learning, Inc. has built a family of authoring tools and methodologies collectively called WinPLATO. Tools are object-oriented and support automated authoring with templates and macros. Methodologies emphasize high-functioning cross-functional teams, rapid prototyping, and a quality management system consistent with ISO 9000. Benefits so far have included improved work flow, productivity gains, improved quality, and reduced maintenance cost when compared to conventional CBT authoring tools and methodologies.

1. INTRODUCTION

The PLATO® library of computer-based instruction contains thousands of hours of courseware. PLATO products make heavy use of graphics and have multimedia production values. They have instructional models ranging from conventional tutorial CBT and tests, to coached instructional simulations and collaborative learning tools. Development projects often produce 100 or more hours of finished product and thousands of files. They require a development effort of a team of specialists measured in many person-years. To meet market windows, development cycle times must be relatively short for products of such size, yet instructional effectiveness, multimedia production values and software integrity must remain consistently high. Product life cycles can reach a decade or more. Products get regular upgrades (including translation, migration to new platforms, and addition of new media, capabilities and content). So, it is likely that more will be invested in maintenance than is spent in building them initially.

Commercial CBT authoring systems do not provide all of the authoring technology needed to succeed in this environment. Instead, the PLATO Product Development group has developed a family of technologies, collectively called WinPLATO, which extend and join together commercially-available tools to form an automated authoring environment capable of yielding superior quality with significantly reduced development cost and time. The vision of WinPLATO uses "technology" in its broad sense to include best practices in methodology as well as software tools.

2. GOALS OF WinPLATO

The goals of the WinPLATO Courseware Development Environment differ substantially from those of most automated development tool projects. To efficiently produce PLATO multi-media curriculum products, WinPLATO has a focus that is, in some ways, narrower than those of a general-purpose tool built for the market. It has the characteristics which are critically important to large-scale, sophisticated development but which are of little importance to market-driven authoring systems optimized for "one-off" production of modest-sized projects on a single work station using a fairly linear development process. By contrast, for example, the WinPLATO environment allows for the "shell" of an activity to be built separately from much of its content. This allows a team to work in parallel shortening development time. This capability, while useful, is not that important for commercial authoring systems.

Our goals in developing WinPLATO technology include:

- *Wide range of standardized product types.* The technology must have strong capabilities to create a wide variety of learning products ranging from tutorials, tests, collaborative tools and instructional simulations, all using multimedia and operating under a common management system. The system must emphasize standardization and reusability of code throughout.
- *State-of-the-art production values.* It must be able to create a contemporary product that is aesthetically pleasing and meets market expectations, at or below industry costs.
- *Inherently sound instructional design.* Superior educational effectiveness must be the natural result of the technology, not an added cost factor.
- *Open and extendable.* The technology must support creation of proprietary components and integration of those licensed from third parties, particularly as new tools and technologies become available.
- *Rapid prototyping.* The technology must facilitate rapid prototyping and iterative, short-cycle development methodologies.
- *Multimedia.* The technology must support a wide variety of media types well, and must be capable of easily incorporating new media formats as they become market standards.
- *Separate structure and content.* Product structure and content are separate, to facilitate maintainability and translation needs.
- *Team development.* The technology must support parallel development by large teams comprising a range of specialized roles, working in a LAN or WAN environment.
- *Automation.* The technology must automate tasks and processes at any point, but particularly in the production stage of development which are the most costly parts

of any project. By comparison, analysis and design are much less costly, and therefore have a smaller return on investment for automation efforts.

- *Quality.* The technology must support appropriate process monitoring and controls at every phase of the methodology to guarantee the consistency and quality of every product in terms of content, production values, instructional effectiveness, and software integrity.
- *Leverage existing tools and technologies.* Wherever possible, use commercially available software tools to leverage internal resources and assure state-of-the-art capabilities.

The WinPLATO technology was a collaborative effort of a chief software designer and programmer who was responsible for writing the software. Development teams, including specialists in instructional design, graphics, programming and testing developed the specifications and constantly reviewed the results, making suggestions for improvement. This is an ongoing evolutionary process.

3. COMPONENTS OF WinPLATO

The WinPLATO technology encompasses four major types of components: a class library of objects, automated authoring tools, automated design templates, and supporting tools and knowledge for the development process and quality assurance.

3.1 Class library

WinPLATO was initially constructed using Asymetrix Learning Systems, Inc., Toolbook II. It consists of a number of systems books which provide WinPLATO authoring and run-time services on top of the capabilities provided by the Toolbook system. Toolbook was selected because it provided:

- a visual authoring environment for productive lesson construction as well as rapid prototyping;
- a means to organize units of system behavior into components which can be built independently of the rest of the activity and integrated into it via well-defined interfaces;
- a fairly complete programming language;
- good built-in support for a variety of media.

The behavior of WinPLATO originates from its class library of proprietary objects (the fundamental "building blocks" of object-oriented programming). This library provides a common set of services such as sending messages, triggering navigation, playing media, saving state and so on. Some of the important WinPLATO classes are:

- *Display Container.* In WinPLATO the scene is the smallest complete unit of instruction. The display container holds the display elements that appear on the scene and orchestrate their behavior. A scene may contain a number of build steps. The display container inherits from the Toolbook page and background (foundation) classes.
- *Display Elements.* WinPLATO display elements inherit from Toolbook foundation classes which include fields, buttons, stages (for playing media), and media clips and provide additional properties and behaviors.
- *Navigation Class.* The navigation class provides the behavior to move from scene to scene. It is a container that combines scenes into larger units but is independent of the scenes themselves.
- Widget Class. Widgets are components which provide interaction and other services. Examples are multiple choice questions, audio enabled text, fill-in questions, video widgets, embedded quizzes and so on. Every WinPLATO widget uses the services of the widget class and contains WinPLATO display elements and programming code to effect its behavior.
- *Message Class.* All messages sent between WinPLATO objects use the message class.

3.2 Automated authoring tools

Every authorable object comes with an editor. This means that programmers must construct the object itself and the editor needed to customize that object. The resulting editors allow non-programmers to author instructional activities by customizing objects through the editors. In addition, the system includes a number of tools to automate repetitive tasks and to check finished lessons. For example, a common problem in CBT development is the accumulation of leftover, unused objects which occur as screens are added or deleted from a lesson as it is built, reviewed and revised. Removing this "dead code" from the finished product reduces the size of the lesson and improves its maintainability. We have built an automated tool to scan and delete unused objects.

Multimedia objects are created using standard production tools. The objects are produced at the component level and placed in a production graphics library which is indexed for easy retrieval during production. In courseware production, the objects are copied from the library and assembled into completed multimedia components. This allows for considerable efficiency by reuse of library objects.

3.3 Design templates and macros

Each PLATO activity is specified in detail via two documents created at the end of the design process. The first is called Programmable Ready Material (PRM). This contains information about the structure and navigation for an activity; graphic and other media specifications; the instruction and directions for other team members. The second document is a storyboard which shows thumbnail graphics for each of the scenes of the activity.

The Instructional Developer creates the PRM starting with a Microsoft Word template. This template contains the Styles the author uses to identify important parts of the PRM such as graphics, text that appears on the screen, audio and video elements and so on. These templates are set up in advance with Style identifiers for important items identified in the product design.

Automation technology is used to read the PRM; extract important information to automatically create the Toolbook file; set up the navigation data and display containers such as chapters, sections and scenes, import text into files readable by the display elements and widgets, and so on.

4. DEVELOPMENT PROCESS AND QUALITY ASSURANCE

A number of features of WinPLATO support our requirements for high quality in instruction, content, software and production values; teamwork; short development cycles for large scale products; and ongoing maintenance and upgrade. The design and development process used with WinPLATO includes the usual processes of requirements definition, project planning, task/content/audience analysis, instructional design and development, production and testing. To this is added rapid prototype development and frequent risk assessment.

4.1 Production team

Development cycle time and project size and complexity require that work flow be parallel and iterative. Project teams include specialized roles such as project managers, subject matter experts, instructional designers and developers, artists, and technical writers. In addition, all of the assets developed for an activity are integrated into a finished lesson by "paste-up" specialists. The team also includes testers from our independent testing department, and a production manager who is in charge of work flow, does configuration management and manages the hand-off of material to the testing group. The WinPLATO editors provide a well-defined interface that coordinates the work of each team member.

A technical manager and staff of programmers is responsible for the development of the WinPLATO technology.

Besides proprietary WinPLATO tools, the software workbench includes a number of off-the-shelf tools for writing, project management, graphic development, programming, testing and configuration management.

Throughout the process, independent reviewers for content and instructional design are employed. Actual learners are also an important project resource, for developmental testing of rapid prototypes as well as Alpha and Beta testing. The WinPLATO design templates and rapid prototyping capabilities provide the materials used for these reviews and tests.

4.2 QSOPS

Quality System Operating Procedure (QSOP) documents detail the development process. Consistent with the philosophy of ISO 9000, the purpose of a QSOP is to describe a major part of the process: when the QSOP should be used, what standards apply, what the outputs are and what reviews and approvals are needed. This quality management system allows us to audit the entire development process for continuous improvement.

4.3 Knowledge base

For each phase of the process there are numerous job aids which document best practice. For example, there are job aids for:

- instructional designers which range from how to decide what content to analyze to how to review completed activities and document defects;
- artists on ways to develop graphics that have a small footprint and yet are visually appealing;
- programmers on how to construct WinPLATO components; and
- project managers and team members on how to use the work flow management tools.

Job aids are contained in a knowledge base on our network, and are accessible to all members of the team. Training occurs when there are new or revised job aids. Refresher training occurs from time to time. The job aids and QSOPs thus define the product development expertise of the group, and they help define training and performance review standards for the development teams.

4.4 Quality plan

Each project includes a quality plan. This document details the development process and work flow, where reviews occur, who has approval, how configuration management is done, how testing will be done and so on. From the quality plan, work flow tracking sheets are devised for each activity specified in the design. The tracking sheet reflects the development process described in the quality plan and includes each of the well over 1,000 components in a given project. Each step on the tracking sheet is assigned an owner. That person completes the task and initials the tracking sheet indicating the work was completed to the best of their ability. Project managers use the tracking sheets to assess status and whether the quality plan is being followed.

4.5 Configuration management

Configuration management is of special interest in large projects with thousands of files in multiple versions. Automated support is critical. The tool we use is the StarTeam configuration management software (by StarBase Corporation). This software allows us to set up projects with a consistent directory structure, with places for project plans, quality plans, media plans, designs, software specifications, activity specifications (PRM), graphics, test plans and so on. The development team checks items into this repository as they are created. Each time an item is checked in, the author writes a short reason for check in such as "New document" or "Revised to include Rob's comments". When an item includes licensed material, the licensor's name is included in the description so we can track permissions. StarTeam tracks different versions of the file in the repository and provides a history for each item. It also provides defect tracking for every item so that reviews can be documented on-line, defects can be examined, cleared systematically and analyzed to improve the process.

StarTeam also provides build management which groups all of the file versions that are used in a particular build of the product.

4.6 Automated software testing

Testing complex software is a daunting task. A diligent tester with a carefully constructed test plan will do a fairly good job the first time when testing, but effectiveness falls off in each successive pass in the iterative development process. To improve testing effectiveness we use automated testing tools when we believe the added cost of developing test code is justified.

In addition test automation includes utilities that scan completed activities looking for common problems we have identified in previous quality audits. Examples include incorrect navigation settings such as unintended looping conditions, graphics which are no longer referenced, inconsistent settings for scoring, questions with no right answer, and so on.

At the end of the process are alpha and beta tests to assess software integrity, content completeness and accuracy, instructional effectiveness and production values. These tests serve to "calibrate" the earlier standards and processes in the development cycle.

5. EFFECTS OF WinPLATO SO FAR

5.1 Teamwork and work flow

Production of Windows courseware with advanced architectures such as those for multimedia coached instructional simulations has required a great deal of teamwork. To accomplish this level of teamwork we have had to change the way we work in three major ways.

First, we could not use a familiar and proven sequential "waterfall" work-flow pattern. Instead, work often proceeds collaboratively and in parallel. Coordination is by "three-way" meetings among instructional developers, artists and programmers to explore ideas and set specifications for the work needed to accomplish instructional ends.

Second, team members, while having a particular specialty, had to understand the work of the rest of the team, use their language and know what is important to them. This was very difficult at first. For example, instructional developers and artists first resisted learning too much about programming. However, it became clear that to accomplish what they wanted, the developers and artists had to learn to be more precise and think in ways that were foreign to them. In time they learned that being able to think in software terms was a very valuable skill enabling more effective communication and better decisions.

Third, the teams had to embrace a rapid prototyping philosophy. Prototypes are usually built to help make difficult decisions. We use a variety of prototyping means: pencil sketches, story boards, interface and interaction prototypes, technology prototypes and full activity prototypes. Every prototype is built with a goal in mind and is deliberately constrained to be constructed and evaluated in a short period of time. We make it clear that most prototypes will be discarded and so there is no point in using more than a minimum effort to build them. The WinPLATO authoring software allows us to build and modify prototypes quickly. In future projects, we plan to make some prototypes available through our web site to collect a broader set of opinions.

5.2 Productivity gains

The biggest gains in productivity with WinPLATO have been in the areas of graphics and programming, where the greatest costs are for any project. We think graphics has been more productive for three reasons. First, the commercial graphics tools we use are getting more and more powerful. Second, we are more experienced in using the tools. And third, our graphics library strategy encourages re-use.

In programming, productivity has increased due to the powerful class of services in WinPLATO. Our programmers use these to build components, which then are used by the production teams. Any class library's integrity is only as good as the discipline of its users, so we have emphasized training about WinPLATO to make sure the programmers do not re-invent services. We also have partitioned the activities to minimize the coupling between the programmers' work, thus allowing for parallel work flows. We think productivity will continue to increase in these areas.

In instructional design, we have also realized some major productivity gains through the use of lesson templates. Each template implements a certain instructional strategy and enabling WinPLATO components. This makes it possible to develop a finished lesson in $^1/_2$ to $^1/_3$ of the time previously needed. However, the templates are far from a complete solution to automating the design and development process. While there are times that templates provide some useful leverage, templates are too monolithic to be a viable solution for any but a narrow range of product features. The challenge of creating productivity tools for instructional designers to use in the early phases of the process remains.

5.3 Quality

An objective measure of quality is the number of defects reported. During a recent project for over 70 hours of courseware, we monitored the number of reported defects (including software, content and instruction) reported during internal reviews. We compared the last modules built to the first, and found defects decreased by a factor of 10. We attribute this to the experience gained by the team in the process and tools during the project. Furthermore, the number of errors reported since product release has been satisfyingly small. A subjective quality measure is the response to the product by our field staff and our customers. Both have been very positive about how the product looks and how it operates. Formal summative evaluation is now in progress.

However, there is always room to improve quality. For example, two current continuous improvement targets are performance and component size (particularly important for Internet delivery).

5.4 Maintenance

The decade-long life cycle of some PLATO products means that maintenance, translation and upgrade are very important. Since WinPLATO products have not yet been through the maintenance cycle, we don't have direct experience to report. However, we believe that the products will be very maintainable because the text and graphics are in a database-like format to make changes easy, and because software behavior is encapsulated either in components or in WinPLATO class services.

6. LESSONS LEARNED

We have learned a number of lessons about large-scale CBT development. Among the lessons are:

- *Good instructional design is worth the cost.* If the product doesn't teach, it doesn't make any difference how good it looks or how well it behaves.
- *Specialists working in high-functioning teams are required.* Large-scale projects require personnel who have mastered their media well enough to communicate with others and devise creative solutions. Our teams include roles not often mentioned in texts, such as testing. We have found it is important to involve the test group early in the process, with as much software as possible.
- *The right process is critical.* The waterfall process doesn't work because it assumes that it is possible to know everything about a project at the beginning, and that nothing will change during the project. Rapid prototyping allows the team to take ideas all the way through implementation and test, including user tests, to see what works and what doesn't. While the waterfall process tries to do everything right the first time, rapid prototyping means that some things won't work and will have to be done over. Staff have to be willing to make mistakes and do re-work. For some, this is very difficult. For a large project, however, the reduction in the risk of discovering a major problem late in the project justifies the change.
- *Standards are essential.* With a large team working on a project, standard ways of doing things are needed to assure that the parts fit together, components are reused, and the end product has a uniform style. Costs of standards development should be amortized over a number of projects.
- *Use the right authoring software.* The tools should be flexible enough to do all of the creative things the team will devise, and yet they should enforce standards. It is also important to separate the structure of the activity from the content.
- *Team members need the right tool workbench.* Capabilities are needed for templating, automated production and object management which exceed those found in current CBT authoring systems. In addition, especially important are tools for project management and work flow tracking, configuration management and defect tracking.
- *Do good software design.* Current CBT authoring systems are well behind the state of the art in software design practice. Without a sound design representing current object-oriented practices, the software is likely to collapse under the weight of the media and all of the changes that normally occur during a large project and in post-release maintenance.
- *Do risk assessment often.* Find out from the team where the risky spots are. Tackle them first. Do prototypes or run tests to answer questions and manage the risks.

While WinPLATO is optimized specifically for PLATO products, we believe the needs it meets are typical of those in any large-scale CBT development organization. WinPLATO also demonstrates that while automating the front-end processes of analysis and design is an important goal, the greatest productivity gains in CBT development will come from automating the end stages of the development and production, and from use of rapid prototyping methodologies. This is especially true in large multimedia projects, and in environments where continuous product maintenance and upgrade is a significant part of life cycle development cost. In settings like these, the substantial costs of developing a technology such as WinPLATO are justified by reduced cost, shorter development cycle time, and improved quality.

WinPLATO illustrates six challenges which face developers of tools for automating CBT authoring. First, the tools must implement current best practices in software design, while recognizing and compensating for the pitfalls of 4th Generation languages and object-oriented programming. Second, the tools must implement current instructional design methodologies, with their emphasis on rapid prototyping, iterative and user-centered design and development. Third, the tools must support use by a high-functioning team of specialists in diverse roles working in parallel. Fourth, the tools must reflect current best practices for software quality management at every phase of design, development, production and implementation. Fifth, the tools must make sound instructional design a natural consequence, rather than an added cost. Sixth, the tools must maximize automation of every step in the process.

Our experience has convinced us of the cost-effectiveness of such tools, at least in environments where quality and rapid development are equal priorities. However, to be cost-effective, we believe such tools must make rather strong assumptions about product architecture, design standards, team roles and structures, and methodology—as has been the case in other types of software engineering. If this is correct, then the goal of building truly general-purpose tools is not realistic, and could cause the tool developers to sacrifice potentially major automation efficiencies in order to achieve generality of purpose. Furthermore, it may be that the approach to automation exemplified by WinPLATO is cost-effective only for large-scale development operations such as PLATO's. Most CBT development is still done in unique and short-lived projects of modest size by small teams which may not even outlive completion of the project for which the team was assembled. In this kind of "cottage industry" environment, the level of investment in tools, methodologies, and team work represented by WinPLATO may not make economic sense. It may be that WinPLATO exemplifies a strategy for "economies of scale" which is not suitable for most current CBT development environments, and thus will not be of interest to developers of commercial CBT authoring systems. The proof will be in the direction taken by the next generation of authoring systems and automation tools.

Chapter 17

Visual Support for Authoring

Rob Wright
Barry Harper
John Hedberg
University of Wollongong

Keywords: Authoring tools, Constructivism

Abstract

Development of educational software has had a long history of use of authoring environments that have enabled instructional technologists rather than programmers to design and develop applications. The advantage of these tools has been that the designer did not need to be highly skilled in high-level languages, but could use a simpler construction set of pre-programmed modules, often supported by a simple scripting language. The disadvantage was that the developer was limited to the pre-programmed modules available and to the underlying assumptions of the structured instructional design models adopted by the tool. The designer also had to work within the visual and procedural structures employed by the tool designers to represent the design process within which learning activities could be constructed.

To simplify terminology amongst authoring environments, in this paper we will refer to the authoring application as a "tool" and the resources, which are collected together for display as a "project". Several authoring tools can construct standalone "projects" (for example, Director creates a "projector") which can be distributed to learners individually. Some tools also focus upon record keeping aspects of the system and create student files that are the student's responses to the embedded questions. Some tools work directly with the visual display, such as Digital Chisel and Oracle Media Objects, all changes and design decisions are represented as visual changes to the display. Other tools provide a more comprehensive meta-view, for example, Authorware, shows the flowchart of the algorithm being employed as well as the runtime display window. Apple Media Tool shows a graphic view of the screens and the associated links, and mTropolis represents a meta-view as an hierarchical "finder"-type structure.

1. VISUAL METAPHORS IN PRACTICE

No matter what authoring tool is used, software designers have to work within the prevailing metaphor of the tool. For example, if a designer employs Director, the prevailing metaphor is a theatrical pageant and is conceived as a time dependent display. Events occur on a stage and are managed by a score, which dictates the depth and movement of each object over time. By contrast, HyperCard employs a card metaphor where links between discrete representations of objects are embedded in an Hypertext relationship. Icon Author and Authorware create a complex flow chart (algorithmic) structure, which is used to design the experience but is not displayed to the learner as they work through the pre-determined paths that are reminiscent of the traditional concepts embodied in programmed instruction.

From these brief examples it is clear that the visual metaphor will create a vehicle in which the instructional designer must work and conceive their project. It follows that there will be a symbiotic relationship between the instructional strategies and the way the tool enables the designer to think about the task. Of the many common tools, the older have taken a more structured approach borne out of behavioural learning theory. The more recent tools have striven to reduce the need for time or procedural structure to create an environment of intelligent objects. To represent this process the tool authors have created different ways of changing and visualising the relationship between creation of a learning environment and using these environments. Thus options which will enable time, hierarchical and spatial display are all possibly needed by designers as the learning tasks and project demand.

2. DEVELOPING AN AUTHORING ENVIRONMENT

The designers of an authoring environment must make assumptions about the instructional design models that the tool will support and the potential end users. Some tools are designed for developers and others for learners to construct their own ideas. Several key writers have called for a reassessment of instructional design models used for the development of technology-supported learning environments that assume constructivist views of learning. Hannafin and Land (1997) have suggested that we should be aiming for open-ended learning environments, Jonassen and Tessmer (1996-1997) argue that we should be aiming at new learning outcomes, and Duffy and Cunningham (1996) have described a range of metaphors which structure how we teach. Additionally, Savery and Duffy (1996) have elucidated several principles that characterise this philosophical view in technology-based learning environments.

Jonassen and Tessmer (1996-1997) have also questioned the commonly-used taxonomies of learning that are the basis of our instructional design models, proposing that engagement with a greater range of learning outcomes is essential for meaningful learning. They have suggested a new framework for specifying the types of learning outcomes modern learning environments should be developing.

In response to these design suggestions and the changes that have occurred in hardware technology options, we have proposed a design model (*Figure 1*) that is cognisant of constructivist approaches to instructional processes and addresses many of the above suggestions for reassessment of instructional design models (Hedberg et al., 1994).

Given the changes to hardware technology options, we wanted to provide design frameworks and visual metaphors that could be employed by highly skilled designers to use the tool for project creation. We also wanted a tool that could support young learners, with access to rich resources, to construct representations of their own ideas using such a tool.

Phase one of the model takes the basic information derived from an assessment of needs and describes the parameters of the Project space. This is the information which is to be included in the materials, how it is structured, what the target audience understands about the information and how it might be structured for the audience. A possible structuring device might be a concept map of the ideas and links that are to be included in the project.

The second phase reviews the basic description and seeks to link the elements through an appropriate instructional or interaction strategy. It also seeks to identify metaphors, which help both the design team and the final presentation of the information structure. The outcome of the second phase would be a formal description such as a design brief. The detail would enable the reader to understand the underlying knowledge structures and the ways it is proposed to link them conceptually and intuitively.

The third phase is a third pass at the same material, this time with the express goal of linking the design ideas into a potential presentation structure. One output of this phase would be an interactive mock-up of the resources using an authoring tool to illustrate not only static display of information but also the graphical and visual metaphors used to create understandable links. The information included in this prototype may include motion and static graphics, sound and data landscapes as appropriate to the concept under development.

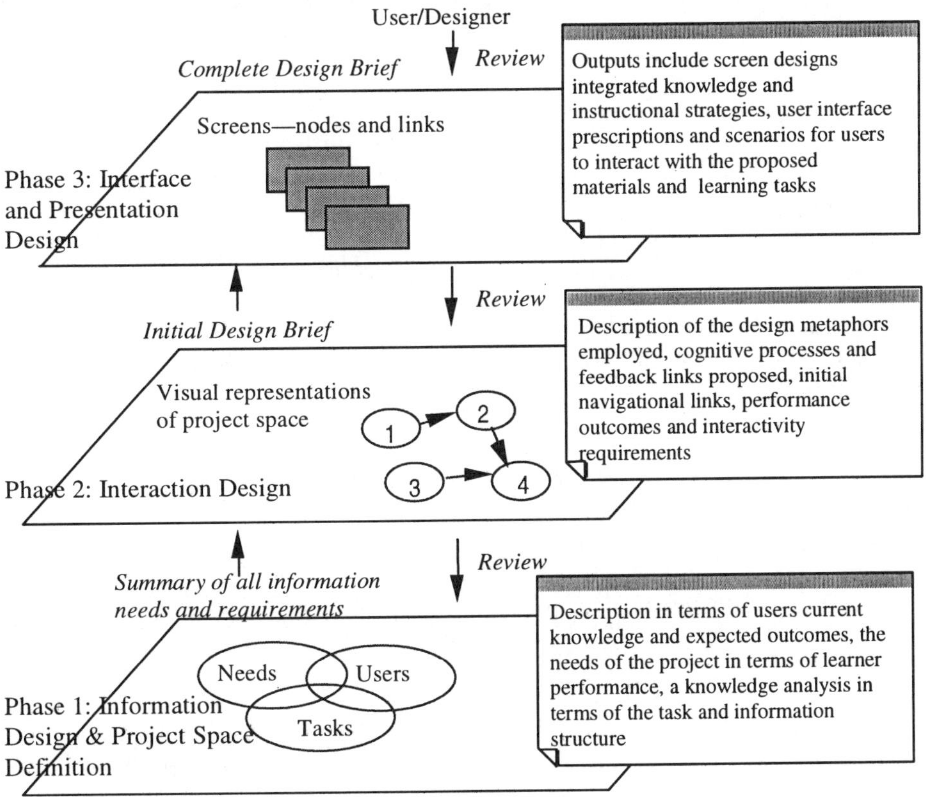

Figure 1. The design process used as the basis for interactive multimedia project development

3. IMPLICATIONS FOR A SOFTWARE ENVIRONMENT

The key implications for an authoring environment that support the design model would be that the environment should include:

- facilities for rapid prototyping of design ideas and restructuring of ideas simply and efficiently through a range of flexible interfaces;
- design elements that can be added through either menu selection, drag and drop, or copy and paste facilities;
- design elements that can be edited, reused, and re-purposed simply and efficiently through a WYSIWYG interface;
- a visual representation of the presentation of the learning strategy options available at all times;
- global highlighting of an object and all of its occurrences in other views of the project (in all design and runtime views);
- extensibility so that new features can be added when necessary through extensions to the authoring tool and bridges to high level languages; and
- networkability so that alternate storage options for graphic, video and audio resources can be used. (For example, data files can be made available on CD-ROMs, hard disks, fileservers, or distributed across all three. Media elements such as pictures and text can be either stored internally or as external files, and in the future, it should be possible to use media and files stored on the Internet).

In addition to these attributes for an authoring tool, the choice of an appropriate visual representation or metaphor will be critical for supporting a range of authoring tool users. This visual metaphor will either support the development of easily learned skills to enable users to express their ideas in multimedia forms, or if unsuccessful will require users to 'learn' how the tool can be used to create projects. This second requirement is typical of most tools currently available. Key to any authoring environment is the way it sets up interactions and supports their simple execution by the tool user.

As we considered the issues we have raised for developing an authoring tool and the attributes that flow from these issues, in developing MediaPlant, we have struggled with questions such as:

- What visual support structures are required to create a learning environment? Is realism enough to make it an authentic context or is there some other way of interacting with the objects so that the task of manipulation is simple? When would a 3D view be required, how should a navigable space be represented and displayed?
- Under a constructivist framework how should the user/learner create the interaction, rather than the learner having the interaction devised for them?
- What useful 'views' are there for project construction? Does it differ for learners, professional designers, or programmers?
- What general support functions should be available in an authoring tool? What cognitive support structures are necessary for learners? (For instance, Ferry et al, 1997 have demonstrated the importance of using cognitive mapping tools to generate understanding of complex knowledge domains.)
- What organisation and support structures (annotation devices, media specific views) are needed to assist in the development of projects that have potentially

thousands of items to track and integrate? Should a level of customisation be available to the project developer so they can 'tailor' the visual organisation of their work to a world more in tune with the content and interactions they are developing?

4. MEDIAPLANT AS AN AUTHORING ENVIRONMENT

With the framework of our design model and the questions we have raised, the Interactive Multimedia Learning Laboratory team set out to develop a software development environment. It was envisaged that this development environment, ‘MediaPlant’, would facilitate the production of complex cross platform learning environments for commonly available configurations, i.e. this would be an authoring tool that offered complete flexibility in the design process and also high-level performance on entry-level machines.

The authoring environment consists of a development tool and a runtime or 'player' program. The development program is used to construct and test the project, which is then distributed with the runtime program. The development and runtime applications enable project construction on both the Macintosh and Wintel compatible platforms, and the project files can be shared across both operating systems. The software environment is based on a C++ application framework tailored for large-scale multimedia development.

The visual representation of the tool consists of a design window which functions as a 'drag and drop' construction space with additional views that include a media window, a tools window and attribute windows for individual screen objects. A meta-view of the initial development program is based upon a tree-file structure. This is being extended to present visual representations of the design by flowchart and Hypertext visual organisation.

The file structure metaphor (*Figure 2*) provides a similar interface to the Macintosh Finder or the Wintel directory hierarchy, allowing easy access to all of the media, screens, screen objects and scripts within a project. By using aliases, files may be accessed wherever they may reside over a network and they may be moved without the link being broken. Projects are constructed using Drag and Drop, as well as copy and paste, which makes adding, editing and reorganisation of elements within a project simple. A designer has considerable freedom in how media is organized, including a choice between using external files, or storing media within MediaPlant's containers. Runtime screens can be displayed whenever desired, allowing easy graphical editing of items on each screen, as well as execution and testing of the product.

The object-oriented nature of the design metaphor offers considerable extensibility for the authoring environment. Cognitive tools for learners, such as note pads, help windows, simulations etc, once developed, can be simply dragged between projects, allowing designers to reuse programmable objects that support learners in their exploration of information and construction of knowledge.

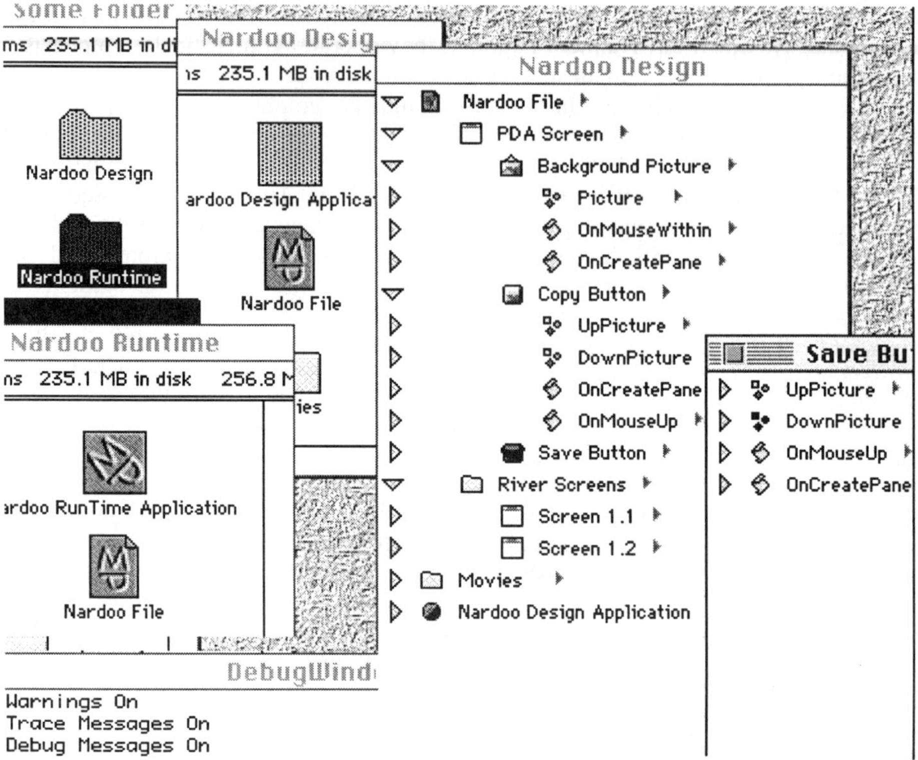

Figure 2. Tree file visual structure used in MediaPlant for the authoring process

5. AN EXAMPLE: EXPLORING THE NARDOO

In exploring the application of the authoring tool, MediaPlant, to a specific project, an interactive multimedia development team used the tool to develop a CD-ROM project, *Exploring the Nardoo.* The design team had experience in developing multimedia products using simpler authoring tools and adopted the design process outlined in *Figure 1.*

The resulting package, *Exploring the Nardoo,* provides a dense information landscape of resources based on general issues in ecology. Its purpose is to engage students in long-term studies using the skills of problem solving, measuring, collating, elaborating and communicating. The package was designed for high school use, for students aged from 14 to 18 years and is based on current Australian state and national curriculum documents for Biology and Geography. A mapping of the achievable learning outcomes for these documents has been included for teachers on the CD-ROM. The package has the potential to provide extensive opportunities for students studying Biology and Geography to extend and consolidate their knowledge base in many of the critical areas of their respective syllabi. In addition they can practice investigative, analytical and communicative skills as well as develop and appreciate different value and attitudes. Specifically the *Exploring the Nardoo* provides extensive support for the study of the interactions between living organisms and the physical and chemical environment in which they operate with particular emphasis on the role and impact of the human species at both a macro and micro level.

The information landscape uses spatial and geographic metaphors: a Water Research Centre (*Figure 3*) and a navigable fictitious river environment. On entry to the environment, users are challenged by three researchers in the Centre to help them solve problems and carry out investigations on the river. The challenge posed by the researchers encourages active learning and supports students to construct their ideas in teams from measurements taken, resources reviewed, maps interpreted and data analysed. By providing a metaphor easily linked to the real world, students are encouraged to apply scientific concepts and techniques in new and relevant situations, throughout the problem-solving process. In so doing, the learner is likely to become more interested in developing questions, ideas and hypotheses about the learning experiences encountered.

Figure 3. The Water Research Centre

A broad range of problems is presented to the user by selecting from the Investigations Notice Board on the back wall of *Figure 3*. The investigations include issues such as fish dying from pollution, weeds infesting the river, and communities discussing farming practice. For example, in one of the investigations weed infestations have affected the Nardoo river ecosystem:

Water Plants and Weeds

Hugh Smythe captains a riverboat that ferries passengers up and down the Nardoo River. As the river is the key to his livelihood, he has become concerned about the rampant growth of weeds and willows that have begun to obstruct parts of the river in recent months.

Your task: Investigate the extent to which the river is effected by the weed and willow growth. Keep a journal of the information that you find and try to re-organize your information to clearly recount the history of the weed problem.

To investigate such issues, students may access a rich set of resources and data that are both embedded in the river environments in situ, through hot buttons, and also in the Water Research Centre through organizing interface metaphors such as reference books and newspaper clippings. Among the resources in the Water Research Centre users will find a book containing descriptions of all of the plants and animals of the river; lists of television and radio news reports and interviews; newspaper cuttings; a filing cabinet containing technical articles and reports on the various investigations. Supportive tools include: a "Presentation Booklet" to support learners in note taking, filing notes and construction of reports and a "Text Tablet" for editing and elaborating the notes and data collected in the Personal Digital Assistant (PDA) (*Figure 4*). Three simulators also support student investigations by helping them to identify, define and explore cause and effect relationships.

Exploring the Nardoo provides the student with a flexible set of tools, some of them cognitive, made available through the personal digital assistant (PDA), to assist in the investigation process. The PDA contains a multimedia notebook for collection of any of the resources in the package, including video, audio, graphics and text; a viewer for viewing the video and graphic resources; a set of measuring tools to take measurements on the river user support through a help file; and navigation tools.

Figure 4. The Personal Digital Assistant (PDA)

6. CONCLUSION

As new theoretical views about technology-supported learning environments develop, we need to be able to explore these ideas using tools that allow us to set up these environments quickly and easily. Tools that assist the task without complex dialogues for handling variables and options. We also need to review the visual and conceptual functioning of tools that we offer both designers to support their efforts in designing high quality software products, and our users as they express their ideas using the full range of media available to them. In the light of changing approaches, it is an opportune time to review our current authoring tools, to suggest options that derive from educational issues, and to address some of the questions raised in this chapter.

MediaPlant is a powerful new authoring tool based on a constructivist framework that may be able to help us achieve these goals. It provides an environment in which ideas can be extended; new options and constructs can be created, augmented and shared by other users/designers. When linked with our understanding about how users work with the more complex environment of computer-based knowledge construction, it also provides supportive components. For example, within *Exploring the Nardoo* the designers have provided a notebook to allow students to collect and annotate what they find as they solve embedded problems. This form of support allows users to undertake the difficult task of multimedia reporting – compiling and editing resources which not only must make a logical argument structure but also will involve thinking visually and in a number of different visual forms. MediaPlant allows designers to declare such pedagogical tools as objects which can then be used in other projects. We hope that we have begun to break away from the constraints of earlier tools that provided one or two ways of representing the underlying ideas of the designer. In this new environment the user can choose the way they wish to view their project in a form that makes 'sense' in terms of the data rather than in terms of what the programmers of the authoring tool chose.

Acknowledgments

We wish to acknowledge the contribution to these ideas by the team who developed the *Exploring the Nardoo* products, in particular, our senior programmer Grant Farr and our programmer, John Orlandi.

REFERENCES

Duffy, T.M. and Cunningham, D.J., 1996, Constructivism: Implications for the design and delivery of instruction, In Jonassen, D.H. (Ed.), *Handbook of research for educational communications and technology*, 170-198, Macmillan, New York.

Ferry, B., Hedberg, J.G. and Harper, B.M., 1997, *Using concept mapping to help pre-service teachers map subject matter knowledge*, Paper presented to the Australian Association for Research in Education 1997 Annual Conference, Brisbane, 30th November to 4th December.

Hannafin, M.J. and Land, S.M., 1997, The foundations and assumptions of technology-enhanced student-centered learning environments, *Instructional Science*, *25*, 167-202.

Hedberg, J.G., Harper, B.M., Brown, C. and Corderoy, R., 1994. Exploring user interfaces to improve learner outcomes. In Beatie, K., McNaught, C. and Wills, S. (Eds.), *Interactive multimedia in university education: Designing for change in teaching and learning*, 15-29, Elsevier, Amsterdam.

Jonassen, D. and Tessmer, M., 1996-1997, An outcomes-based taxonomy for instructional systems design, evaluation and research, *Training Research Journal*, *2*, 11-46.

Savery, J.R. and Duffy, T.M., 1996, Problem based learning: An instructional model and its constructivist framework, In Wilson, B.G. (Ed.), *Constructivist learning environments: Case studies in instructional design,* 135-14, Educational Technology Publications, Englewood Cliffs, NJ.

Chapter 18

Cognitive Tools to Support the Instructional Design of Simulation-based Discovery Learning Environments: The SIMQUEST Authoring System

Ton de Jong
Renate Limbach
Mark Gellevij
University of Twente

Michiel Kuyper
Unilever Research

Jules Pieters
University of Twente

W. van Joolingen
CIBIT

Keywords: Discovery learning, Computer simulations, Authoring environments

Abstract

IMQUEST is an authoring environment for creating learning environments that combine simulations with instructional support. SIMQUEST *learning environments* are based on a computer simulation of, for example, a process, a procedure, or equipment. Learners can engage in a process of discovery by manipulating values of variables and observing the outcomes of their actions. Since discovery learning is not easy, SIMQUEST learning environments provide learners with instructional support that helps them in the process of discovery learning. For example, while working with the simulation, learners can ask for extra explanations of phenomena, or can ask for assignments to guide their interaction with the simulation. The SIMQUEST *authoring environment* offers authors (teachers) the possibility of creating learning environments without the need for programming knowledge or specific pedagogical knowledge. An author creates a learning environment by, first, selecting 'building blocks' from a library, and, second, by adapting the selected elements and combining them into a complete learning environment. Building blocks include templates for simulation models, interfaces, and instructional support. In addition, authors receive 'cognitive tools' that help them in the design and production process. These tools comprise an on-line help system, a wizard, and an advice tool.

1. INTRODUCTION

Computer simulations are a promising teaching device. Computer simulations can be used to instigate a process of discovery learning in which learners through experimentation induce characteristics of the model underlying the simulation (cf. de Jong & van Joolingen, 1998). While schools generally have access to simulations, these simulations have an industrial background and as a consequence a) the content is not tailored to the specific learning situation, b) the simulations are not embedded in instruction, and c) teachers have no opportunity to change even simple parts of the simulation. We, therefore, saw a clear need for an authoring environment that offers developers an easy way to create (or adapt) simulations and to embed them in a learning environment.

In the market, we find an abundant number of tools, both for creating simulation languages and for creating instruction. There are particularly large numbers and varieties of simulation languages and tools. Several of them are even (partly) created for the education and training market. Some examples of these dedicated tools are STELLA, RAPID (from EMUTEK), and Powersim. The simulations delivered by this group of tools are similar in that they are suitable for use in a classroom or in training situations, for instance because the simulations created with it have a learner friendly (i.e., less complicated) interface. However, none of these tools produce simulations *integrated* in a learning environment. This means that simulations created with these tools operate as stand-alone products with no embedded instructional support for the learner.

Tools for generating computer-based instruction, mostly general authoring languages such as Toolbook and Authorware (see e.g., Rieber, 1997) include facilities for making simulations, but offer no genuine integration of simulation and instructional support. The two examples of tools we know of that integrate support and simulations are RAPIDS (Munro, 1995) and ID2 Instructional Simulator (Merrill, 1996). The disadvantage of these approaches is that the modelling mechanisms used and the type of support provided is rather limited. Another paramount issue in this context is that the reuse of existing material is not sufficiently taken care of in existing products (see De Jong & Sarti, 1994). Also, development of more complex learning environments, in which for example non-linear learning behaviour will occur (as in simulation), requires development methods and tools that go beyond the stage-based approaches and the more commonly used off the shelf authoring tools.

In the SimQuest authoring system,[1] we have overcome these limitations by enabling a full integration of simulation and a range of powerful learning support mechanisms.

[1] SimQuest was developed with support from the European Commission in its Telematics program in the **SERVIVE** project (ET1020).

2. THE SIMQUEST ENVIRONMENT

2.1 The learning part

The central part of any SIMQUEST learning environment (application) is a simulation of a system or a process. The SIMQUEST modelling system can be used for creating static models and for dynamic models based on first order differential equations. Currently we have learning environments under development in the physics topics of motion, oscillation, collision, electricity, and thermodynamics, the mechanical engineering topic of stress and strain, and two broad environments (one on water purification and one on greenhouses) that cover topics from physics, biology, and chemistry.

In SIMQUEST applications, simulations are embedded in instructional support which is aimed at supporting learners in the discovery process. Currently, the SIMQUEST authoring environment provides four types of instructional support for learners (see also De Jong et al., 1998):

- *Model progression.* A learning environment created with SIMQUEST may contain a number of different simulation models, ordered along dimensions such as difficulty, 'order' (qualitative vs. quantitative), or perspective on the domain.
- *Assignments.* Assignments provide the learner with short-term goals, like finding a specified relation, predicting the behaviour of the simulation or achieving a specified simulation state. In co-operation with model progression, assignments decompose the overall learning goal of a simulation into a number of subgoals. Learners can answer assignments and then receive feedback on their answer.
- *Explanations.* In the SIMQUEST system, the author can define textual, graphical, and multimedia explanations. These explanations can be used to provide for extra information on variables, relations, or events in the simulation.
- *Monitoring.* The monitoring tool helps learners monitor, compare, and replay the experiments they have been doing, and that, in relation to answers to investigation assignments, can provide feedback on the relation between the experiments and answers chosen.

Figure 1 gives an example of a part of a learner interface created with SIMQUEST. This example shows a simulation on characteristics of alloys. Students can change the temperature and mixture of materials (in this case metals) and see the effects on the characteristics of the alloy. The output is given in numerical values and in an animation that shows how the two materials combine. The assignment in the example asks the students to find the relation between the proportion of eutectic material within materials A and B. The eutectic composition is the alloy composition which solidifies with a solidification point which is lower than the solidification point of the two materials in the alloy.

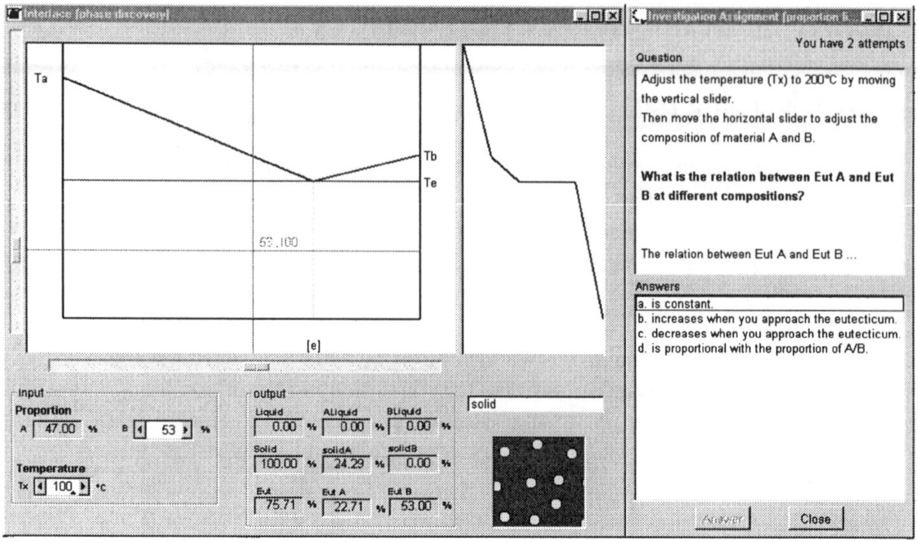

Figure 1. Example of a simulation with an assignment for a learner (designed by Jan van der Meij, CINOP)

Instructional support (e.g., assignments) and the simulation are fully integrated in SIMQUEST. The simulation can be set in a certain state by the assignment and can be controlled from the assignment by the learner. Also, values from the assignment are used for generating adequate feedback.

We have now done several empirical studies in which several hundreds of learners have worked with different configurations of learning environments created with SIMQUEST. The overall conclusions are that SIMQUEST environments are especially helpful in obtaining what we have called 'intuitive knowledge' (Swaak & de Jong, 1996), and that assignments are very helpful for learners. We have strong indications that model progression helps learners. We cannot draw conclusions about explanations, as we thought it necessary to have them present in all our environments.

2.2 The authoring part

In the SIMQUEST authoring system, we have chosen an object-oriented approach. This means that a SIMQUEST author will *select* elements to work with from a central resource, the *library*. Once an author has selected the library element, it can be *instantiated,* which means that the element is placed in the learning environment under development. Then, *specialising* (or *editing*) the object is done by opening the appropriate editor on the element. Through the editor the author changes specific characteristics of the element and fills in the domain content (see *Figure 2*). SIMQUEST was developed through continuous interaction with authors from the field of vocational training. Pilot users used an intermediate version of the system and posted bugs and requirements on dedicated web-site. Mediated through a coordinating team this information was used to make patches and enlarge the functionality of the system (see Kuyper, de Hoog & de Jong, 1996, for a description of this process). In addition, controlled studies with authors of different expertise were performed (see e.g., Chapter 9 of this book, and Gellevij et al., 1998).

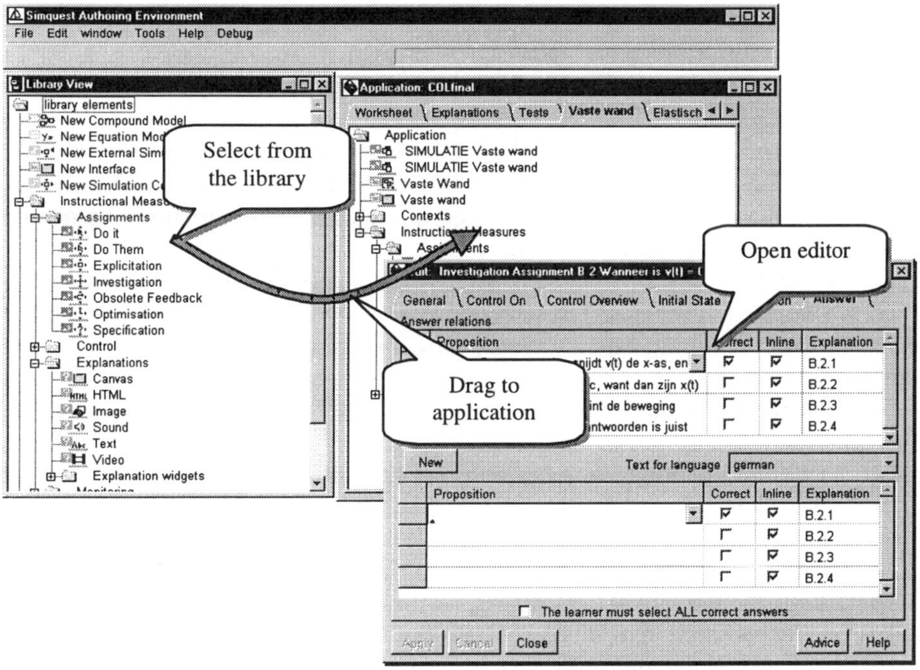

Figure 2. The authoring process in SIMQUEST

3. AUTHOR SUPPORT

In the SIMQUEST authoring process, conceptual support for the author is available in the form of templates offered in the building blocks. For example, authors do not need to design assignments from scratch but are offered templates for the design of different types of assignment. In addition, we provide authors with explicit support in the authoring process. For designing this support, our target user was a domain expert, not necessarily an expert in programming and/or in the didactics of discovery learning. The domain expert can be a teacher from a school, or an expert in an enterprise who may work in co-operation with an instructional designer. In the process of creating the elements of a learning environment, we distinguished three main aspects: design, implementation, and methodology.

The SIMQUEST system is constructed to support the author in these tasks. A specific feature of SIMQUEST is that it tries to support authors by including *implicit* support in the system itself. This is realised differently for the three different aspects.

- *Design.* Templates are included in the building blocks. These templates tell the author how specific elements need to be designed (e.g., specification assignments). They concern the model, the interface, the instructional support (measures), and the instructional control.
- *Implementation.* The author interface is created in such a way that an optimal transparent system results. A consistent look and feel has been given to all aspects of the authoring system.
- *Methodology.* Constraints in the system unobtrusively guide the author through the different methodological stages. For example, the way the author interface is set-up provides the author with a basic overview of the application organised following (in principle) model progression levels.

Despite the emphasis on including as much support as possible within the core system, SIMQUEST provides the author with *explicit* author support. *Figure 3* summarises the different aspects of support. This support covers the three areas mentioned above:

- *Design support.* On-line *advice* on several pedagogical aspects of the to be designed *product.*
- *Implementation support.* Implementation (or operational) support concerns the actual *implementation* of ideas and can be found in SIMQUEST in *on-line help* and a user *manual* (for the manual see Gellevij, et al., 1998).
- *Methodological support.* Specific support in the process of *authoring* an application. The basic structure is a *wizard* that helps the author navigate through all the tasks that need to be done and that may help the author with very specific hints. The SIMQUEST wizard automatically performs specific task the author wants to perform.

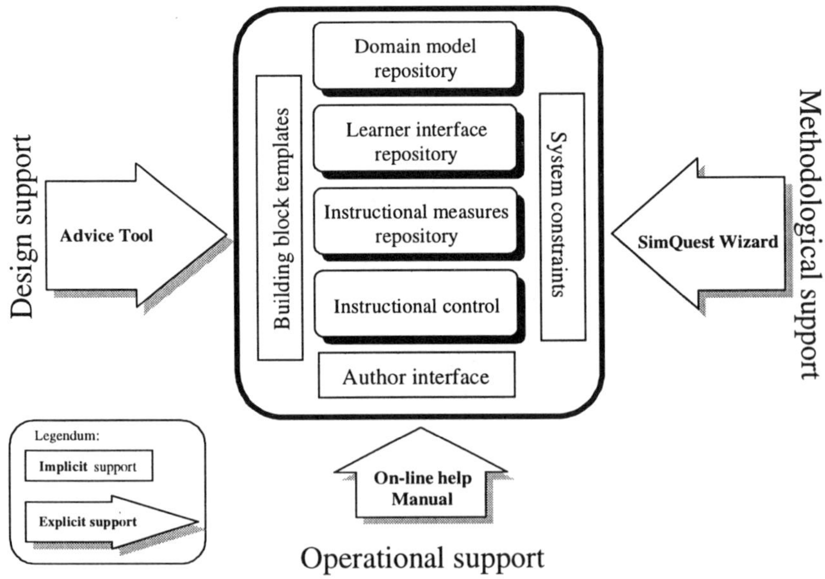

Figure 3. Overview of author support

3.1 The SIMQUEST advice tool

Because of its novelty, our prospective authors usually lack experience and detailed knowledge on the specific learning process that we want to induce in learners, namely discovery learning. Since the design process could be seen as a process of knowledge acquisition (Tripp & Bichelmeyer, 1990), we decided to include an information tool in SIMQUEST. Instead of supporting what designers *do or should do*, information systems support what the designers *think or could think of.* Information systems provide an environment that often requires designers to think harder about the subject matter domain being implemented in a learning environment while generating thoughts that would be difficult without the system. In the design process, the information system should function as a catalyst for adapting or extending the designer's knowledge base.

For the development of an information system, three broad types of design decisions have to be considered (de Vries & de Jong, 1997), namely the issues regarding structure (relations between parts of the material), content, and means of navigation (types of access). We decided to create the information system as a two-dimensional card-tray. On the one hand, different topics on which information is available are presented, and, on the other hand, each topic is discussed in multiple tabsheets. Through this flexibility, it should be possible for designers to find information quickly.

The classification of the information is based on four categories:

- What is?: provides information of a definition or explanation of a (sub) design aspect.
- Example: provides (a) textual or visual example(s) of a (sub) design aspect. Including examples is very important as became apparent from the reactions by designers in the study we performed (de Jong et al., 1996). The examples provide ideas about how to design a part of the learning environment e.g., ideas for how to design an assignment that supports the discovery learning process of the learner.
- Considerations: the designer is provided with useful considerations about a (sub) design aspect.
- Background: provides all kinds of background knowledge e.g., information about studies from the literature.

For the designer to understand the information it is important to stay as close as possible to the 'world' of the designer and to avoid using scientific and typical educational concepts. The information should be short and goal-directed. The content of the information system is based on the results of studies (see Limbach, de Jong & Pieters, 1998a; 1998b) and questions the designers may have during the process of designing (see the results of the study in de Jong et al., 1996). In Chapter 9 of this book, we give a more detailed account of the advice tool.

3.2 On-line help

Operational support offers the user detailed steps to implement the design of the learning environment in SIMQUEST. The system can be divided into three parts: help access, help topic selection, and help content. A fourth part of the system contains navigational aspects, which are interwoven through the whole system.

Help access. The help system provides access via:

- a help menu, which can be considered standard over various computer programs and leads to an index in which the user selects the appropriate topic;
- a help option in the background menu of library-elements leading to specific information on that element;
- help buttons in editors that give information on how to use that specific editor; and
- help buttons in feedback windows that open specific information to support overcoming interruptions of the workflow (e.g. caused by feedback windows or error messages).

Help topic selection. The design of the topic selection is guided by two basic thoughts. First, it should be easy to choose a topic, and, second, the user should have to take as few steps possible to select one (3 steps maximum). The help system offers a short index with clearly decomposed tasks and short explanations.

Help Content. Procedures for, and explanations about, the six basic user design tasks in SimQuest - designing levels, models, simulations, assignments, explanations, and tests - form the help content. Procedures are the core of the system because it is believed that users consult the online help mainly for "How to do...." information. To clarify complex procedures and objects, explanations are added that answer "What is..." questions.

Help Navigation. Providing easy navigation means providing answers to three user questions. The question *"Where am I now?"* is answered by the topic title, explaining the content of the help window, and the text in the top window bar, providing information on the type of information (index, procedure, or explanation), the topic, and the place of the information in the hierarchy of the help system. *"Where did I come from?"* is answered in help topic selection by leaving the previous choices visible, emphasising the choice made. A 'back' button leads to the help window used last, and a 'history' button presents all help windows used so far. The question *"Where can I go?"* is answered by links to associated procedures and links to more detailed procedures or explanations. From the latter, only links back to the main procedure are available to prevent the user from getting lost in the system.

3.3 The SIMQUEST wizard

Methodological support in SIMQUEST focuses on novice authors. By means of a wizard-like tool, these authors can generate a default application structure step by step. This structure can later be modified and extended in the free, advanced editing mode to fit application-specific requirements. Wizard support helps the author in the production process:

- in selecting the right element from the library at the right time;
- in providing the minimally required subject matter knowledge at the right time;
- by establishing default component links, providing default computation values, and by generating default test interfaces automatically; and
- in making the transition towards advanced ways of working.

The wizard support also helps the author in the design process:

- in organising the elements of the application in an integrated part-whole structure, that reflects the didactical decomposition of the application; and
- by providing didactical decision alternatives that govern the choice of assignments.

In this last respect, the SIMQUEST wizard is different from known wizards. It provides the author with design knowledge and also takes over part of the instructional design process. This is illustrated in *Figure 4* where the wizard, based on the learning goal and previous choices, suggest the author add a specific type of assignment to the learning environment (in this case an 'explicitation' assignment).

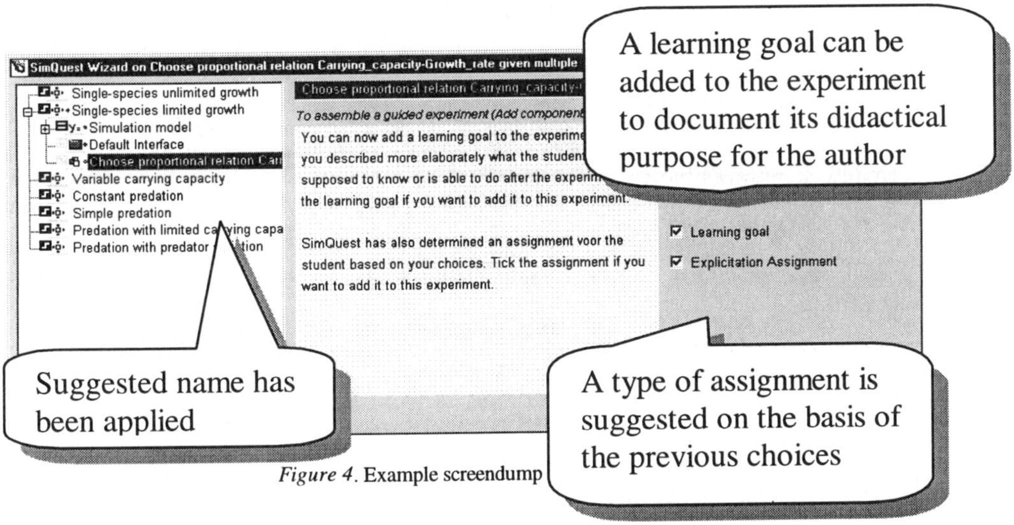

Figure 4. Example screendump

4. CONCLUSION

The SIMQUEST system is an authoring system for designing integrated simulation learning environments. SIMQUEST is both a production and a design environment (see de Jong & Sarti, 1994). SIMQUEST is a production system since it delivers actual learning environments. The on-line help tool further helps in this production process. The design support is first reflected in the elements in the library offered to authors. Elements are pre-designed templates that authors only have to adapt and fill with content (a process we have called 'specialise' or 'edit'). Templates, as such, contain instructional knowledge. In addition to this, specific tools are supplied to further help the author in the design process. The *advice tool* concentrates on optimising the *product* by providing the authors a broad range of tips and background information on all aspects of the instructional design of discovery learning environments. The SIMQUEST *wizard* combines design and productions support and concentrates on providing *process support* by helping the author take the needed steps by providing pre-selected sets of choices and making suggestions. These three tools are all directly accessible from a main pull down menu. The advice and help can also be accessed from within an editor and in this case are context sensitive by bringing the author directly to the appropriate part in the tool. Also, the tools point to each other, for example from the wizard, the on-line help and the advice are accessible in a context sensitive way.

REFERENCES

Gellevij, M., Meij, H. van der, Jong, T. de and Pieters, J., 1998, Operational support study, In Jong, T. de (Ed.), *Evaluation of the SIMQUEST authoring system and SIMQUEST applications,* pp. 41-71, University of Twente, Enschede

Jong, T. de and Joolingen, W.R. van, 1998, Scientific discovery learning with computer simulations of conceptual domains, *Review of Educational Research, 68,* 179-202.

Jong, T. de, Joolingen, W.R. van, Swaak, J., Veermans, K., Limbach, R., King, S. and Gureghian, D., 1998, Self-directed learning in simulation-based discovery environments, *Journal of Computer Assisted Learning, 14,* 235-246.

Jong, T. de, Kuyper, M., Hoog, R. de, Gellevij, M., Limbach, R., Veldhuizen, B. van, Wieberdink, B. and Horst, M. ter, 1996, Author support in SIMQUEST, In Jong, T. de (Ed.), *Designing integrated computer simulation environments for discovery learning: First project progress report,* pp. 97-139, SERVIVE project (ET 1020), University of Twente, Enschede

Jong, T. de and Sarti, L. (Eds.), 1994, *Design and* production *of multimedia and simulation based learning material,* Kluwer Academic Publishers, Dordrecht

Kuyper, M., Hoog, R. de and Jong, T. de, 1996, Evaluation studies – first requirements for SIMQUEST, In Jong, T. de (Ed.), *Designing integrated computer simulation learning environments,* pp. 27-71, University of Twente, Enschede.

Limbach, R., Jong, T. de and Pieters, J., 1998a, Supporting instructional design with an information system: the SimQuest advice tool, In Jong, T. de and Joolingen W. van (Eds.), *SIMQUEST: an authoring system for integrated simulation learning environments,* pp. 75-95, SERVIVE deliverables D7.2 and D8, University of Twente, Enschede.

Limbach, R., Jong, T. de and Pieters, J., 1998b, The design of instructional simulation-based discovery learning environments: an exploratory study, In Jong, T. de (Ed.), *Evaluation of the SIMQUEST authoring system and SIMQUEST applications,* pp. 17-40, SERVIVE deliverable D6, University of Twente, Enschede.

Merrill, M.D., 1996, Instructional design tools based on knowledge objects, In Frasson, C. and Gauthier, G. (Eds.), *Intelligent tutoring systems,* p. 29, Springer, Heidelberg.

Munro, A., 1995, *RIDES QuickStart*, Behavioral Technology Laboratories, University of Southern California, Los Angeles.

Rieber, L., 1997, *Getting interactive with authorware. Building simulations and games,* University of Georgia, Available through http://itech1.coe.uga.edu/ Faculty/lprieber/authorwareinfo.html

Swaak, J. and & Jong, T. de, 1996, Measuring intuitive knowledge in science: the development of the what-if test, *Studies in Educational Evaluation, 22,* 341-362.

Tripp, S.T. and Bichelmeyer, B., 1990, Rapid Prototyping: An Alternative Instructional Design Strategy, *Educational Technology, Research and Development, 38,* 31-44.

Vries, E. de and Jong, T. de, 1997, Using information systems while performing complex tasks an example from architectural design, *International Journal of Human-Computer Studies, 46,* 31-54.

Chapter 19

CASCADE – SEA: Computer Assisted Curriculum Analysis, Design & Evaluation for Science Education in Africa

Susan McKenney
University of Twente

Keywords: Curriculum development, Professional development, Electronic Performance Support Systems, Science education

Abstract

The CASCADE-SEA program aims to support curriculum development within the context of secondary level science and mathematics education in sub-Saharan Africa. This project focuses on the iterative design of a computer-based curriculum development support system for the creation of classroom materials. Through this process, developers hope to gain a better understanding of curriculum development within this context, in addition to creating a product that is valid, practical and effective. The CASCADE-SEA endeavor is rooted in the notion that teacher development and curriculum development occur simultaneously, even synergistically. Thus, the prototyping process takes place through inter-related study of teacher professional development and curriculum development in the arenas of mathematics and science education in southern Africa. This chapter discusses the origins of CASCADE-SEA as well as the first three episodes of its prototyping process.

1. BACKGROUND

The CASCADE-SEA research builds on a previous study that aimed to explore the potential of the computer's supportive role in curriculum development. This previous study was a design-oriented research project initiated by the Department of Curriculum at the University of Twente together with the Dutch National Institute for Curriculum Development (SLO). Among other results, this earlier research has yielded a computer program called CASCADE: Computer Assisted Curriculum Analysis, Design and Evaluation concentrating on formative evaluation for the SLO. Evaluation of this program has indicated that such a tool may offer much to the world of curriculum development, particularly with regard to the formative evaluation of classroom materials (Nieveen, 1997). Findings indicate that use of the CASCADE program saved time and improved consistency of formative evaluation plans and activities, while motivating users to carry out an evaluation that otherwise might have been omitted. The CASCADE-SEA study uses the previous CASCADE findings, (both in the forms of knowledge and product) as a launching pad for continued investigation into computer supported curriculum development.

The previous CASCADE study suggested that additional fruitful applications might lie in contexts outside that of the Dutch SLO, where curriculum materials are also being produced. Educational reform projects have been in operation in southern Africa for over 20 years; and the creation of locally relevant classroom materials has been a recurring theme among many. Despite concerted efforts, many problems arise in the local implementation of such reform projects (Verspoor, 1989), and many of the conditions for effective schooling are not met. While some authors argue that lack of success is primarily a function of a "lack of fit" of innovations (Guthrie, 1986), others stress shortcomings in the planning of the implementation phase (London, 1993; Verspoor, 1989; 1992). Yet many will agree that sustainable development can only happen when investments are made into (local) human resources, or capacity building. In terms of curriculum development, this implies investments into the professional development of teachers, teacher educators and curriculum developers. According to Ware (1992) such investments (often realized in the form of inservice) should occur in coordination with other major change components, such as curriculum redefinition and materials development, reform of the examination system and improvement of pre-service education.

Various on-going reform efforts stress the notion that teacher development is directly linked to curriculum development. The Vrije Universiteit Amsterdam (VUA) and the University of Twente (UT) coordinate several projects in sub-Saharan Africa which are designed around this philosophy. The VUA-supported Science Education Inservice Teacher Training project (SEITT) in Zimbabwe and the joint UT/VUA-coordinated Teacher Education Assistance for Mathematics and Science project (TEAMS) in Tanzania are two examples. Both exemplify the notion that a significant synergism exists between teacher development and curriculum development. They do so by incorporating teacher and curriculum development needs into their pre and inservice programs.

Within the latter half of the 20th century, many southern African countries achieved independence. Among other changes, this has often brought about a new curriculum for basic education, new subject syllabi, and reform with regard to teaching methodologies (such as a call for more learner-centered teaching). Most of the countries in this region already suffer from a serious shortage of qualified teachers (Caillods et al., 1996); and such changes have, for many, brought on additional challenges, such as the inclusion of subject matter completely new to them and the adoption of a teaching methodology they had never experienced/been taught. Many curriculum development efforts have been initiated throughout the region to help secondary level mathematics and science teachers cope with post-independence changes. Three recurring themes within such innovations include improvements in pre-service teacher education, the establishment/improvement of (sustainable) inservice education, and the more general aim of capacity building among key players. Within the arenas of inservice and capacity building, the creation of classroom materials has been deployed to help teachers: (1) update their subject matter knowledge, (2) strengthen their basic teaching skills and (3) begin to understand and implement more innovative teaching methods. At the same time, sharing the materials among other teachers (regional or national colleagues) can assist in implementing the new curriculum. The notion of engaging teachers in materials development as an effective form of inservice is widely advocated (Ball & Cohen, 1996; Ben-Peretz, 1990; de Feiter, Vonk & van den Akker, 1995). So, in view of the stimulating nature of the activities involved in materials production, the creation of classroom resources has been introduced into various inservice programs as a teacher development strategy. Simultaneously, this fulfills the needs in the curriculum development arena for new materials. Thus, a support system for the creation of lesson materials possesses the potential to contribute significantly to these development processes.

Many countries in this region (Zimbabwe, Tanzania, Namibia, and the Republic of South Africa, to name a few) employ the use of regional resource centers to achieve inservice education. In most cases, such centers are staffed with specially trained teachers who are responsible for inservice activities. Often, these facilitator teachers coordinate regional projects, including the development of lesson plans and classroom materials. Both the aforementioned projects, SEITT and TEAMS, work with regional facilitator teachers on materials production. The CASCADE-SEA research project aims to support such facilitator teachers through a computer-based system that offers relevant advice, information and tools to the materials developer.

2. RATIONALE

Three key fields of study fuse together to form the foundation of the CASCADE-SEA research endeavor. These are the notions of Electronic Performance Support Systems (EPSSs), curriculum development and (teacher) professional development. In this exploration, the former is seen as a vehicle to enhance the relationship between the two latter themes. An EPSS is an integrated computerized environment that supports task performance by providing information, advice and learning opportunities. Both advocates of the concept of EPSS as well as developers of various support systems for curriculum development presume several advantages of providing computer support (Gery, 1991; Stevens & Stevens, 1995 and Nieveen & van den Akker, 1996). An effectively designed and implemented support system would succeed in promoting improved task performance (better quality materials), improved curriculum design and development knowledge (teacher professional development) and organizational learning (among resource teachers). These factors could then contribute to additional benefits: those associated with good quality materials and teacher inservice. Goodlad, et al. (1979) looked at curriculum from varied representations: ideal, formal, perceived, operational and experiential. High quality materials can help in closing the gap between the ideal and experiential curricula, as implementation research has shown (Snyder et al., 1992; de Feiter, Vonk & van den Akker, 1995; Ottevanger & van den Akker, 1996). At the same time, the learnings acquired from creating as well as using good quality materials may serve to bolster confidence in teaching and, therefore improve overall classroom performance (McKenney, 1995).

3. DEVELOPMENT PROCESSES

While exploring the potentials of computer-based curriculum development in this context, the CASCADE-SEA study employs a developmental research approach. Through a cyclic process of design, formative evaluation and revision, EPSS prototypes are used as a vehicle to learn about building an effective tool for this context. This process is guided by the following main research question:

What are the characteristics of a valid, practical and effective computer-based support tool for curriculum development in the arena of secondary level science and mathematics in southern Africa?

The approach used to answer the main research question may best be characterized by the term 'evolutionary prototyping.' These prototyping activities are separated into three basic phases: (1) needs and context analysis, (2) design and formative evaluation of prototypes and (3) more summative evaluation of a final prototype. These phases differ not only in activities, but also in focus, with regard to the kinds of characteristics (of a computer-based support tool for curriculum development in the arena of secondary level science and mathematics in southern Africa) being examined. In the earlier phases of research, more attention has been given to the validity of the EPSS prototypes. That is, the quality and relevance of the ideas embedded in such a system as well as their cohesiveness. As the validity becomes clearer, the role of practicality enters into focus. Here, developers begin to concentrate on making the EPSS friendly and usable. Once these aspects have been elaborated, the effectiveness may be examined. Using a final prototype, developers will look to see if the tool actually does what it intends to do. *Figure 1* illustrates the plan for evolutionary prototyping in the CASCADE-SEA research at the time of this publication.

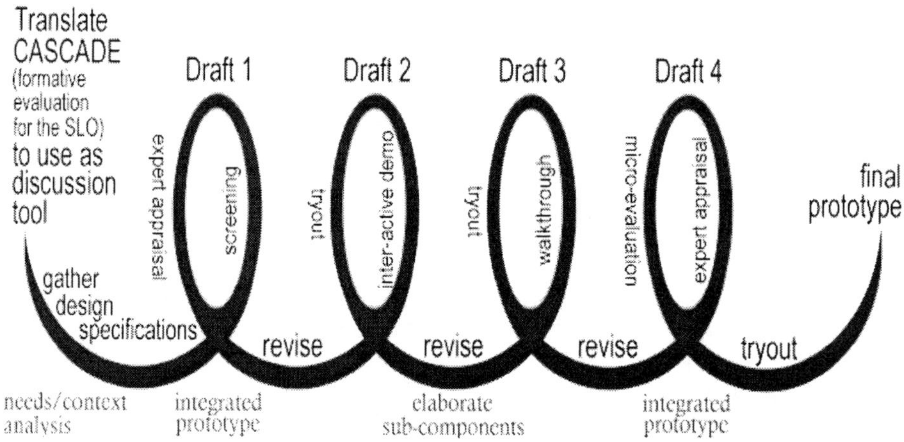

Figure 1. CASCADE-SEA overall prototyping plan

4. DESIGN PRINCIPLES AND MAJOR CHOICES

Details regarding design decisions may best be divided into three main areas, relating to the various prototypes that have been created and revised thus far: needs and context analysis, integrated prototype and elaborated sub-components.

4.1 Needs and context analysis

The needs and context analysis yielded information about various kinds of design specifications, including technical aspects (such as an authoring system and user interface), user profile information, data on the target setting and ideas about how to approach the elaboration of program content. During the actual creation of prototypes, various questions were posed such as: who should be responsible for the programming of prototypes; what platform(s) will the prototypes run on; and what languages (and software) will be used for programming?

Numerous options were surveyed and weighed, and Authorware Professional was selected as the main vehicle for prototype development, since it spoke to most of the aforementioned concerns:

- *Multi-platform.* Programs developed with Authorware Professional may be run on Windows, Macintosh or Internet systems. Since the target users vary greatly in platform use, this feature allows for additional flexibility and options with regard to testing during prototype development.
- *Easy to use.* Authorware Professional is an extremely user-friendly, icon-based system which is still powerful enough to support complex product development. Following a small initial investment in training, this has eliminated the need for a professional programmer in the project. Because Authorware is so easy to use, the developers/researchers are able to create prototypes independently; the potential for truly rapid prototyping is hereby greatly increased.
- *Flowline.* Authorware professional uses a built-in form of structure (called a flowline) that forces the user to work in (inter-related) branches. As a result, pieces of a program may be altered without forcing major updates in other parts of the system. This also aids significantly in facilitating the evolution of prototypes, as it allows the developer to "cut" and "paste" sub-components in and out of a larger whole.

Information collected during the needs and context analysis phase of study also yielded design specifications related to the users and, as a result, the user interface of a support tool. The establishment of regional teacher resource centers and specially trained teachers to staff them was discovered to be a trend that many organizations are adopting for themselves. Such a user group, while large, still possesses many similarities, such that the definition of a user profile is a realistic goal. Resource teachers tend to be selected because they are experienced and motivated, although most of them are also new to computers.

While the interface might evolve through various forms, an effort would be made to satisfy three main qualifications:

- *Easy.* Most of the resource teachers involved in the creation of lesson materials have little or no experience with computers; however, experience has shown that they are very willing and eager to learn. Therefore, an effort must be made to reward that enthusiasm with (relatively) easy-to-achieve success. That is, this program must not abuse the users' willingness to learn by bogging them down with complexities; rather, it should be easy to learn. This means building in, for instance, redundancies, so that the user may be able to access certain kinds of information in more than one way.
- *Inviting.* Even though this group is, on the whole, motivated and willing to learn, a dull, uninviting program might serve to dampen that enthusiasm. Steps should be

taken to promote user self-confidence, as well as to increase user interest in using the program. One way to do this is to make the practical applications of its use abundantly clear, so that the teachers see the immediate benefit(s).

- *Safe.* Of the few users having experience with computers, the overwhelming majority have used DOS machines. Besides being command (as opposed to menu) driven and just plain difficult to learn, earlier programs were far less safe-guarded than those operating under, for example, Windows. Users need to know that they aren't going to break the computer or ruin files if they push the wrong button. Further, this program must aid and encourage them to explore, while taking responsibility for ensuring that experimentation yields positive (not negative) results.

4.2 Integrated prototype

Following the needs and context analysis, the cyclic process of prototype design, evaluation and revision was initiated. This began with the construction of a relatively shallow prototype that integrated the key phases of materials development: analysis, design and evaluation. The main function of this version was to provide a vehicle through which the total configuration (of a CASCADE-SEA program) could be examined. *Figure 2* illustrates the main tasks supported in this integrated version, following the first round of evaluation and revision.

After the integrated prototype was revised, developers then began to elaborate individual sub-components. Determining which area(s) would be targeted for elaboration revision was based on the anticipated needs of users who would be using (testing) the program at the next available research opportunity. Various components of CASCADE-SEA were considered as possible areas to elaborate in the next development cycle. After much deliberation, the component selected for immediate revision was that of building individual lessons. The lesson builder sub-component was created and tested in two field settings (workshops with teachers in Zimbabwe and Tanzania). For testing purposes, this prototype was designed to run independently from the integrated prototype, but with the intention of being able to merge the two, later. In order to understand the depth and breadth of this elaboration, it may be useful to describe the "Lesson Builder" sub-component in detail.

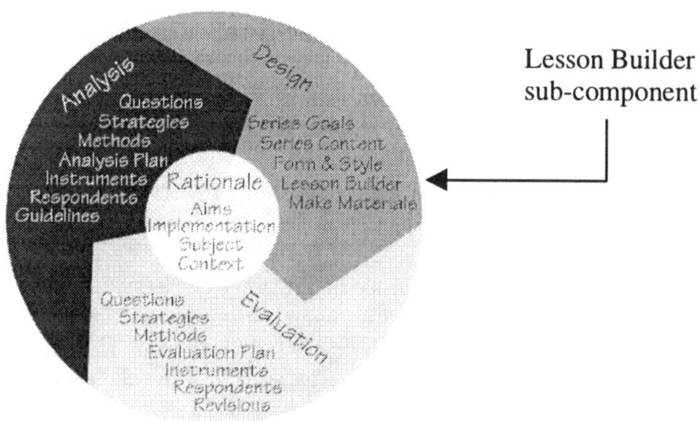

Figure 2. Revised task map of integrated prototype

4.3 Sub-components: the Lesson Builder

The "Lesson Builder" component has been designed to help teachers make paper-based exemplary lesson materials (such as teacher guides) that can be photocopied and shared. The way in which lessons are built depends on how individual users wish to work. For those desiring support throughout the building process, they can respond to all of Lesson Builder's prompts and then let the program generate an initial draft. Support is offered in various forms, such as giving the user multiple choice options on how pupils should present their work. Another type of support features the sharing of ideas. The body of the lesson is constructed of various building blocks, called activities. All of the activities are stored in a database and divided into the following categories: demonstration, experiment, explanation, field trip, game, homework, practice and project. Each entry in the database indicates the subject area, grade level, title, objectives, materials needed, instructions, expectations, and is cross-referenced when applicable (for example, some experiments are also suitable for demonstrations). Users may search or browse through the database until finding an activity (or a few) appropriate for their lesson. These activities may be added to the lesson plan, and users are also encouraged to contribute their ideas (activities) to the database. For those preferring to get off to a quick start without much support from Lesson Builder, they can simply go to the "Generate" folder and click the 'Do it for me' button. A template lesson will then be opened. All of the fields will still have to be filled in, but a general structure will get the user off to a good start.

Figure 3 illustrates the various 'main pages' within the lesson builder program. Bearing in mind that this component has been built to function independently, it then makes sense that it comes complete with additional functions (such as help) that will later be merged with those inside the 'mother' program.

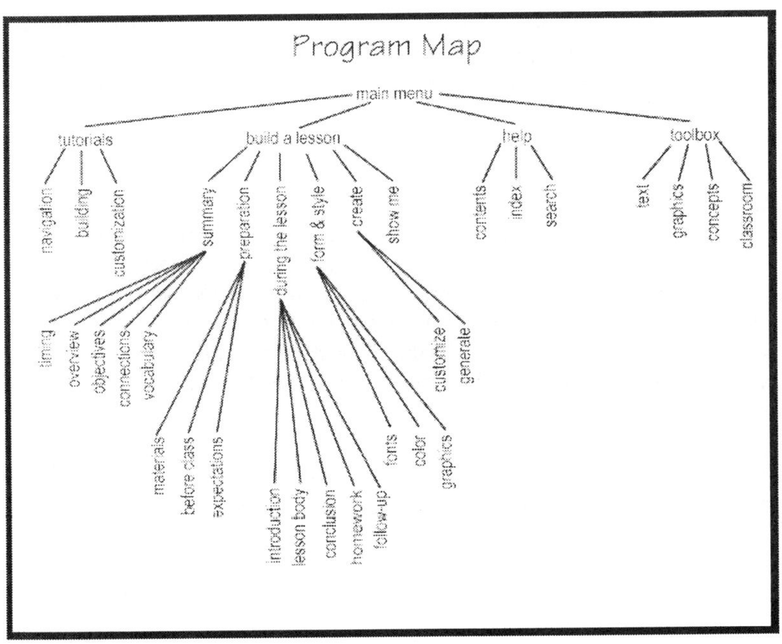

Figure 3. **Lesson Builder program map**

Becoming acquainted with any new program can be challenging, even frustrating. Lesson builder tries to help users become acquainted with this program by offering tutorials on three main subjects: navigating, building and customization. The navigation tutorial helps one understand how to get around in the program. The building tutorial helps one understand what to do on each page while building a lesson. And the customization tutorial aims to help users enhance their draft lesson plans generated by lesson builder.

In the toolbox, users are offered tools to help customize and elaborate the draft lesson plans created by the Lesson Builder. The tools are divided into four categories: text, graphics, concepts and classroom. The text page contains links to two text editing programs, the graphics page contains links to two drawing programs and to a clip art gallery, the concepts page contains a link to a concept mapping tool, and the classroom page contains a link to an idea book (containing suggestions for ways to improvise when classroom resources are scarce).

Help is offered in three forms within the lesson builder: a table of contents attempts to give answers to frequently asked questions (this area is arranged by topic); an index helps users find answers by keyword; and via the find function, the user can ask the program to search through all the help files and find those pages relating to whichever word(s) they have entered.

5. FUTURE DIRECTIONS

Developers have struggled with many difficult issues, such as how to strike a useful balance between generic and content-specific support. In CASCADE-SEA, general guidelines and heuristics (independent of subject matter) are offered throughout the program, while all examples are subject-specific. For instance, users creating biology lessons are presented with generic guidelines (such as a list of action verbs to get a start on writing objectives) while all samples and examples relate to the field of biology. Such an approach yields both practical advantages for the development of the system, as well as increased potential for learning as a result of using the tool. From a practical standpoint, this helps the developers to maintain focus on offering support for good materials design in general. With an eye toward the possibility of ever creating such a tool for other subject areas, this approach keeps more options open for future elaboration. Keeping in mind the overall aims of this EPSS, such an approach is also consistent with the goal of helping teachers learn through the process of creating exemplary lesson materials by offering design principles, explaining what they mean/why they are important and then giving examples that are meaningful and relevant to the user.

As this project evolves, additional sub-components will be elaborated and trailled, before being merged into the (revised) integrated version. Through the creation, evaluation and revision of prototypes, developers hope to identify other important attributes (such as generic guidelines coupled with specific examples) that should be employed in such a system. The rationale behind each design decision strives to fill a current void in research literature. While much has been written with regard to components of good quality lesson materials, strikingly little information is available about how to support teachers in creating such materials themselves. And while the current body of literature decidedly supports the notion of embedding lesson material creation into professional development scenarios, it lacks much information on how such an approach should be implemented. By carefully documenting the reasoning behind each design decision throughout this iterative process, the CASCADE-SEA development team hopes to begin filling a gap in educational research literature that will be useful toward both theory building and improvement of practice.

REFERENCES

Ball, D. and Cohen, D., 1996, Reform by the book: what is - or might be - the role of curriculum materials in teacher learning and instructional reform?, *Educational Researcher (25)* 9, 6-8, 14.

Ben-Peretz, M., 1990, *The teacher-curriculum encounter,* State University of New York Press, Albany, NY.

Caillods, F., Göttelmann-Duret, G. and Lewin, K., 1996, *Science education and development: Planning and policy issues at secondary level,* Pergamon, Paris.

Feiter, L. de, Vonk, H. and Akker, J. van den, 1995, *Towards more effective science teacher development in southern Africa,* VU University Press, Amsterdam.

Gery, G., 1991, *Electronic performance support systems: How and why to remake the workplace through strategic application of technology,* Weingarten, Boston.

Goodlad, J., Klein, M. and Tye, K., 1979, The domains of curriculum and their study, In Goodlad, J. and Associates (Eds.), *Curriculum inquiry: The study of curriculum practice,* 43-76, McGraw-Hill, New York.

Guthrie, G., 1986, Current research in developing countries: the impact of curriculum reform on teaching. *Teaching and Teacher Education, 2,* 81-89.

London, N., 1993, Why education projects in developing countries fail: a case study. *International Journal of Educational Development, 13,* 265-275.

McKenney, S., 1995, *The formative evaluation of a Namibian physical science teacher guide,* Master's thesis, University of Twente, Enschede

Nieveen, N. and Akker, J. van den, 1996, Computer supported curriculum development, In Plomp, Tj. and Ely, D. (Eds.), *International encyclopedia of educational technology,* 153-158, Pergamon, Oxford.

Nieveen, N., 1997, *Computer support for curriculum developers,* Doctoral dissertation, University of Twente, Enschede.

Ottevanger, W. and & Akker, J. van den, 1996, *Supporting curriculum changes in Namibian classrooms,* Paper presented at the AERA annual meeting, April 8-12, New York.

Snyder, J., Bolin, F. and Zumwalt, K., 1992, Curriculum implementation, In Jackson, P. (Ed.), *Handbook of research on curriculum,* 402-435, Macmillan, New York.

Stevens, G. and Stevens, E., 1995, *Designing electronic performance support tools: improving workplace performance with hypertext, hypermedia and multimedia,* Engelwood Cliffs, NJ, Educational Technology Publications.

Verspoor, A., 1989, Using textbooks to improve the quality of education, In Farrell, J.P. and Heyneman, S.P. (Eds.), *Textbooks in the developing world: Economic and educational choices,* 52-60, World Bank. Washington, DC.

Verspoor, A., 1992, Planning of education: Where do we go? *International Journal of Educational Development, 12,* 233-244.

Ware, S., 1992), *Secondary school science in developing countries.* PHREE background paper series, World Bank, Washington, DC.

Chapter 20
The TeleTOP Decision Support Tool (DST)

Betty Collis
Wim de Boer
University of Twente

Key words: Implementation, Design Methodology, Rapid prototyping

Abstract

Instructional design models have been classified (Gustafson & Branch, 1997) into three general categories: classroom orientation, product orientation, and system orientation. A combination of the models is taking place at the University of Twente in the Netherlands via the TeleTOP Project of the Faculty of Educational Science and Technology. In this chapter we show how the strengths of each of these methods can be captured in a new hybrid instructional design model and illustrate this through the process taking place at the Twente faculty. In particular, we illustrate how the WWW-based *TeleTOP Decision Support Tool* is a key resource in this new approach.

1. FACULTY-WIDE COURSE RE-DESIGN AND THE NEED FOR A NEW INSTRUCTIONAL DEVELOPMENT MODEL

In their taxonomy of instructional design models, Gustafson and Branch (1997, p. 30) identify three general classes of models (classroom orientation, product orientation, and system orientation). What they label as the classroom orientation reflects the typical process of course design in traditional higher-education settings: the individual instructor using little or no explicit design methodology is at one time the designer, developer, and deliverer of his or her course; the content expert and the person responsible for student organisation, student support, and student evaluation. Despite the potential weaknesses of the idiosyncratic course-development method, it is by far the predominate way that courses are designed and delivered in schools and universities. And its weaknesses can also be its strengths: the instructor as content expert fully responsible for the course can mentor, stimulate, scaffold, and personally interact with his or her students so that the course is much more than a systemic way to meet pre-defined objectives but also can be a framework for an apprenticeship-type mentoring relationship between instructor and learner (Sfard, 1998). Instructors can also monitor and adapt during the instruction; tasks that are often difficult to accomplish with technology based instruction.

In contrast, the product-orientation and system-orientation instructional-design models assume professional development teams using an explicit design methodology, and sophisticated analysis, design, and development tools. In contrast to the instructor-rooted classroom-orientation models, the product- and system models implicitly or explicitly assume that it is not likely that members of the development team will actually have a role in the product's ultimate implementation or use.

Each class of instructional-design models has its strengths, and its limitations. A combination of the models, in a way that optimises the strengths and reduces the effects of the limitations, seems desirable. Such a model could retain the skills and insights of the personal instructor and at the same time make use of the time involvement, specialisation skills, sophisticated tools, and task division possible in the professional development team.

Such a synergy is not a new idea; many educational-software development projects have called for close involvement of representative classroom teachers, for example. However, such a synergy does not yet much occur in practice. What can we expect then with respect to the design of WWW-based course-support environments, and the redesign of existing courses related to such course-support environments? Will the instructional design of these take place via the individual instructor, or via professional development teams? Can a synergy occur? If it can, what methodology and tools can help promote the synergy?

2. WHAT SORT OF INSTRUCTIONAL-DESIGN METHOD FOR OUR OWN FACULTY SETTING?

We work in a faculty in a traditional university setting, where course design and delivery takes place predominately in the classroom orientation. In this context, a number of instructors in our faculty have been pioneers in the re-design of their courses involving new technologies. For example, members of the faculty were among the first to use the WWW as a collaborative-learning environment for course assignments, in March 1994, so that experts in different places in the world could interact with the students in the course on the collaborative writing of course materials using the WWW as a common dissemination environment (Collis, 1997b). By the end of the 1996-1997 academic year, our faculty could be characterised as having moved from a *support-the-pioneers* stage with respect to the use of information and communication technology (ICT) in instruction, to a *1,000 flowers blooming stage* (Collis, 1997a).

In the momentum of this experience, our faculty has decided to move from the *1,000 flowers blooming stage* to a stage of *managed change* in our instructional practice. We have made this decision on the basis of not only the naturally evolving interest and momentum for course re-design taking advantage of the potential of ICT for increased interactivity and communication within courses, but also because of the strategic decision to offer our education program in a more-flexible way. In particular, the decision was made in mid 1997 that by September 1998 students entering our program could participate as local students, as students meeting together in a satellite campus on the other side of our country, or as part-time mature students, already in the workplace and maintaining their jobs and home situations while participating in our program. All three of these cohorts will come together to our campus one day every two weeks for various common sessions and opportunities for face-to-face interaction, but after that each course in the first year (16 courses) as well as a variety of others (a total of 30) is being re-designed so that all three of these cohorts can have their particular needs met within a shared course experience, with no extra instructional staff.

In order to steer and manage this complex implementation project, an instructional development team, called *TeleTOP* has been formed. The task of the TeleTOP team is to lead and carry out a systematic and integrated course re-design initiative, what Gustafson and Branch (1997) would categorise as a system-orientation approach, but working within the reality and strengths of the local instructor-specific, classroom-orientation culture. A key question for us was: what instructional-design model should we us? None of the 13 models described in Gustafson and Branch's 1997 taxonomy integrated the strengths of a system orientation with a focus on the central involvement of the individual instructor in the re-design of existing courses for more-flexible delivery. Thus we have evolved our own instructional-design model and tools to support the model in practice (see *http://teletop.to.utwente.nl*). But we would emphasise that virtually all the existing ID Models seem to assume new design rather than redesign which is a major limitation in our view.

3. THE TELETOP COURSE RE-DESIGN MODEL

There is not space in this chapter to describe the overall TeleTOP Course Re-Design Model in detail; various papers, documents, and presentations can be accessed from the *Publications* link of the TeleTOP homesite (*http://teletop.to.utwente.nl*). The main features of the Model are: (a) an emphasis on instructor engagement, on extending the strengths of the instructor via technology, not reducing the instructor's involvement or personal impact; (b) an awareness of predictable problems and phases in the diffusion of an innovation in an educational setting (Fullan, 1991), and in particular of factors that affect the instructor's likelihood of changing his or her instructional practice (Norman, 1997; Northrup, 1997); (c) a pedagogical model to guide the instructor's decision making about course re-design and technology integration (Collis, in press; Laurillard, 1993); (d) the use of rapid prototyping of the evolving WWW environments to support each re-design course as a key instructor-involvement strategy; and (e) a common goal and challenge--to provide quality instruction to the different cohorts of students in a way that is both individualised to the needs of each cohort but also recognises the limits on time and energy that the instructor can invest in the aspects of his or her work relating to instruction. A key to integrating these features is our *TeleTOP Decision Support Tool* (the *TeleTOP DST*; Collis and de Boer, in press; de Boer, 1998; de Boer, Fisser, Strijker, Peters, Verheij, and Collis, in press). In the remainder of this chapter we discuss the *DST* in more detail.

4. THE TELETOP DST

In the following subsections we indicate the educational principles on which the *TeleTOP DST* is based (Section 4.1), outline briefly how the *TeleTOP DST* evolved over three versions during 1997 during its own rapid prototyping and evaluation cycles (Section 4.2), describe the version used in the 1997-98 cycle of the *TeleTOP Project* (Section 4.3), and describe how the *TeleTOP DST* is used as a major tool in our course re-design methodology (Section 4.4).

4.1 Pedagogical concepts underlying the TeleTOP DST

In our approach to *pedagogical re-engineering* (Collis, 1997b; Collis and Fisser, 1998), we take as a starting point the current organisation, general instructional approach and objectives of the individual course, and then ask the instructor to think of his or her course in terms of its decomposition into six major generic components, each of which presents a variety of tasks for the instructor:

1. Organisational aspects of the course, including planning, developing course outlines, setting and revising objectives, choosing study materials, communicating with the students about procedures, maintaining records on student attendance and work, determining the final grades of the students and reporting these, etc.
2. Preparing and delivering class presentations (usually lectures).
3. Setting up the self-study expectations of the course; choosing study materials, organising and handling on-going small assignments, dealing with questions relating to the course study materials usually in preparation for examinations; organising and perhaps supervising laboratory or practicum sessions.
4. Setting up, guiding, and evaluating the major assignment(s) in the course.
5. Developing, monitoring, and grading the course examination(s) and discussing problems encountered on the examinations with individual students.
6. Providing opportunities for general individual contact and other forms of communication (establishing office hours, indicating how and in what ways students should make appointments for help, etc.).

Instructors are shown examples of how a variety of different courses can each be profiled in terms of the relative amount of time and effort the instructor spends on each of the above categories of tasks. Then the instructors are asked to think of their own courses in terms of these general categories of tasks, and for each category, consider ways in which some of the tasks could be adapted for more-flexible and more-communication oriented delivery, via telematics support. *Table 1* shows examples related to the categories *General Course Organisation* and *General Communication Support.*

Examples of how such pedagogical re-design has already occurred in courses within our faculty, with help of ICT, are made available, to help instructors better picture possibilities for their own courses. These examples were linked to a WWW page containing a simple matrix (see *Figure 1*). Over a series of one-hour workshops, the instructors were shown these examples and encouraged to study them further in their own offices, via a WWW browser.

In this way, instructors begin with their existing courses, get familiar with a common way of decomposing and describing those courses in terms of six sets of familiar tasks, and only consider technology use when it serves an educational purpose—to make a task more flexible to better serve the needs of different types of students or to enrich the communication and interaction aspects of the learning process.

Course aspect	**Current practice, strengths**	**What to extend, make more flexible? How?**
1. General course organisation	-Faculty bureau makes on-going information available to students via display boards, personnel available for questions, bulletins in the student newspaper, telephone availability, putting messages in students' individual mailboxes, sending email -Instructors provide course outline with all details of course organisation; give updates and changes during lecture sessions, put messages on student notice board or in student post-boxes (physical and email)	1. General course organisation -Faculty bureau makes on-going information available to students via display boards, personnel available for questions, bulletins in the student newspaper, telephone availability, putting messages in students' individual mailboxes, sending email 2. All courses are supported by an integrated WWW site; the courses themselves are integrated in larger WWW sites with faculty-wide information 3. Administrative information (names of students, email addresses, scores to date, etc.,) are available through the WWW site 4. "Newsflash" areas on the WWW sites present announcements, changes, updates 5. Faculty bureau maintains channels of personal contact and availability, printed material still available on courses, timetable, etc; Email and the WWW used for student contact, updates in information, sign-ups for elective courses, etc.
2. Communication	-Between instructor and students: Takes place during lectures, informally in the canteen and hallways, via personal visits to the instructor in his office, via notes between the instructor and students via their physical mailboxes or email -Among students, organised or informal, in the hallways and other locations around campus, in the computer laboratories, via their mailboxes or email	-Maintain all of these, but add: -convenient communication through an e-mail centre in the WWW sites, where not only individuals can be messaged but also groups within the course (via aliases). Group members can have their own private communication areas within shared workspaces -the opportunity for guided reflective discussions via a WWW board (a form of computer conferencing within a WWW site) -tools for scheduling and carrying out real-time communication at a distance including chat, Internet telephony, desktop meeting tools (such as *NetMeeting*)

Table 1. Pedagogical re-engineering, TeleTOP approach (Collis and Fisser, 1998)

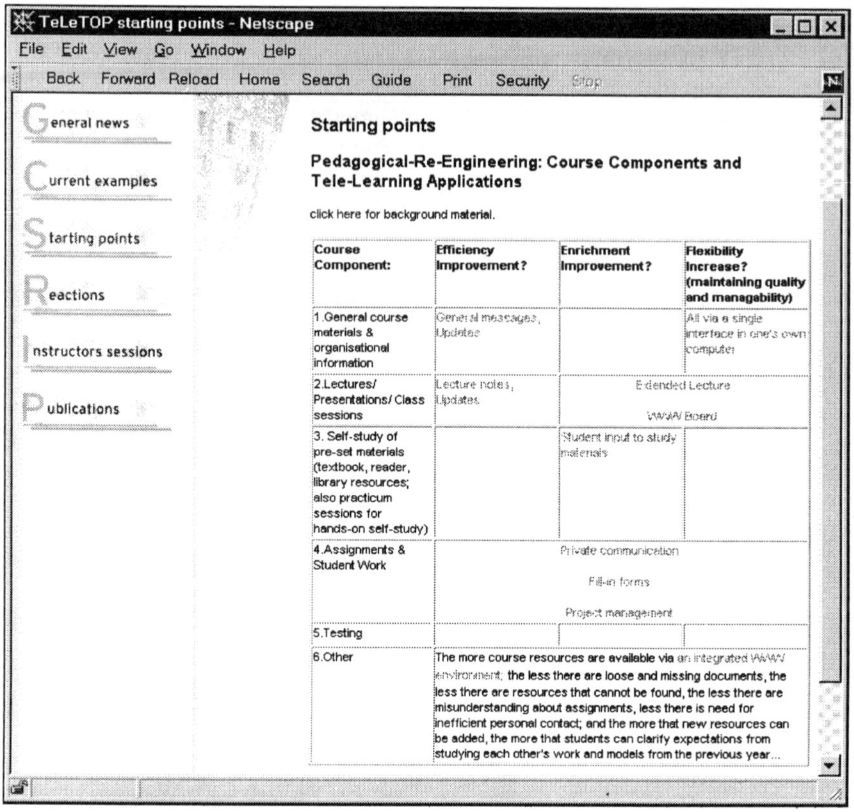

Figure 1. Screen dump of the "Starting Points" matrix from the TeleTOP site (http://teletop.to.utwente.nl/). Links are available in the cells, to show examples from currently operating courses in the faculty

4.2 The evolution of the *TeleTOP DST*

During this same time, one of authors, now an instructional designer on the TeleTOP team, was working on a project with a similar sort of context as that of *TeleTOP* (de Boer and Hamel, 1998). In this earlier project, instructional designers of language courses to be offered with telematics support on a Europe-wide basis needed to make decisions about how to re-design existing courses to serve the needs of heterogeneous student groups over distances and how to choose appropriate technological options for their courses to meet this challenge. It evolved that the course designers not only needed a systematic way to become familiar with technological possibilities, but also needed help in seeing how these possibilities might be useful within an actual language-learning course. Thus the idea of developing a decision-support tool that would not only help the instructor become aware of technical possibilities for his or her course but also help see how these could be integrated in an educationally useful way in a course to be offered partially or totally at a distance, emerged. A paper prototype of the decision-support tool was designed and evaluated by instructors and educational technologists involved with the language courses.

After revision, a first electronic version of the decision-support tool was made in html, but now focused on the broader situation of courses in our own faculty rather than just language courses. This electronic version was continually adapted during several months of rapid prototyping with feedback from a number of different TeleTOP instructors and team members. This second cycle of rapid prototyping concluded with a structured formative evaluation of the decision-support tool, involving not only a guided walkthrough but also a post-use interview and questionnaire, with five TeleTOP instructors (de Boer and Hamel, 1998).

Many changes to the DST were made on the basis of these results, particularly with regard to the level of detail in the DST; the examples chosen; the order in which components and examples appeared; the sorts of questions asked of the instructor during his or her use of the DST; the user-interface layout and navigation; and the flexibility available to the user in terms of skipping options, and reviewing and changing choices when using the DST. At the same time, the underlying technical architecture of the tool was also improved. This architecture will be described in the next section.

4.3 Functional and technical specifications of the *TeleTOP DST*

The TeleTOP DST is a WWW-based environment consisting of a series of more than 65 questions related to the six major generic components that are our basis for pedagogical re-engineering (see Section 4.1).

The questions in the DST are very short and deal with the "normal questions" that an instructor has to ask him- or herself when setting up a course, such as: "What kind of presentation formats will I use in my lectures?; How, when, and to what detail will I provide feedback on assignments?", etc.. Each such consideration was presented in terms of a question. The questions are of the following forms:

- *For course overview.* Do you want to have the possibility through a sort of "newsflash" on the homepage of a course-support WWW site to give short updates and announcement?
- *For communication.* Do you want your students to participate in group discussions, where they can enter their reflections at a time convenient to them?
- *For lectures.* Do you want to make additional information available to your students, relating to material in the lectures?
- *For self-study.* Do you want to give occasional personal feedback to the students based on their progress with self-study questions available through the WWW site?
- *For major assignments.* Do your students work in groups on a major assignment, and if they do, do they sometimes have problems managing themselves and staying on tempo?
- *For testing.* Do you want to give students access to previous test questions, along with feedback and overviews of previous students' answers?

In order to help the users (instructors) to better consider their answers to these questions, and at the same time relate the answers to the range of new instrumental and technical possibilities, examples can be chosen and studied relating to each question. The choices can be made for the new WWW-based learning environments which the instructors will be using to support their courses. The examples are all taken from existing on-line WWW-based support environments from courses in the Faculty of Educational Science and Technology, and sometimes from other parts of the world. Important however is the fact that they are all examples from an educational context. This makes it much easier to "understand" the examples which go with the question, in order to answer it in a more-considered way. *Figure 2* shows a part of the *DST* with the question window and with an example window, in this case both referring to a WWW-based course roster.

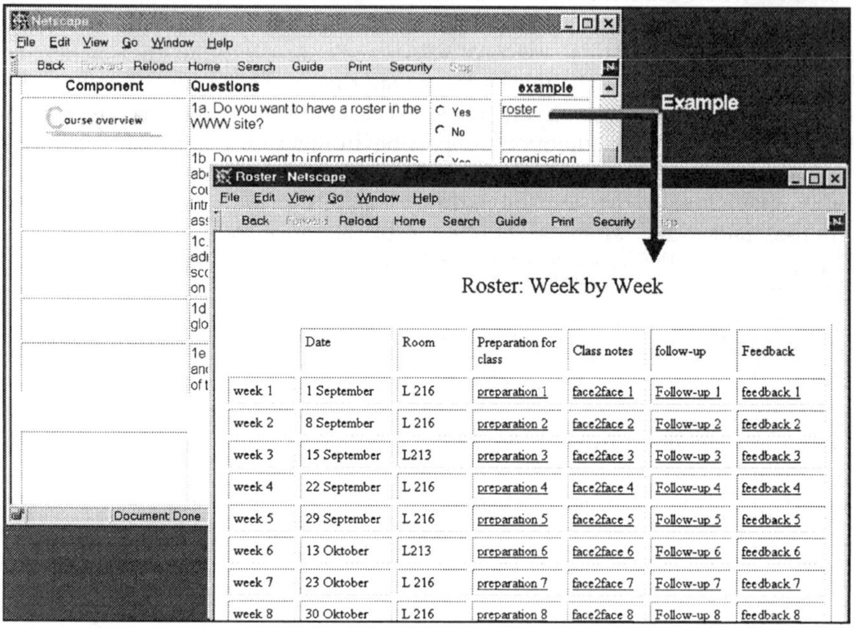

Figure 2. The Decision Support Tool with an "example" window showing how a course roster can be operationalised in the form of a hyperlinked html page in a WWW environment

Each of the questions in the DST is answered by entering a yes or no choice. Instructors are encouraged to answer "yes" if they are potentially interested in an option; changes can be easily made in later rounds of prototyping. In some cases, a "no" answer jumps the user over a subsequent series of questions and examples; for example, if the instructor has no interest in group-based project work, then a range of questions about how to support such project work especially when students are not always at the same place at the same time is omitted. After completion of responses to all pertinent questions in the DST, the database which underlies the DST automatically generates a unique WWW page with the specific answers and the chosen options made by the user. In this way, users immediately can see via the WWW browser or a printout what they have chosen, and can continue to examine the examples in their own offices, simply by loading the WWW page which was generated for them.

The *DST* is build upon a database, which organizes the components, the questions, and the examples. The questionnaire is designed in a table, with in the left column the components (e.g. *Communication*), then the questions with yes-no response options, and in the right-hand column, the hyperlinks to the examples. Examples are always displayed in a new browser, so that the *DST* itself will not disappear from the user interface. The use of a database makes it possible to generate and store the new forms which are generated every time the *DST* is used in a structured way. The stored information can be later retrieved or edited and restored by the users. The use of a database can be seen as "write once, read many, change if necessary" actions in an already-structured environment. The design of the database is essential for the storing and retrieval of information. The information can only be stored and retrieved in the way that is defined in the designed structure of the database.

4.4 Using the *TeleTOP DST*

The instructor who is going to use a WWW-based course-support environment needs to decide what features he or she thinks will be appropriate for his or her course. But at the same time, the instructional designer of the TeleTOP team responsible for designing and building the WWW-based course-support environment needs to be well informed about the instructor's wishes and decisions. Therefore, an appointment for one hour was made with all instructors participating in the TeleTOP project. Two TeleTOP team members took part in the interview with the instructor physically present. Because the *DST* is available on the Internet (password protected), a computer with an Internet connection was used during the interview. The screen display was projected on the wall of the meeting room, thus facilitating a more-natural discussion environment rather than that of a group of persons huddled around a computer screen. First, the team members explained to the instructor what the goal of the session was. Second, the *DST* was introduced as a tool designed for support of a structured interview and to enable the instructor as well as the team members to make decisions in a structured and organised way.

To illustrate the approach, a session will be briefly described.

In January the instructor of the course Instructional Design, Dr. Gulmans, was invited to come to a DST session and an appointment at his convenience was made. The TeleTOP team knew how his course was presently being organised, through studying the course organisation, goals, etc., found in the current study guide. In the course Instructional Design the students are divided into groups of four. At the end of the course the groups deliver a report, which will be marked. The short interview that was held at the start of the DST session made clear that Gulmans organises his course around six lectures. The TeleTOP team members explained that the new groups of part-time and distant students that will be participating in his course in the following year should have the possibility to participate in the course as easily as the regular students and that the DST could help him find ways to make this flexibility feasible. Furthermore it was emphasised that the role of the instructor was not going to be eliminated or reduced and that face-to-face sessions were still going to be very important. It was explained that every student would come every other week to the University to have face-to-face lectures or other forms of contact sessions. The regular students could have even more face-to-face sessions, but a consequence would be that the other cohorts should be able to participate in one or another way in these sessions as well.

Subsequently the Decision Support Tool and the goals of working with the tool were introduced. The name of the course and of the instructor were filled into the WWW-based form and the process of responding to the questionnaire was started.

Taking the first component of a course considered in the DST, the "Course Overview" component, Dr. Gulmans was asked first if he prefers to have a roster in his WWW-based course-support environment. Dr. Gulmans did not have a clear picture of what was being asked, and agreed that it would help him to see the example which the DST provided (see Figure 2). After the example Dr. Gulmans had a clearer picture of what was meant by a WWW-based roster and the functionalities it could support. This enabled Dr. Gulmans to make a more considered response to the question " Do you want a roster in your WWW-based course-support environment ?"

In a similar way a series of questions related to each of the six course components were discussed by the instructor (Dr. Gulmans) and the TeleTOP team members. The choices were made and indicated via the DST. After answering all questions and discussing the components and examples, the DST automatically generated a "decisions made" form. This form is available in the DST environment and it was printed out for Dr. Gulmans.

Via the homepage of the *DST* all instructors can find their courses and the pages that were generated containing the choices made by the instructors during the *DST* sessions. The instructors are able to compare the choices they have made with the choices of their colleagues. By comparing their own courses and the chosen options of the other courses, the instructors could get more ideas for their own courses. Each instructor has the possibility to reconsider his or her own choices for quite a while during the subsequent rapid prototyping cycles. The option of changing the decisions is always there, so that instructors do not need to feel locked into choices made when they work with the *DST*.

The "decisions made" form, which is unique for every individual instructor, is also the base for the next step of the TeleTOP rapid-prototyping method. Within a few days after the *DST* session, the TeleTOP team makes a first prototype of the specific course-support environment, based on the decisions made by the instructor. After the development of this prototype two members of the TeleTOP team visit the instructor in his or her office, and conduct a walk-through of the first prototype of the course WWW site, further discussing the ideas and reactions of the instructor, and adjusting the prototype.

5. EVALUATING THE EFFECTIVENESS OF THE *TELETOP DST*

The *TeleTOP DST* has been evaluated formatively by the TeleTOP team in terms of both its usability and its utility during its use with the 30 courses being adapted in the 1997-98 phase of TeleTOP. Usability relates to how easily the user can interact with the system; utility relates to how well the system does the task it is meant to do (Nielsen, 1994; Sweeney, Maquire and Schakel, 1993). Some results follow:

In terms of usability, the *DST* has been evaluated as being easy to use in the interview setting and then as a follow-up tool for self-study as well as for further design-oriented discussions between an instructor and TeleTOP team members. In terms of Nielsen's major attributes of usability of computer-based systems (1994) (learnability, consistency, ease of remembering how to use over time, and subjective appeal), we have no evidence of any instructor having problems or dissatisfaction with the use of the tool. Consistency is a particularly strong feature of the tool, not only among the various sections of the *DST* but also, between the DST interface and the interface of the site generated directly as a result of use of the *DST*. This consistency is further maintained in the first prototype of the WWW-based course environment based upon the instructor's decisions made with the *DST*, and in the final course-support environment eventually produced. All of these interfaces seem to the instructor as if all the products have just blended from one to another. *Figure 3* shows the interfaces for one course re-designed for TeleTOP: the *DST* environment, the output from the *DST*, the first prototype site, and the site as in use with a full group of students.

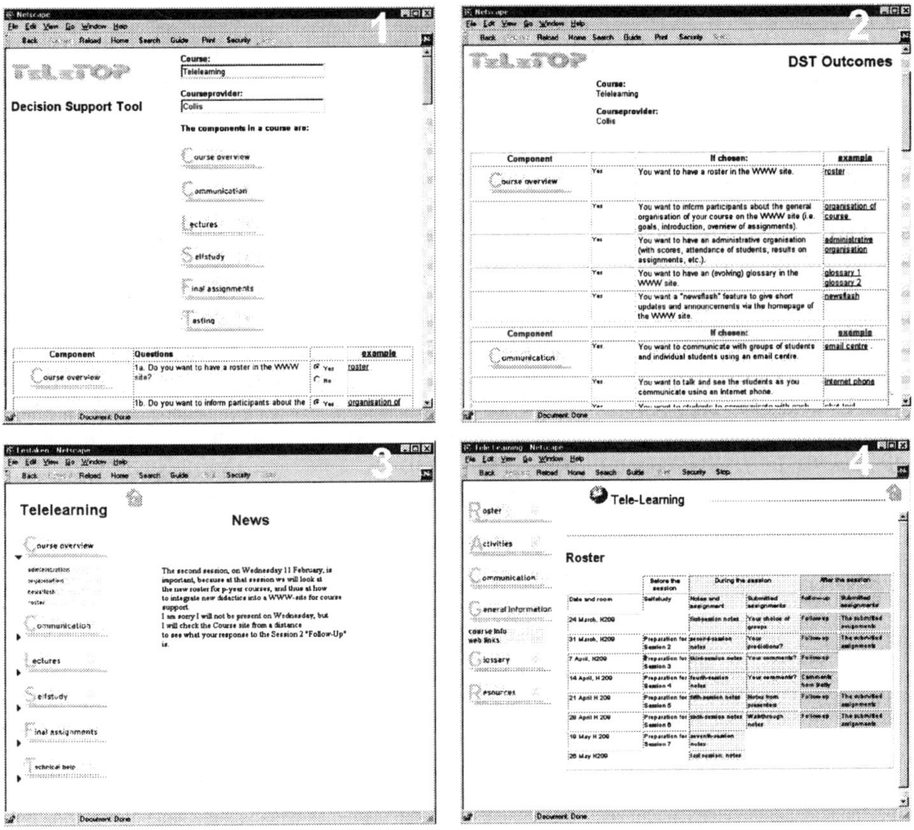

Figure 3. Interfaces of the four stages (1.DST session; 2.DST output; 3.Prototype of the course; 4.Final course environment) of realising a WWW-based course-support environment using the TeleTOP instructional-design Model, illustrated with the course named Tele-Learning.

In terms of usability, it must be remembered however that the *DST* tool is not designed to be used independently, but rather in an interview setting under the leadership of an instructional designer deeply familiar with the tool, as well as with some personal knowledge of the instructor involved and the sorts of course re-design features that are likely to be of interest to his or her own style of teaching and type of course. When these conditions are not in place, it would be necessary to add more explanatory materials to the *DST* and perhaps a guided tour through an example session using the tool. This guided tour would be best if it combined with short video segments synchronised with selected screens of the *DST*, so that the potential user could picture the intended use of the *DST* as a tool to support a structured interview.

As a mixture of both usability and utility considerations, instructors appreciated very much that immediately after their interviews they could be handed a printout of their choices made during the *DST* interaction, and also that their choices were immediately incorporated into a WWW page linked to their names and course titles, so that they could review their deliberations in their own offices. All persons involved appreciated the fact that the *DST* made the interview process time-efficient; each interview was completed in the one hour scheduled.

As another aspect of usability, system robustness, there were no problems with the use of the *DST* or the subsequent access of the WWW pages generated by the *DST* containing the individual instructor's choices. All works well, and easily; just as any WWW environment. The *DST* and the pages generated by the interviews remain available to all the instructors as part of the (password-protected) *TeleTOP WWW* site; extending the value of the *DST* yet further, in that this collection of resources has been used repeatedly in the on-going interaction with the instructors during the subsequent cycles of the course re-design process.

In terms of system utility, the *TeleTOP DST* has been invaluable to the re-design team. It served as a comfortable but businesslike way to carry out 30 structured interviews, each staying within the one hour scheduled for it. It provided a clear and common focal point for discussions with the instructors, and the access to examples was critical to raise the instructors' awareness of what different WWW-based functionalities could look like integrated into a larger course-support site. Thus it also provided an important staff-development function at the same time as serving to professionally facilitate what could have been a complicated or even contentious interview (for the several instructors who were not initially convinced that they wanted anything to do with a re-design of their courses or WWW environments to support the re-design). From a practical perspective for the TeleTOP team, the *DST* provided an immediate archive to record the decisions made by the instructors, in a common WWW environment, without any need for further record keeping and transcription. This is a considerable time savings. Being able to use the results of the *DST* as a discussion focus with the individual instructors gave the design team a conveniently accessible and common way to work with any instructor, in his or her own office as well as in the team's offices and computer rooms. It also gave the team a way to ascertain the level of familiarity each instructor had with using a WWW environment, and the level of upgrading that would be necessary for the instructor's own desktop computer, in that it was quite natural to visit the instructor in his or her office, and just ask him or her to start up a browser and go to the *DST* results. By watching how comfortably or uncomfortably the instructor responded to this chore, the team got adequate information to determine how much WWW/Internet literacy might be needed to provide to various instructors, without asking the instructors to self-describe themselves on a questionnaire. The team also was able to see for itself which instructors needed computer upgrades or new versions of a browser or plug-ins, and provide these for the instructors without them being bothered or even needing to know what they needed.

6. DISCUSSION AND FURTHER DEVELOPMENTS

Critically, the *DST* provided the instructions with a conceptual link to the first prototypes of the instructors' course-support environments. The choices the instructors made via the *DST* reappeared in their first prototypes in a way that looked and operated as if the first prototype was simply a direct result of the use of the *DST*. (It was, in terms of substance, but some handwork had to be done to actually set the first prototypes up over the database environment). From these first prototypes, the instructors have continued their explorations and filling-in activities as they continued the rapid prototyping process over weekly instructor sessions during 1998. The migration to the final fully functioning WWW-support environments will seem to the instructors as just the continued evolution of the their first experiences with the *DST*. Because all output environments--the DST results, the first prototype of the course environment, and the final version of the course environment--are accessed through the same WWW browser, with the same look and feel of the interface of the *DST*, we believe that the *DST* is a critical and successful tool in our instructional-design process. We also believe it is an innovative tool: combining strategic, staff-development, and rapid-prototyping functionalities during our TeleTOP mixture of systemic- and classroom-orientation instructional-design models (Gustafson & Branch, 1997).

However, we are continuing to evolve the DST, based on the reactions of the instructors. Several issues have emerged:

- Should instructors begin by being led through a minimal subset of possibilities, and only be exposed to a broader range when they are likely to be comfortable with them? We used the approach of beginning with a broader range, on the assumption that many instructors had no real idea of possibilities, and thus would make better choices, if better informed. However, the risk is that the amount of examples may be an overload.
- To what extent should the design team attempt to influence the instructor, by recommending for or against certain options, even if the instructor has other ideas? We tried not to influence the instructors during their DST sessions, but later did make suggestions, for features to include that they may not have originally chosen (four of these: a roster, a newsflash, an e-mail centre, and an overview of their course), and to omit features that we predicted would not be feasible to handle (for example, desktop videoconferencing for tutorial sessions). But it might be a compromise to identify a core set at the start, and then allow time and opportunity for each instructor to add to this set in different ways.
- To what extent is the consistency of interface more important than individual instructor's interface preferences? For several of our instructors, already experienced with the use of a WWW-based course support environment, the elegance of our approach in terms of interface consistency (see *Figure 3*) was counter to their own preferences of the colour and layout and interface organisation that they wished to have for their own courses.

These and other issues will not have definitive answers, even as our experience grows. We plan a variety of evaluation strategies as the new course-support sites come into use, asking students as well as instructors if they feel that the options in the sites fit well with the "classroom approaches" of the instructors, and perhaps add new aspects that the instructors had not been able to realise before. More generally, we will continue to define and validate our hybrid instructional-design method, integrating classroom, product and system orientations.

REFERENCES

Boer, W.F. de, 1998, 13 May, *A rapid prototyping method: The decision support tool and the first prototype,* Paper presented at ORD '98 [National Educational Research Days], University of Twente, Enschede.

Boer, W.F. de, Fisser, P., Strijker, A., Peters, O., Verheij, G.J. and Collis, B., 1998, Implementing tele-learning: Decision support for instructors in the TeleTOP Project, In Banks, S., Graebner, C. and McConnell, D. (Eds.), *Networking lifelong learning,* pp. 2-10 - 1-14, Sheffield, UK: The University of Sheffield Centre for the Study of Networked Learning.

Boer, W.F. de and Hamel, Q., 1998, *Designing courses for the WWW,* Masters thesis, Faculty of Educational Science and Technology, University of Twente, Enschede.

Collis, B., 1997a, Implementing ICT in the faculty: Letting a 1,000 flowers bloom or managing change?, In Miranda, M., Riemersma, J. and Veen, W. (Eds.), *De digitale leeromgeving* [The digital learning environment], pp. 121-136, Wolters-Noordhoff, Groningen

Collis, B., 1997b, Pedagogical re-engineering: A pedagogical approach to course enrichment and redesign with the WWW, *Educational Technology Review, 8*(3), 11-15.

Collis, B., in press, New didactics for university instruction: Why and how?, *Computers & Education*.

Collis, B. and Boer, W.F. de, in press, Rapid prototyping as a faculty-wide activity: An innovative approach to the redesign of courses and instructional methods at the University of Twente, *Educational Media International* (EMI).

Collis, B. and Fisser, P., 1998, *TeleTOP: Telelearning at the University of Twente,* Paper at Tele-Teaching, '98, IFIP World Congress, Vienna, August.

Fullan, M., 1991, *The new meaning of educational change,* Cassel, London.

Gustafson, K.L. and Branch, R.M., 1997, *Survey of instructional design models (3rd* edition), Syracuse, NY, ERIC Clearinghouse on Information & Technology, Syracuse University.

Laurillard, D., 1993, *Rethinking university teaching: A framework for the effective use of educational technology,* Routledge, London.

Nielsen, J., 1994, *Usability engineering,* Academic Press, San Diego, CA.

Norman, N., 1997, Communication technologies and education: Lessons in the potential of innovation, *Journal of Advanced Learning Technologies (ALT-J), 5*(3), 43-53.

Northrup, P., 1997, Faculty perceptions of distance education: Factors influencing utilisation, *International Journal of Educational Telecommunications, 3*(4), 343-358.

Sfard, A., 1998, On two metaphors for learning and the dangers of choosing just one, *Educational Researcher, 27*(2), 4-13.

Sweeney, M., Maquire, M. and Schakel, B., 1993, Evaluating user-computer interaction: A framework, *Journal of Man-Machine Studies, 38,* 689-711.

Chapter 21

Enabling Technologies to Design, Produce and Exploit Flexible, Electronic Learning Materials

Martin Valcke
University of Gent

Paul Kirschner
Evelijn Bos
Open University

Key words: Tools, Distance education, Flexible learning materials

Abstract

The base for this chapter is a large scale project, set-up at the Dutch Open University (OUNL) to re-design the current approach to design, produce and exploit distance education learning materials. The project has been influenced by two major paradigm shifts in the educational approach of the OUNL. These paradigm shifts are being described in the chapter of Kirschner, Valcke and Sluijsmans (see chapter 7 of this book).

The first paradigm shift was influenced by the results of a strengths-weakness analysis of the 'traditional' approach oriented towards the production of printed books with some extra multimedia (called the 'second generation of distance education). This revealed shortcomings at four levels: economical, educational, technical and scientific. The analysis resulted in a re-engineering of the overall approach, founding a first level of the 'third generation of distance education'. This new approach implies the availability of tools that help to design, produce and exploit learning materials in a completely innovative way: electronic (multimedia), indexed with user (student) profiles, based on models (content model, didactic model, learning path, etc.), continuous updating and actualisation, etc. As such, large repositories of learning materials are available from which, just-in-time, on-demand and tailor-made subsets can be selected and published-on-demand (internet/intranet, CD-Rom and/or in print). Students receive materials that are in line with their needs or characteristics.

In this chapter we give an outline of the problem analysis, the design approaches of 'generations' of tools, and reflection upon research results involving students using the flexible electronic learning materials.

1. THE CONTEXT OF DISTANCE EDUCATION AND OPEN LEARNING: BASIC CHARACTERISTICS

The Open University in The Netherlands (OUNL) develops and provides higher distance education with a high degree of freedom as to where and when to study. Everyone from 18 years on can subscribe. There are no specific admissions criteria or requirements with respect to prior education. The modular system of 'guided independent study' courses and credit points allows students a wide range of options; for instance to study single courses, short diploma lines or university degrees. Most students (> 70%) are in the work force. Their professional environments are very different. As a consequence, this openness results in a very heterogeneous student population when we consider study intentions, prior knowledge levels, personal interests, prior experiences, educational background, etc. This heterogeneous nature of the student population and other factors bring about a set of specific problems that questioned the specific approach to course design, production and utilization. It is in line with these problems and the potential of information and communication technologies (ICT) that we have elaborated this chapter.

2. A FIRST PARADIGM SHIFT: AWAY FROM A SECOND GENERATION APPROACH TOWARDS DISTANCE EDUCATION

Although the OUNL is a relatively young educational institute, its approach towards designing, producing and exploiting materials resulted - after about 14 years - in a routine-based industrial approach. A thorough rethinking of the mission of the OUNL in view of the changing societal context, as exemplified in the chapter of Kirschner et al. (this volume) was felt as necessary. This process was set up along the following lines:

- Problem definition and analysis.
- Design and production of a prototype solution:
  - iterative design of new tools in collaboration with all problem owners (students, content specialists, tutors, educationists, technologists, etc.);
  - production of prototype tools;
  - production of a prototype organisation and procedures (scripts, flows of activities);
  - developmental evaluations.
- Design and development of a final system.
- Development of an implementation plan, featuring:
  - organisational changes;
  - procedural changes;
  - training/re-training;
  - transition from old approach to new approach.

In this article, we will especially focus on the first two phases.

3. PHASE ONE: PROBLEM DEFINITION AND ANALYSIS

At the start of the process, a thorough analysis of the current approach towards the design, production and exploitation of learning materials at the OUNL was carried out (Valcke, 1994). This strengths/weaknesses analysis focused on the following features:

- General:
  - administrative processes;
  - course model development;
  - student modelling (market analysis).
- Content design:
  - content model design;
  - analysis of and taking account of target population (market focus);
  - elaboration process of basic content (scientific content).
- Didactic-Educational design:
  - definition of didactic/educational models;
  - learning paths;
  - tutoring;
  - analysis of and taking account of target population (market focus);
  - elaboration of didactic components in learning materials (examples, summaries, schemes, cases, advance organisers, etc.).
- Developmental evaluation of learning materials:
  - approaches to evaluation during development process;
  - execution of evaluation.
- Production processes:
  - graphical design and lay-out;
  - editing process;
  - reproduction process;
  - distribution process.
- Management features:
  - legal issues;
  - developers team (composition, managing of team, communication, etc.);
  - financial aspects;
  - timing/planning.

A protocol was designed, covering the above issues. Based upon this protocol all 'problem owners' that play a role during the current design, production and exploitation process of learning materials at the OUNL were interviewed: course team managers, domain managers, authors, tutors, head-tutors, media-developers, educational technologists, students, editors, referees, lay-out people, assessment specialists.

Analysis of the interview data helped to define problems with and the flexibility of the current approaches and the final aims: demand-driven education and tailor-made courses. The results of this analysis can be summarised as follows.

Current OUNL-learning materials are mainly based on 'printed' text books. Despite the high quality of the materials, this heavy focus on printed materials results in a number of problems:

- The efficiency of the current process of designing, developing, producing and exploiting the learning materials is low, in terms of speed of the process, costs, consistency of the materials being developed, their flexibility. The life cycle of the materials is too long considering the need to update the scientific content (e.g., a new theory has been elaborated) or to add new examples (actual data, examples, names, references, etc.). The books get - as a consequence - outdated after 2 to 3 years. Their revision implies reiteration of the design, development and production process. Next to text processing tools, no advanced computer tools were available and being used to support the entire or specific parts of the process.
- The effectivity of the learning materials, delivered as 'books' and the efficiency of the production process are low when we see that they are not suited to supporting just-in-time learning and tailor-made courses. The OUNL failed to meet the heterogeneous and quickly changing demands of part of the learning 'market'.

The first problem is of an educational-economic nature. Questions are asked in relation to the speed of material development, the model-based approach when designing sets of materials, continuous quality control, the degree of flexibility when exploiting sets of learning materials, the possibilities of re-using existing materials, the possibilities to re-use the same materials for different purposes (for instance, commercial exploitation, adaptation to professional contexts), etc.

A second major problem is linked to an educational argumentation. Traditional printed courses are based on a priori decisions and selections of basic content (scientific domain knowledge, theories, principles, formulas) and a priori decisions about the (embedded) support needed by students. In the distance education context of the OUNL, there is barely face to face contact between students and teachers/tutors, so the learning materials have to be sufficiently supportive. Therefore an essential part (about 40%) of the learning materials consists of embedded support devices (ESD's), such as pre- and post questions, schemes, illustrations, content pages, indexes, tasks, summaries, advance organisers, objectives, examples, prior knowledge references and study guidelines. ESD's aim at scaffolding and focusing the study process. Course developers assume these ESD's have certain functions and effects. We can distinguish three subsets of functions and/or effects:

- helping to get access to the new content, for instance, by means of overviews, content pages, margin texts;
- supporting processing the content, for instance, by using advance organisers, activities, tasks, pre-questions; and
- facilitation of the testing of mastery of the content/objectives, for instance, adjunct questions, test-items with feedback, self-tests.

Next to these rather 'cognitive' functions and effects, ESD's also substitute for teachers by motivating students, invoking interest in new topics and structuring the material.

Research has confirmed the functions and effects of ESD's: it was found that students use ESD's to a high extent, students appreciate them and using ESD's has a positive effect on study results. But the studies also revealed that the adaptation of ESD's to the individual characteristics of the students, such as prior knowledge, educational background, is needed (Valcke, Martens, Poelmans & Daal, 1993; Martens, Valcke, Poelmans & Daal, 1996; Martens, Valcke & Portier, 1997).

In the existing OUNL-approach, all students get the same book, with the same examples, exercises, etc. This is in sharp conflict with the demand for more demand-driven education. The era of supply driven education is considered to be ineffective and inefficient. The need for tailoring courses to the specific needs of student is a world-wide issue (Kirschner & Valcke, 1994). In this way one optimizes the degree to which students get materials that concur with their prior knowledge and that are at the level of their didactic elaboration adapted to their specific needs.

Besides the economic and educational problems, technical problems also emerge. The delivery and distribution of educational materials is increasingly being done electronically (e-mail, Internet, news groups). The current design-approach of learning materials is not suited to be delivered in these environments; long expository texts, too little structure, no built-in possibilities to choose parts of the materials, lack of integration of printed media and more dynamic media (for instance audio, video, CAI, tools, etc.). These problems urge us to find a new design and development approach to the learning materials.

4. PHASE TWO: DESIGN AND DEVELOPMENT OF A PROTOTYPE SOLUTION

As indicated earlier, the analysis was followed by research and development (R&D) activities in the following fields:

- iterative design of tools in collaboration with all problem owners;
- production of a prototype tool;
- production of a prototype organisation and procedures (scripts, flows of activities); and
- developmental evaluations.

4.1 Iterative design of the tools

The iterative design of a solution to cope with the problems observed in the first phase and with the scope to realise demand-driven education and tailor-made courses and the tools was dealt with in a very specific way. Literature on the development and implementation of innovations (including ICT-solutions) explicitly indicates the importance of involving the 'problem owners' and future users of innovative approaches and tools in the initial design, development, production and evaluation processes. In our approach we adopted an iterative design and production approach that did involve future users in iterative cycles. The first versions of the tools were called ILCE (Interactive Learning and Course development Environment) and ILE (Interactive Learning Environment). At each stage prototype versions of procedures and systems were discussed, presented and used by these users to gather as much feedback as possible to direct the further R&D activities. The final - now commercially available - system is as a consequence the outcome of a collaborative operation that took two and a half years.

4.2 Production of prototype tools

Figure 1 depicts the modular structure of this system, called Mercator. Each module could be considered as a distinctive tool that interacts with a shared database of resources or models.

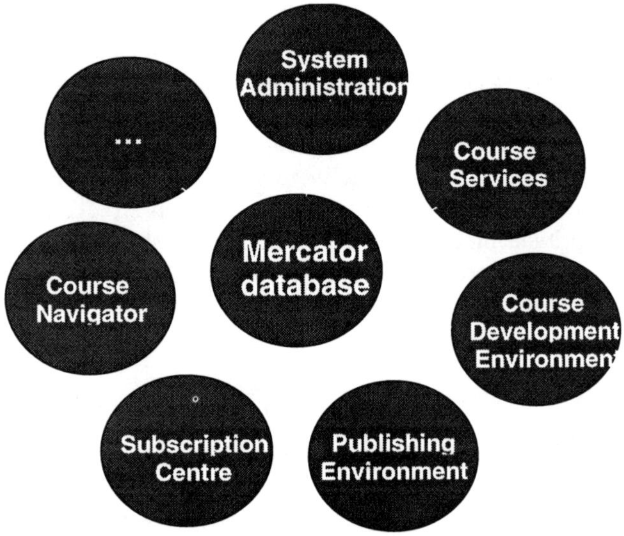

Figure 1. Modular structure of the Mercator-system

The MERCATOR-system is based on an architecture that reflects the interests of all parties: modelling, development and publishing modules versus a subscription and a navigator module. The first set of modules helps to design and produce the material; the second sets helps to select specific materials for the student and that supports the actual delivery in a printed (printing on demand) and/or electronic (learning from screen) mode.

In describing the modules we will attempt to include a large set of procedural and organisational specifications. In this way, we will discuss topics that will further be dealt with in 4.3.

The central part of the system is a database that contains user data, modelling info, and the learning materials.

Module 1: System administration

This module is used by the system administrator and defines the user identification and the password for Mercator users. As a consequence, users only get an authorised access to specific modules, depending on their roles and responsibilities (content specialist, author, educationalist, media specialist, student, etc.).

Module 2: Course services

Course Services are of prime importance. Within this module basic models are defined that are to be used in the development environment and that influence the overall flexibility of the learning materials being developed.

First, the model concentrates on the structuring principles to build up the database of the knowledge elements (cf. *Figure 2*). Users define and structure such structuring principles, such as:

- main theme – theme – sub-theme;
- case – problem;
- main task – task – problem; and
- competency – skill.

Of course, users can also decide to combine structuring principles. *Figure 2* depicts such a model.

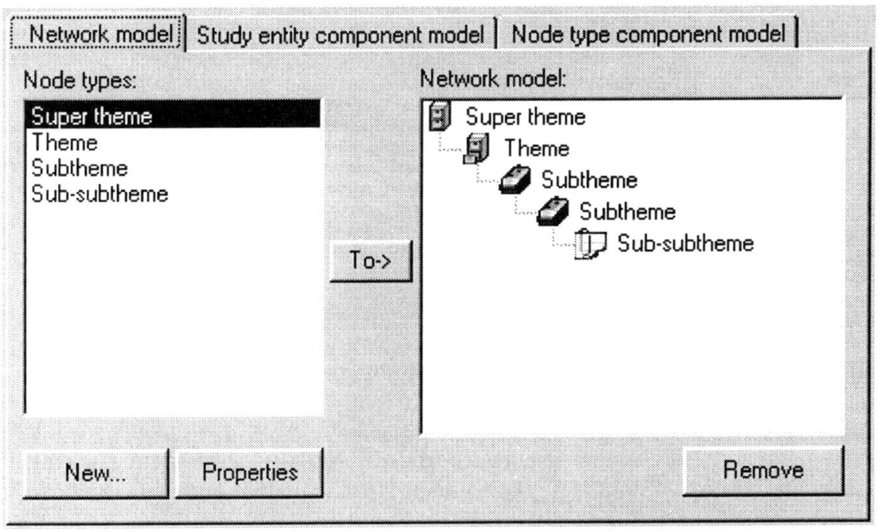

Figure 2. A choice for thematic structuring principles

Second, users of the Course Services module have to decide what the didactic elaboration will be. In other words, how students will be activated when they receive knowledge in relation to for instance thematic knowledge component: example, introduction, task, feedback on the task, self test, scheme, summary, group work, internet search task, simulation, objectives, etc.

The user can assign specific subsets of such didactic elements to a specific structuring principle.

These didactic elements are also called Embedded Support Devices (ESD) since they replace the tutor/professor in the interaction of the student with the learning materials. They help to have access to the materials, to process them and to help students to test their mastery of the objectives.

Module 3: Development environment

Once models have been defined, they can be used in the development environment where the developers (authors, content specialists) enter concrete materials in relation to their content domain. *Figure 3* depicts how a user can use the structuring principles to build up the content model of the database (at the left side) and how the user can use these elements to build a learning path for students (at the right hand).

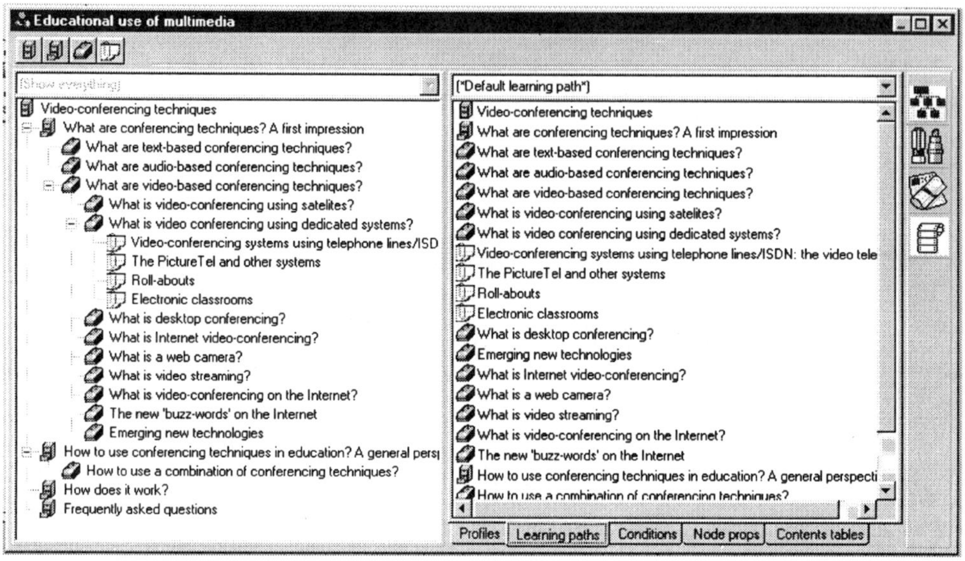

Figure 3. Building the content model and the learning path in the development environment

Opening one of the themes in the content model enables the developers to enter learning materials by making use of the pre-defined set of didactic elements. All kinds of data can be entered: text, audio, video, links to other software packages, links to communication environments (e.g. e-mail, discussion groups, conferencing tools), etc. *Figure 4* shows for instance how video is being embedded.

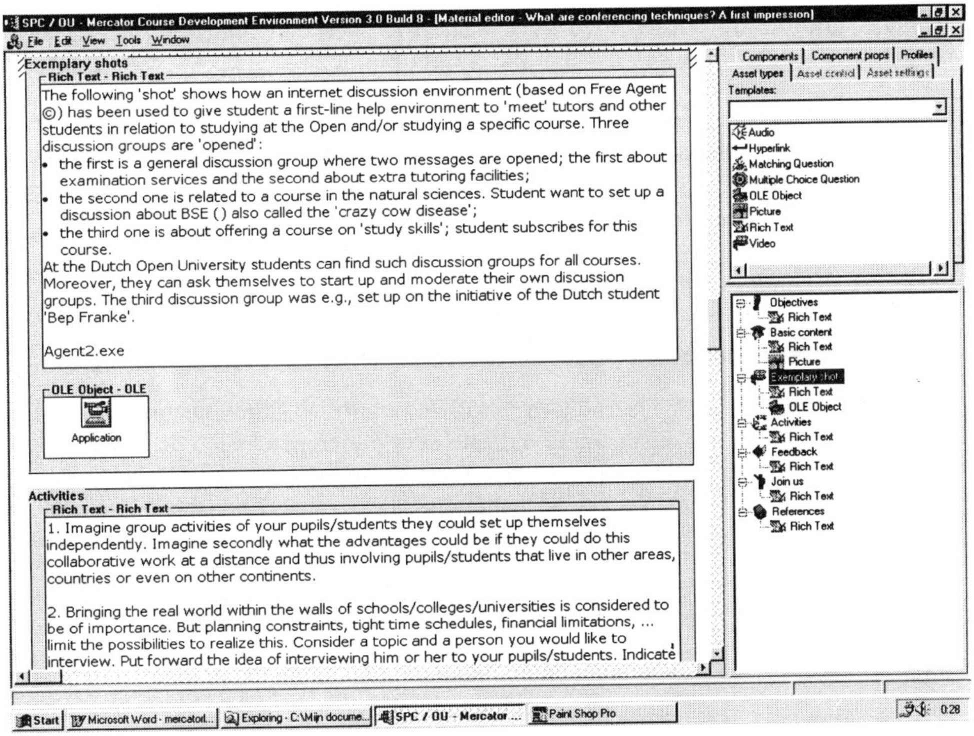

Figure 4. Embedding video in the didactic component 'Basic content'

In this environment, the developer can also construct student profiles. These are of prime importance, since they can be related to specific themes, sub themes (the structuring principles) and/or to a specific didactic elaboration (for instance, other examples, cases, shorter/longer introductions, different tests, alternative self tests, etc.). *Figure 5* shows a variety of profiles that has been used to adapt the content of an electronic course to specific student variables in a student profile.

- [x] Technician
- [x] Primary school teacher
- [x] Secondary school teacher
- [x] Teacher higher education
- [x] Professional trainer

Figure 5. Five student profiles applied to a database of learning materials; in this example all the student profiles are active and as a consequence the student will receive all the materials in line with all these profiles

Module 4: Publishing environment

Once the developers consider the set of materials 'ready', they can be published. Three alternatives are available, even in a specific combination: printing-on-demand, CD-ROM and/or internet-delivery (on-line courses). Moreover, in this module the selection criteria are chosen to deliver a specific subset of materials that as such constitute a 'course' by making use of the student profile.

Module5: Subscription centre

In this module students are assigned to a specific 'published course'. This module helps at the one hand to register students as users of the Mercator Navigator and at the other hand to indicate what published course will be accessible for specific students.

Module 6: Mercator course navigator

This module is used to study the materials that are delivered on a CD-ROM. *Figure 6* depicts a typical screen. The example shows text and a button to start a video-clip.

If this 'course' includes learning materials in relation to different student profiles, students can take an intake session to indicate their preferences, choices, to take a test and consequently to ask the system to give them those materials that are most in line with their characteristics. Of course materials can be annotated, there is access to interactive tools (e-mail), etc.

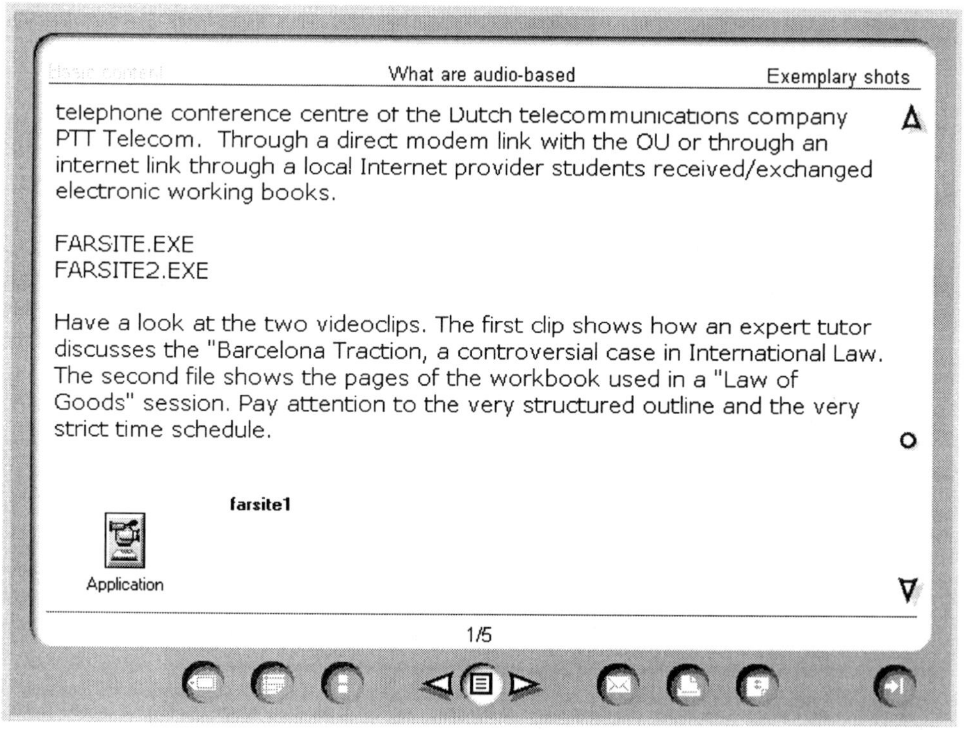

Figure 6. CD-ROM delivery of a course

4.3 Production of a prototype organisation and procedure (scripts, activity flows)

As stated in the former section, the R&D activities also included:

- the design and production of a prototype for the organisation that has to be set up in order to be able adopt the design and development approach, including the tools developed; and
- the design of a scripts, procedures for a variety of users of the system.

Brok and Martens (1997) describe these aspects in detail. They did research to check the relevance, effectivity and efficiency of the procedures and organisational designs. This research was set up in line with the developmental evaluations discussed in the next sections. Their general conclusions reflect a moderately positive attitude towards the changes in their work flow and the new procedural/organisational features.

Figure 7 represents a possible organisational scheme that is proposed for the OUNL. In the representation it is clear that decisions at the micro-level depend upon decisions taken at a higher level (domain- or OUNL-level).

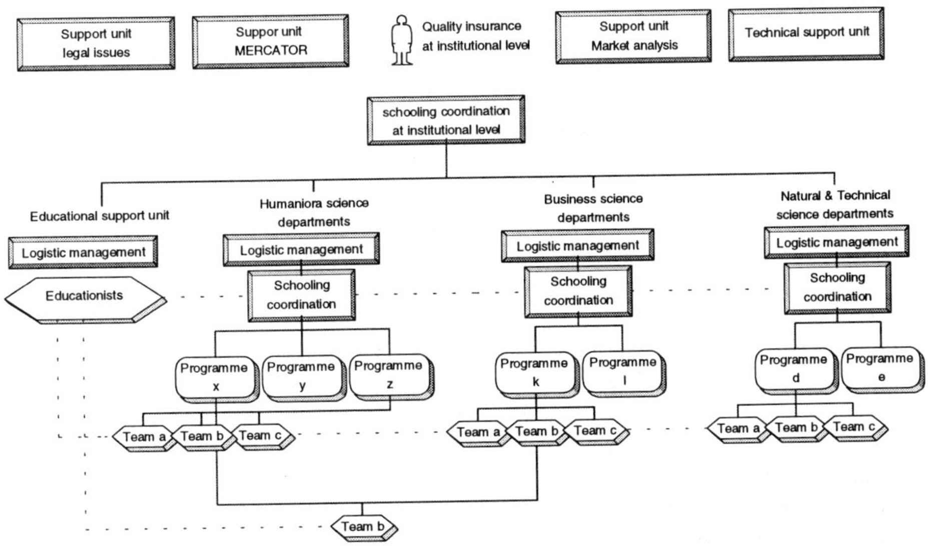

Figure 7. Possible organisational restructuring of the OUNL in view of the re-engineering of the processes involved

In *Figure 7*, the design and development of schooling programmes is no longer a responsibility of teams at micro-level, but will be influenced by the co-ordination teams at meso-level (cluster of departments) and at macro-level (institutional co-ordination).

4.4 Developmental evaluation of prototype versions of the tools

Whereas the iterative cycles in designing and developing the system primarily involved actors from an institutional perspective, the student perspective was taken during a large evaluative set-up with courses from three different content domains (social, technical and law). In this developmental evaluation, the focus was therefore on testing the 'delivery' subsystem (the studies are fully documented in Martens, Portier, Valcke, Van Buuren, Dekeyser, Goeminne & Schuyten, 1995 and Martens, Valcke, Portier, Weges & Poelmans, 1997).

The main research question dealt with was "How do students cope with electronic courses that had been developed in line with the approach described in the former sections? We wanted to know whether they think that the electronic delivery subsystem (Navigator) is a suitable learning environment and if it generates learning materials that are well suited for self-study. Moreover, considering the broader aims of the innovation processes, the research also focused on topics going beyond the mere micro-level (organisation, overall study approach).

The students studied flexible demand-driven courses at home, as they do in the regular situation of studying distance education materials at the OUNL. The experimental courses also contained 'regular' OUNL learning material but with adaptations in line with the increased interactivity and flexibility. The materials also required substantial study time of the students. Furthermore, students were informed that they would be tested on their mastery level at the end of the course. In order to enhance external validity, comparisons were foreseen with other student populations (regular OUNL students, regular students from a non-distance university and students in other experimental settings) and across several content domains.

The results of the researches indicate that students find the interactive learning environment suitable for their study. They do not question the learning access level of the materials. Learning from a computer screen does not seem to be the problem. Their opinions about the electronic learning materials do not differ very much from students studying regular OUNL courses or studying at more 'traditional' universities. In case of differences, these differences are in favour of the electronic learning environment. The results can be generalised across content domains.

An important conclusion is that students want more interactive support; for instance, interactive tests, animation, dynamic links with other computer programs, more advanced navigation tools and communication tools.

The findings from the evaluation studies have been used to improve the design of the system in order to start the production of the commercially available Mercator-system. Mercator®, version 3.0, has been developed in close collaboration with a private enterprise[1]. Next to this, the OUNL continues to set up evaluation activities to document the use of the system in different settings. Moreover, the system is being further developed to be integrated in environments for evaluation and assessment, internet-services, administrative modules, etc. The Mercator tool was positioned in relation to other tools such as the StudieNet-system of the OUNL[2], an automated testing system and groupware provisions. An integration of the functions of all these

---

[1] SPC/Group; e-mail j.bijlstra@spcgroup.nl or s.portier@spcgroup.nl, http:// www.spcgroup.nl

[2] http://WWW.OU.NL

tools was sought for. This new tool is called the Electronic Learning Environment (ELO). This ELO can as such be viewed as the instrumental elaboration of the innovative knowledge and skills of the OUNL. This generic systems is currently under construction. Kirschner et al. (this volume) describe in more detail this ELO in line with a second paradigm shift at the OUNL.

5. HOW DOES MERCATOR WORK?

To describe in more detail how the Mercator-system works, we can rely on the frame of reference, discussed earlier in this book to describe development tools.

As to the "purpose and evidence of benefits", we can refer to a recent evaluation study in which the Mercator-tool was evaluated as a curriculum-development tool (Krabbe, 1998). She focused on the modules "Course services" and "Development Environment". She reports an increase in efficiency and effectivity when using the Development Environment to produce actual learning materials. As to the Course Services, she reported that the users found less support in this tool to develop 'blueprints' for curriculum development. In her research set-up Krabbe did not present an initial set of didactical models that could serve as inspiring "blueprints" for the developers. Such an approach is recommended by the developers of the system. Nevertheless, the study of Krabbe clearly indicates that the purpose of the system is to develop generic (content and didactic) models that next are being used for the development of large repositories of learning materials. These repositories can next be used to extract subsets (based on student profiles and learning paths) and be published in alternative formats (print, CD-ROM and/or internet).

When we consider the "development process and underlying theory", we can conclude that the Mercator-system supports the design, production and delivery process at macro-level (knowledge domain, e.g., economics) and at the micro-level (e.g., a specific electronic workbook for students in the field of micro-economics). The tool offers designers and developers of learning material 'blueprints' and "templates" and at the same time generic mechanismes to design such blueprints and templates. The target group of the output is at tool-level (blueprints, templates and the working environment) for developers of learning material and those involved in managing the delivery of learning materials. At the level of the learning materials the target group consists of teachers (selection of learning materials) and students (learning materials and a study environment).

The theoretical base for the Mercator-tool is in part based on Martens (1998) who researched the role of embedded support devices (important when designing models/blueprints) and a set of approaches discussed in more detail in Vuist and Valcke (1995).

As to the "support" users get when using the system, the most recent version of Mercator presents step-by-step procedures that help users (designers and developers) to elaborate models and learning materials. Next to the system the designers and developers receive a manual that presents a typical exemplary procedure to use the system. When the system cannot proceed, because of inconsistent or inadequate use

by the end-user, there are automatic system-prompts that indicate what step or information is missing.
Students receive (CD-ROM-delivery) an electronic environment that resembles an internet-browser with limited help-facilities (button-labels and comment-fields).

Research with users indicates that the use by students is a straightforward thing. But for designers and developers this complex system implies an initial training session and a thorough introduction based on a production-task taken from their own environment/practice. Evaluation by Martens (1997) indicates that users are positive, but it is necessary that the system contains examples (e.g., blueprints and repositories with learning materials).

6. CONCLUSIONS

This chapter described in detail the design and development of new tools for the OUNL to cope with a major paradigm shift. The driving forces behind this process are based on a rethinking, reconsideration and positioning of educational activities at the micro-, meso- and macro-level.

As such, the Mercator-tool is but one element in a major innovation process and should always be approached like that. Moreover, further research is needed to found the basic assumptions underlying the paradigm shifts described. This underpins the validity of the new OUNL-research programme that focuses upon competency-based education, self-, peer- and co-assessment, flexible structuring of large repositories of knowledge, alternative curriculum structures, etc.

REFERENCES

Brok, E. and Martens, R., 1997, Authoring modular courses: exploratory experiences, *Distance Education*, 18(1), 44-58.

Kirschner, P. and Valcke, M., 1994, From supply driven to demand driven education: New conceptions and the role of information technology therein, *Computers in Human Services*, 10(4), 31-53.

Krabbe, E., 1998, Tussen intuïtie en rationaliteit - Een onderzoek naar mogelijkheden tot standaardisatie voor leerplanontwikkeling [Between intuition and rationality. Research into the possibilities to standardize the process of curriculum development], Doctoral dissertation, University of Twente, Enschede

Martens, R., 1997, *Evaluatie Mercator-workshops, Mercator-benadering en Mercator-software* [Evaluation of Mercator-workshops, -approach and -software], Open University, Heerlen

Martens, R., 1998, *The use and effects of embedded support devices in independent learning,* Lemma, Utrecht.

Martens, R.L., Portier, S.J., Valcke, M.M.A., Buuren, van, J.A., Dekeyser, H., Goeminne, K. and Schuyten, G., 1995, *The use of embedded support devices in interactive learning environments: the impact of student characteristics. Research results*, Open University, Heerlen.

Martens, R., Valcke, M., Poelmans, P. and Daal, M., 1996, Functions, use and effects of embedded support devices in printed distance learning materials, *Learning and Instruction*, 6, 77-93.

Martens, R., Valcke, M., Portier, S., Weges, H. and Poelmans, P., 1997, Research with interactive learning envrionments in three content domains: descriptive statistics, continuous mathematics and substantive criminal law, *Distance Education,* 18(1), 44-58.

Martens, R., Valcke, M. and Portier, S., 1997, Interactive learning environments to support independent learning: the impact of discernability of embedded support devices, *Computers & Education,* 28(3), 185-197.

Valcke, M., 1994, *Het cursusontwikkel- en productieproces van de Open universiteit onder de loep* [Analysis of the course design and development process of the Open university - Views, problems and expectations of problem owners], Open universiteit, Heerlen.

Valcke, M. and Martens, R., 1997, An interactive learning and course development environment: Context, theoretical and empirical considerations, *Distance Education*, 18(1), 7-23.

Valcke, M.M.A., Martens, R.L., Poelmans, P.H.A.G. and Daal, M.M., 1993, The actual use of embedded support devices in self-study materials by students in a distance education setting, *Distance Education, 14*, (1), 55-84.

Vuist, G. and Valcke, M., 1995, A model-based design approach for the flexibilisation of courses, F. Lockwood (ed.), *Open and distance learning today*, pp.185-196, Routledge, London.

Weges, H. and Portier, S., 1997, Demand driven education through the delivery of interactive learning materials within the domain of environmental law, *Distance Education*, 18(1), 76-92.

Chapter 22

The IDXelerator™: Learning-centered Instructional Design

David Merrill[1]
Bennet Thompson
Utah State University

Key words: Instructional design, Computer-based instruction, Authoring tools

Abstract

IDXelerator™ is an authoring system implementing the notion of learning-centered instruction. Instructional design is built-into the system via built-in lessons of different types. It is based on the assumption that different instructional goals required different types of instructional presentation and practice. The system includes four built-in lessons each complete with presentation, practice, and learner guidance: information about, parts of, kinds of, and how to. In addition the system includes a text module and slide show function. IDXelerator™ is a ToolBook™ plug-in adding an IDX item to the authoring menu. Any IDX component can be linked to any ToolBook object. IDXelerator™ increases authoring efficiency by at least 50% and simultaneously increases the effectiveness of the instruction by using scientifically verified instructional strategies consistent with instructional outcomes.

1. INTRODUCTION

In this paper we describe the IDXelerator[2]. This is a learning-oriented instructional development tool (Merrill, 1997a). A learning-oriented development tool is one that includes built-in instructional strategies based on scientifically verified principles of instruction (Gagné, 1985; Merrill et. al., 1996; Merrill, 1997b). Each of these built-in

[1] The authors wish to acknowledge their associates in the ID_2 Research Group who made major contributions to the design and programming of IDXelerator. Mark Lacy assisted in the design and programmed the Electronic Trainer. Ben Thompson assisted in the design and programmed IDXelerator. Building this tool would not have been possible without their significant contribution to this project. Other members of the ID_2 Research Group at Utah State University also contributed to the design and development of this tool.

[2] An earlier version of the IDXelerator™ has been previously distributed under the name Electronic Trainer™. IDXelelerator™ is available as a commercial product from Mindware Creative, Inc. For more information see their web site www.mindwaresystems.com

strategies is a complete set of all of the interactions required for the student to acquire a specific kind of knowledge or skill. This system contains a library of preprogrammed instructional strategies that have been carefully designed to include the presentation, practice, learner guidance, and knowledge structure appropriate for achieving a particular learning goal. The user selects a goal and the appropriate instructional strategy. The authoring system then prompts the user to provide the multimedia resources necessary to implement the entire instructional strategy. The system automatically generates the instructional strategies, including practice, learner guidance, and knowledge structure. No programming is necessary: The instructional strategy is built-in and has been previously programmed.

The advantage of such a learning-oriented system is that novice instructional designers and subject matter experts are much more likely to develop effective instruction than is likely with other instructional design tools that require considerably more instructional design expertise. A user can learn to link multimedia resources into this system in a few hours instead of days or weeks. The 30 to 60 percent programming effort required by authoring systems is reduced to one or two percent of the total effort. The design effort is also reduced since the user is not required to design every interaction but merely to select appropriate strategies. Since maintenance merely requires replacing a multimedia resource it can be accomplished in a matter of a few minutes instead of days or weeks. Learning-oriented tools such as IDXelerator™ enable subject matter experts, with only a minimum of instructional design training, to develop quality instruction that will really teach.

Learning-oriented instructional development tools such as IDXelerator™ do have a few limitations. First, these tools are not *theory-neutral*. If the user disagrees with the instructional design theory built-into the system, he or she will not be comfortable with the system. Therefore, the instructional strategies built-into IDXelerator™ have been carefully constructed to implement the best that is known about scientific instructional strategies. (Some of these strategies have been briefly outlined in Merrill, 1997b).

The IDXelerator™ is also limited by the set of instructional strategies included in its library. There are a limited number of classes of effective instructional strategies, nevertheless, there is an infinite number of ways to implement the principles required for a given type of instructional outcome. An authoring system has considerably more flexibility for building a variety of instructional strategies. Therefore, the version of IDXelerator™ described here is a *plug-in* to Asymetrix ToolBook™. When the built-in strategies of the system are appropriate the advantages listed above are available. When a given instructional goal requires some other form of interaction the user can use ToolBook™ to program this strategy. In this way the user has the best of both worlds.

2. THE STUDENT[3] VIEW OF IDXELERATOR™

IDXelerator™ is designed to run within Multimedia ToolBook™[4] and requires the ToolBook runtime system to deliver the instruction to the student. This runtime system can

---

[3] We refer to the learner as *student* and the author of the material to be learned as *user*.

[4] Current versions run in either ToolBook Instructor version 5.0 or 6.0.

be distributed free of charge with ToolBook lessons. The system includes the following components: a text module, a slide-show, an information-about lesson, a parts-of lesson, a kinds-of lesson and a how-to lesson, a quiz module, and student registration with performance record. Each of these components are briefly described. A student view of the kinds-of lesson is included below (see *Figure 1*).

2.1 Text module

The *text module* can be linked to any ToolBook object so that when a student clicks on that object the text module is called. The text module, consisting of one or more pages of text, is a view window that overlays the current screen display. It displays approximately 16 lines of text on each page. At the bottom of the text window are 5 navigation buttons: help, find-phrase, close-text, page-up and page-down. The *help* button invokes a context sensitive windows help file which directs the student in the use of the text page. The *find-phrase* button enables the student, by typing in any word or phrase, to immediately jump to the page containing this phrase and highlight the requested phrase. The *close-text* button returns the student to the ToolBook page from which the text module was called. Location on a page is an important learning cue, therefore, the text module does not scroll but rather contains individual pages which can be configured by the user to facilitate these visual cues. The *page-up* and *page-down* buttons enable the student to navigate from one page to the next.

The built-in lessons of IDXelerator can be linked to any word or phrase in the text module. A given phrase can be linked to several different kinds of lessons. The name of the lesson is highlighted in the text and when the student clicks on this highlighted phrase a menu pops containing one or more of the following: learn more about, learn the kinds of, learn the parts of, or learn how to. Clicking on the menu item displays the corresponding built-in lesson as described below.

2.2 Slide show

The slide show can be linked to any ToolBook object so that when a student clicks on that object the slide show is called. The slide show, consisting of one or more graphic images, is a view window that overlays the current screen display. Each graphic image can be accompanied by a text caption, an audio caption, or both. The text caption appears in a scrolling text box beneath the graphic. The audio caption plays when the graphic is shown. The student clicks a next button to see the next slide.

2.3 Information-about lesson

An information lesson is called when the student clicks on its name appearing as a highlighted word in a text module (see above) or by clicking on any ToolBook object on a regular ToolBook page to which the lesson is linked.

The information-about lesson consists of a view window divided into two side-by-side panels. The portrayal for the information appears in the left panel. This information can be any combination of multimedia: a text display, a graphic with or without a written or audio caption, a video display or animation with or without audio. When the portrayal includes audio or video, control buttons appear beneath the left panel that allow the student to stop, or replay the audio or video material. The right panel contains the written information to be learned. Beneath the right panel are optional buttons linking to more information. This *more information* can be any combination of multimedia information similar in format to the portrayal in the left panel. Additional buttons can also link to any other executable program. When the student has finished with the additional information or executable program they are returned to the information-about display. Beneath the left panel is an *OK* button that returns the student to the text module or ToolBook page from which the lesson was called.

2.4 Quiz module

The quiz module is automatically generated by IDXelelerator. This module provides either a multiple choice or short answer quiz for each of the information-about items selected by the user. The multiple choice version displays the descriptive information (right panel information) followed by 4 alternative answers. The alternatives include the name of the information described and a random selection of the names of 3 other information items from the ToolBook course under consideration. The student clicks on the correct answer and is given right/wrong feedback. The short answer version also displays the descriptive information (right panel information) followed by an answer box in which the student types the name corresponding to the information. The quiz has provision for either exact match or key word matching to the name.

The quiz module randomly orders the information items each time it is requested. In the multiple- choice version the alternatives are also randomly ordered each time an item is displayed. The quiz module uses sampling with replacement meaning that when a student misses an item it is put back into the item pool and tested again later in the quiz. Items are replaced until the quiz is mastered. The bottom of the view window includes a *Cancel* button that allows the student to abort the quiz before finishing, an *OK* button used by the student to indicate that their answer is complete and should be evaluated, and a display showing the item number and the number of items in the quiz (e.g. 3 of 10).

The student record system records date/time the student began the quiz, the number of tries required to get all of the quiz items correct, and the date/time the student finished or aborted the quiz.

2.5 Parts-of lesson

A Parts-of lesson is called when the student clicks on its name appearing as a highlighted word in a text module (see above) or by clicking on any ToolBook object on a regular ToolBook page to which the lesson is linked.

The parts-of lesson consists of a view window divided into six parts. The main display occupies about ¾ of the viewer in a panel to the left. This panel contains a graphic on which the parts have been identified. The student explores this display by moving the cursor over the graphic. When the cursor passes over an identified part it is highlighted and its name appears in a text field below the primary display panel. The right side of the view window contains three sections. The top section contains three buttons: *naming quiz*, *locating quiz*, and *cancel quiz*. The center section contains the following directions during the explore phase: "Explore the graphic. Left-click an image for a description, right-click for a demonstration. After exploring the graphic select a practice mode." The bottom third of the right hand section is used for the naming and locating quiz. At the conclusion of the executable program the student is returned to the parts-of lesson. The bottom of the right section contains an *OK* button. Pressing the *OK* button returns the student to the ToolBook page or text module from which the lesson was called.

The naming quiz highlights a part in the graphic display and displays the names of the parts in the lower section of the right panel. The center right section directs the student to click on the name that corresponds to the part. The student is given right/wrong feedback and the next part is highlighted. This is repeated until the student has correctly named each of the parts.

The locating quiz displays a name in the lower section of the right panel. The center right section directs the student to click on the location of the part on the graphic display. The student is given right/wrong feedback and the next name is presented. This is repeated until the student has correctly located each of the parts.

The quizzes use sampling with replacement so a missed item is put back into the pool for representation to the student. The record system records quiz start and end times and the number of tries required to name and/or locate all of the parts.

2.6 Kinds-of lesson

A kinds-of lesson is called when the student clicks on its name appearing as a highlighted word in a text module (see above) or by clicking on any ToolBook object on a regular ToolBook page to which the lesson is linked.

The kinds-of lesson consists of a view window divided into five panels. The upper left panel (about ¼ the display) consists of a portrayal for a reference example (*Bitmap File Format* in *Figure 1*). This portrayal can be any combination of multimedia objects: a text display, a graphic with or without a text or audio caption, or video with or without sound. Audio and video are accompanied by appropriate stop/replay controls. The name of the concept being taught appears above the portrayal panel (*Example of: Bitmap File Format*). The right hand panel (about ¼ of the display) contains a list of the essential or key characteristics of the concept being taught. The lower left ¼ of the display contains two panels: the upper contains a list of examples of the concept and the lower panel contains a list of non-examples of the concept. When the student clicks on an example name the system displays a multimedia portrayal of the example using any combination of media as described for the reference example. A similar display is presented when the student clicks on the name of a non-example. The example or non-example display can be dragged about the screen to allow the student to compare the example or non-example with the reference example or the essential characteristics. The system records the presentation start and stop times.

The lower left panel of the display contains two buttons: *OK* and *Take Quiz*.

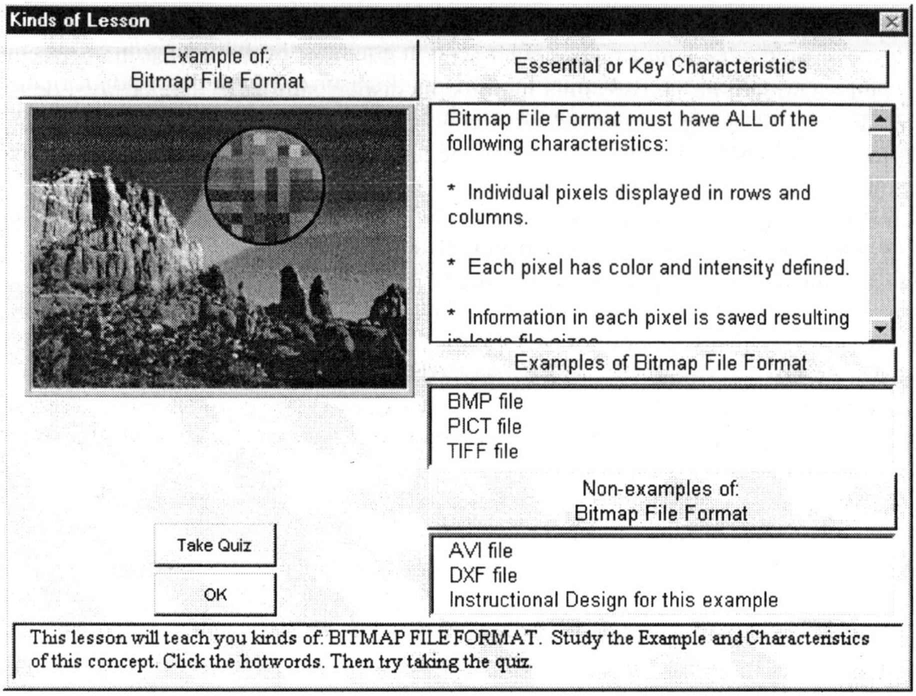

Figure 1. Student view of Kinds-of lesson

For the kinds-of quiz the system presents examples and non-examples using the same format as above. The student is directed to click on a *TRUE* button if the display is an example of the concept being taught or the *FALSE* button if the display is a non-example of the concept being taught. The quiz uses sampling with replacement and requires the student to get all of the instances correct. The quiz also includes a *Cancel* button allowing the student to abort the quiz. The system records quiz start and stop/abort and the number of tries required to get all of the instances correct.

A narrow text box across the lower part of the display provides directions to the student as the lesson/quiz progresses.

2.7 How-to lesson

A How-to lesson is called when the student clicks on its name appearing as a highlighted word in a text module (see above) or by clicking on any ToolBook object on a regular ToolBook page to which the lesson is linked.

The how-to lesson consists of a view window divided into two panels. The left panel contains a list of the steps in the procedure being taught. The right hand panel contains a portrayal of the step highlighted at the left. The portrayal can be any combination of multimedia objects including text, graphic with or without text or audio caption, or video with or without sound. Audio and video portrayals are accompanies by appropriate stop/replay controls. The bottom of the display contains three buttons: *Consequence*, *OK*, and *QUIZ*. The *Consequence* button is optional and is used to show what the consequence of a given step might be. It calls up a multimedia display similar to the portrayal described above. The *QUIZ* button is used to start the how-to quiz. The *OK* button returns the text module page or ToolBook page from which the lesson was called. The system records start and stop times for the lesson presentation.

The how-to quiz randomly orders the list of steps in the left panel. The student is directed to drag-and-drop the names of the steps until they are in the correct order. The demonstration of each step is available in the right panel as with the presentation. A *Cancel* button allows the student to abort the quiz. An *Evaluate* button caused the system to compare the student's order of steps with the correct order. If a step is not in the correct order the student is given feedback to this effect and asked to try again. The quiz is finished when the student has arranged all of the steps in the correct order. The system records the start and stop/abort time and the number of tries required to get all of the steps in the correct order.

2.8 Student registration and records

When a student signs on to the run time version of the ToolBook course they are asked to register. A registration form requests the student's name, ID number, course/period and instructor's name. If the student has previously registered the student's record is located and his new course start time recorded. If the student has not previously registered a new student record is created. The registration form also contains an *instructor login* button that allows the instructor, after providing a pass word, to review all of the student records.

3. THE AUTHOR VIEW OF IDXELERATOR™

IDXelerator adds an *IDX* menu item to the ToolBook authoring page. This menu contains the following items: *Edit Text Module, Launch Text Module, Link Text Module* (divider), *Edit Slide Show, Launch Slide Show, Link Slide Show* (divider), *Edit Lesson, Launch Lesson, Link Lesson* (divider), *Edit Info Quiz, Launch Info Quiz, Link Info Quiz* (divider), *About IDXelerator.* Each of these functions is described in the following paragraphs.

3.1 Front-end analysis

Like all instructional development creating a course using IDXelerator requires an analysis phase and an instructional design phase prior to beginning development. In the analysis phase the instructional design must first determine audience requirements, goals of instruction, need for the instruction, etc. In the design phase the designer must decide what kind of lesson is required for each goal and whether the lessons built-in to IDXelerator are sufficient for the goals of the instruction. IDXelerator does not yet contain a wizard to help decide what kind of lesson is required for each goal. The designer must then decide what multimedia objects are required by each of the built-in lessons of IDXelerator.

3.2 Development phase

The development phase requires three activities: The user must first build a course interface using ToolBook Instructor. Such an interface consists of menus, course structure, overview materials, etc. The built-in functions and lessons of IDXelerator are called by linking its components to ToolBook objects in this interface.

Second, the user must obtain or prepare the multimedia objects required to implement the built-in strategies of IDXelerator. The folder containing the ToolBook Course must contain folders with the following names: *text, graphics, audio, video*. The corresponding multimedia objects are placed in these files.

Third, the user must link multimedia objects into appropriate slots of template forms corresponding to the various modules or lessons of the system. At this phase in the development process authoring IDXelerator components is relatively easy. The only computer skill needed is the ability to use the Windows™ file manager.

Clicking on one of the editors from the IDX menu calls up a list of all of the components (text modules, slide-shows, info lessons, parts of lessons, kinds of lessons, how to lessons, quiz modules) of this type already created for this ToolBook course. A given course can have as many of each type of component as required. Clicking on the name of a component on the list provides the user with the following options: *Add a new* <component>, *Edit* <component> <name>, *Delete* <component> <name>, *Select* <component> <name>. *<Component>* gives the name of the IDX component: Text Module, Slide Show, Info Lesson, Parts of Lesson, Kinds of Lesson, How to Lesson, Quiz Module. *<Name>* provides the specific name of the component selected.

For example suppose we have a text module called *"The Four Stroke Engine"*. Selecting the IXD menu item *Edit Text Module* would display a list of all the text modules in the course. Among the list would appear the name *"The Four Stroke Engine"*. Selecting this item would pop-up a menu consisting of the following items: Add a new text module, Edit text module "The Four Stroke Engine", Delete text module "The Four Stroke Engine", Select text module "The Four Stroke Engine." Selecting the item Edit text module "The Four Stroke Engine" opens the text module editor.

Each of the IDX editors works the same. They consist of a series of slots labeled with the information required. If the slot requires only a text name a dialog box appears into which the user types the name. It the slot requires a resource file clicking on the slot calls up a menu consisting of the following items: *Select video resource file, Select graphics resource file, Select text resource file, Edit resource file, Deselect resource file pointers*. If the slot only supports text then the *Select graphics* and *Select video* items do not appear, etc. Clicking on a *Select graphics* or *Select video* or *Select text* item causes the Windows file manager to list all the resources in the appropriate multimedia file (text, graphics, audio, video). The user merely clicks on the name of the file. The file is then linked to the slot in the authoring template and the name of the file appears in the slot in the IDX editor (See *Figure 2*).

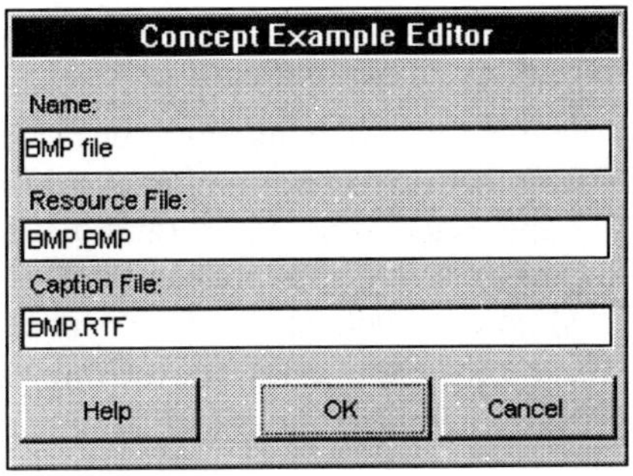

Figure 2. Typical IDX editor for linking multimedia resources

Selecting the *Edit resource file* item on the pop-up menu opens the appropriate text editor, audio editor or video editor. The user can then view or modify the multimedia resource in the context of authoring.

Selecting the *Deselect resource file pointers* item on the pop-up menu unlinks the resource from the slot in the authoring dialog.

Each IDX editor also contains the following buttons: *Help*, *OK* and *Cancel*. *Help* provides context sensitive help for authoring this editor, *OK* accepts the links made in the editor. *Cancel* discards the links made in the editor.

Selecting a component enables the *Link* <component> and *Launch* <component> items on the IDX menu. The *Launch* command allows the user to see the student view of the selected component in authoring mode. The *Link* command allows the user to link the component to a selected ToolBook object on the current ToolBook page. When a student clicks on this object in runtime the linked IDX component will be launched. For example, after having selected the item "The Four Stroke Engine" the Launch text module ... and Link text module ... items would pop-up the name "The Four Stroke Engine" when they are selected.

The *Link* command assigns Toolbook script to the selected object. This script uses a *to handle buttonclick* command structure to initialize lessons. Users can edit this script to better fit their programming strategy. For example, they can edit an object's script to initialize the lesson on a *doubleclick.* This allows the author to take full advantage of Toolbook functionality.

3.3 Text module editor

The text module editor contains the following slots: Text Module Name, Text File, Info Lessons: (a list), Kinds of Lessons: (a list), Parts of Lessons: (a list) and How to Lessons (a list). Clicking in the lesson lists pops-up the following menu items: *Select* <lesson type> *lesson*, and *Deselect* <lesson type>: <lesson name>. The *select* item displays the list of all lessons of the type under consideration. Clicking on a lesson name adds it to the list. *Deselect* cuts the lesson from the list but does not delete the lesson from the course.

3.4 Slide show editor

The slide show editor contains the following slots: Overview Slide Presentation: (name of slide show), a left hand panel listing the name of each slide, a right hand panel showing the file name of the slide graphic and slide audio (audio file). Clicking in the slide list panel pops up the following menu: Build new slide, Edit Slide: <slide name>, Move slide up: <slide name>, Move slide down <slide name>, Deselect slide: <slide name>.

Build new slide pops up a slide show editor consisting of three slots: Slide Caption: (text), Slide Graphics File: (link to graphic file), Slide Audio File: (link to audio file).

3.5 Info lesson editor

The Info Lesson Editor has two tabs: Info Lesson Data and Info Lesson Extras. The Info Lesson Data tab contains the following slots: Info Lesson Name: (text), Info Keywords: (text), Info Representation (graphic or video file), and Info Description (text file).

The Info Lesson Extras tab contains 4 panels. The top left panel is a list of Additional Information names that appear on the student view of the Info Lesson. The right top panel shows the multimedia files linked to these hot word names. Clicking in the Additional Information panel pops up an Additional Info Editor dialog with the following slots: Hotword text (text), Resource File (graphics, video, or audio file), and Caption File (text or video). In the student view clicking on a hotword displays the resource file and caption to the student.

The bottom left panel is a list of names that Link to Other programs. The lower right panel shows the execute files associated with these names. Clicking in the *Other Programs* panel pops-up a Learning Applications Editor with two slots: Program Introduction (text) and Program file (name of executable file). When the student clicks on one of these names the associated program will execute.

3.6 Quiz module editor

The Quiz module editor provides a way to practice/text the information presented by the Info lessons. The Info Quiz Editor contains a Quiz name slot (text) and a Quiz Item list panel. The Quiz Items are the names of Info lessons. Clicking in the *Quiz Items panel* pops up a menu with two items: Select Info Quiz items, Deselect Info Quiz <name>. Selecting Select Info Quiz items presents a list of all the Info Lesson in the course. Clicking on the name of the Info Lesson adds it to the quiz.

The Info Quiz Editor also includes two buttons: *Multiple Choice* and *Short Answer*. This determines the format of the Quiz items. Either or both options can be selected.

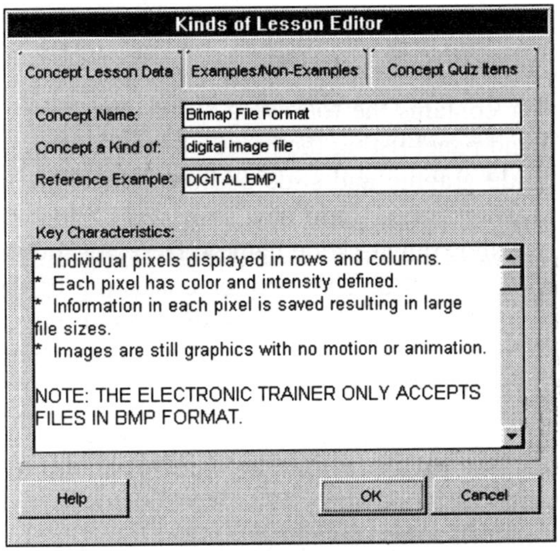

Figure 3. Concept lesson authoring dialog

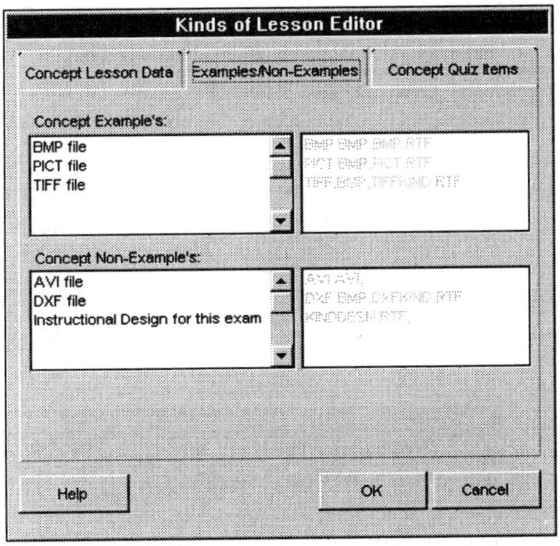

Figure 4. Examples/Non-Examples dialog for kinds-of lessons

3.7 Kinds-of lesson editor

The Kinds of lesson editor has three tabs: Concept Lesson Data, Examples/Non-Examples, and Concept Quiz Items (See *Figure 3*). The *Concept Lesson Data* tab contains the following slots: Concept name (text), Concept a Kind of (text), Reference Example (link to text, graphic, audio, or video file and text or audio caption file), Key Characteristics (text). A portion of the *Help* file for the Kinds-of authoring dialog is illustrated in Figures 5 and 6. This is context sensitive imperative help that directs the user in the next steps of the authoring process.

The *Examples/Non-Examples* tab contains four panels (See *Figure 4*). The upper left is a list of Concept Examples names. The upper right panel lists the linked text, graphic, audio, or video file and the text or audio caption file associated with the example name. The lower left is a list of the Concept Non-Examples. The lower right panel lists the linked text, graphic, audio, or video file and the text or audio caption file associated with the non-example name.

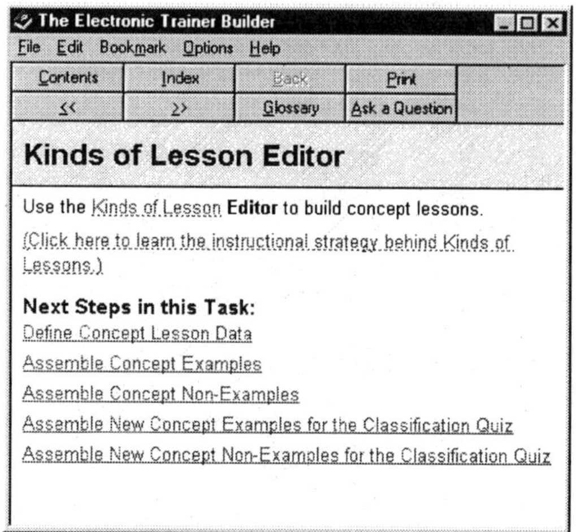

Figure 5. Help screen for kinds-of lesson editor

The *Concept Quiz Items* tab also contains four panels. The upper panels listing the Concept Examples and associated multimedia files, the lower panels listing the Concept Non-Examples and associated multimedia files. In the quiz to the student these items are presented to the student in random order. Linking files into these slots is as described for other editors.

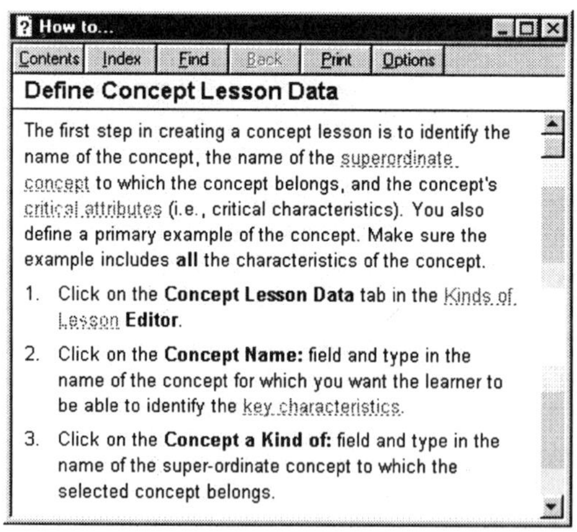

Figure 6. Help screen for *Define Concept Lesson Data*

3.8 Parts of lesson editor

The Parts of lesson editor consists of the following slots: Object's Name (text), Objects Image (graphic file), three list panels Object's Component Parts. The name of each part appears in the left panel, the multiple media resource files associated with each part is listed in the center panel, the location of the part on the screen is listed in the right panel. Each part can have two pieces of information a description (text file) and a resource (graphic, video file) with an audio or text caption file. Clicking in the left panel pops-up a menu with the following items: Add New Component, Edit Component <name>, Position Component <name>, and Delete Component <name>. Linking files is as previously described. Positioning a component consists of outlining the location of the part on the Object image.

3.9 How to lesson editor

The How to lesson editor consists of the following slots: Lesson Name (text), and three panels. The left panel is a list of the Events in the lesson. The middle panel lists the multimedia files which demonstrate the event (video, graphic, text) with audio or text caption. The right panel lists the multimedia files that show the consequence of the event (video, graphic, text) with audio or text caption. Authoring is the same as previously described. Clicking of the left panel pops up the following menu: Add new event, Edit event <name>, Move up event <name>, Move down event <name>, Delete event <name>. The quiz for the How to lesson consists of ordering the events into the sequence shown in the left panel.

REFERENCES

Merrill, M.D., 1997a, Learning-oriented instructional development tools, *Performance Improvement Quarterly*, 36 (12), 51-55.

Merrill, M.D., 1997b, Instructional strategies that teach, *CBT Solutions*, November/December, 1, 4, 8, 10-11.

Merrill, M.D., Drake, L., Lacy, M.J., Pratt, J. and ID_2 Research Group, 1996, Reclaiming instructional design, *Educational Technology*, 36(5), 5-7.

Gagné, R.M., 1985, *The conditions of learning and theory of instruction*, Holt, Rinehart and Winston, New York.

Chapter 23

Intelligent Support for Instructional Development: Approaches and Limits

Michael Spector
University of Bergen

Keywords: Instructional development, Intelligent performance support

Abstract

Systems which provide intelligent performance support for instructional development are the focus of this chapter. The discussion will focus on two of the more successful efforts. One effort (AGD) supports the upstream analysis and planning tasks associated with designing distance learning curricula and courses. AGD adopts a strategy intended to extend the capabilities of less experienced designers. The other effort (GTE) supports the downstream development and implementation tasks associated with the construction of intelligent tutoring systems. GTE adopts a strategy intended to replace the capabilities of an experienced courseware developer. Because the differences in these two systems are quite significant, and because each of these systems has been succesful within its targeted domain of application, it is interesting to consider their characteristics and contributions. What will become clear from these and other efforts is that the activity of attempting to automate complex processes provides deep insight into those processes.

1. INTRODUCTION

Computer-based instructional materials have now been in use for more than twenty years. Early systems were hosted on mainframes and were primarily text-based (e.g., Plato). Some of these instructional systems are still in use, partly because the planning that went into courseware development was quite thorough, and the courses developed were both robust and effective. Today's learning environments most often involve multimedia and microcomputers, often making use of networks and the world wide web. The developer's watchwords of those early systems might have been to plan well, to present simply, and to practice often. Today's developers might emphasize the engagement of learners, encouragement of reflection, and support for exploration. These characterizations are admittedly oversimplifications, and they are only meant to indicate that the times and our expectations have changed. Some argue that the notion of learning has changed (Twigg, 1994). Such a radical conclusion, however, is not central to this discussion, which concerns the application of artificial intelligence to the domain of planning and implementing learning environments.

For this purpose, traditional but broadly stated definitions are adequate. Learning involves stable, persisting changes in human attitudes, beliefs, knowledge, and skills. Learning environments are those settings intentionally designed to facilitate learning. Instructional development involves the planning and implementation of learning environments. Courseware engineering involves the development of computer-supported learning environments. Artificial intelligence involves techniques and technologies to get computers to exhibit selected but constrained attributes of human intelligence.

The attempt to automate some aspects of courseware engineering has led to deep insights into the more general processes of instructional design and development. Indeed, the theme of this chapter is that we have only begun to witness the many ways in which intelligent support can be provided to instructional designers and developers; however, we would be wrong to expect computers to replace human expertise in key aspects of instructional design and development processes. The following discussion will focus on two systems which provide automated and intelligent support for the creation of computer-based learning environments. The analysis and remarks are pertinent to the more general domain of instructional design and development. It has been a natural development to provide computer-based support for the development of computer-based learning environments, but such support can and should be envisioned for more traditional settings as well. The instructional design considerations and difficulties overlap significantly with substantial divergence mainly arising in the implementation and deployment phases of an instructional development effort.

In order to generate a reasonable framework for speculation concerning reasonable approaches and expectations, two systems will be presented in some detail: AGD (L'Atelier de Génie Didactique) and GTE (Generic Tutoring Environment). Both of these tools address knowledge modeling and representation issues relevant to the development of learning environments, and both involve object-oriented design methodologies; these shared aspects will provide a basis for speculation about future possibilities. AGD, however, is intended to provide a general purpose design framework, whereas GTE targets the development of intelligent tutoring systems.

2. TWO INTELLIGENT SUPPORT SYSTEMS

2.1 AGD (L'Atelier de Génie Didactique)

The AGD project began in 1992, partly as an attempt to provide a pedagogical design assistant to content experts acting as instructional designers at DMR, a computer consulting company, and at Télé-université, the distance learning institution at the University of Québec (Paquette, 1992; Paquette, Aubin & Crevier, 1994). It is well established in the instructional science community that instructional quality increases when appropriate attention is given to analysis and design details in the beginning phases of a project. The entire set of processes involved in developing a learning environment has been called instructional systems development (ISD). At least four generations of ISD methodologies have been identified with more than one hundred variations (Tennyson, 1994a, 1994b).

The ISD model implemented in AGD is recent, up-to-date, comprehensive, and quite robust. AGD can be considered a computer-based instantiation of Tennyson's ISD^4 (1994a), although it was not designed with that in mind. Tennyson's ISD^4 model can be roughly characterized by the following qualities:

- includes a variety of evaluations pertinent to development tasks at hand;
- makes explicit an appropriate situational evaluation for wherever the designer happens to be in the process;
- represents the reality that instructional development processes begin at different points depending on the type of effort (e.g., new development vs. revision, etc.);
- clearly subscribes to an iterative design approach with tasks, persons, and phases interacting in many dynamic ways; and
- integrates recent cognitive learning theory and knowledge modeling methods.

In short, ISD^4 provides both flexibility and detail about many activities involved in different kinds of instructional developments. These qualities allow ISD^4 to be a reasonably descriptive and adaptable model, suitable for use in a general purpose instructional design support system such as AGD. While AGD does not support all aspects of ISD (e.g., there is no direct performance support for courseware production, as can found in GTE, for example), those aspects of ISD which are supported, mainly the earlier phases of analysis and design, appear consistent with Tennyson's model.

In addition, AGD provides the user-designer with a critique at various stages in the analysis and design phases. This feature was designed into the system to insure coherence and consistency of learning objectives with types of knowledge, targeted learners, and identified learning needs. Such a design critic was first suggested in the research literature by Duchastel (1990). AGD can be considered the first such implementation of Duchastel's notion of a *Design Critic.*

AGD provides a knowledge modeling tool along with curriculum/course design tools. The knowledge modeling component has since been modularized and made available as an independent and stand-along tool called MOT (modélisation par objets typés). This particular tool allows a subject matter expert to create a semantic network of a selected subject domain to a level of detail believed to be appropriate for instructional purposes. The semantic network created by AGD (and also by MOT) has two interesting features: (1) it is designed specifically for instructional purposes (e.g., there are links to indicate relationships which have instructional significance); and, (2) the objects in the network are part of an object-oriented network (e.g., they can be instantiated at various points in a curriculum/course and retain relevant aspects).

It is also worth drawing attention to the fact that such a knowledge modeling tool can be used as a way to assess how learners are progressing toward the general goal of acquisition of expertise. If several such semantic models of a domain are created by experts and they happen to exhibit recognizable patterns, as reported by researchers at Télé-université (Paquette, Aubin & Crevier, 1994), then learners can be given a similar task before and after the learning experiences. If the learning environment has been effective, then there should appear changes in the learners' construction, and the changes should be in the direction of the models created by experts.

The inputs to the curriculum and course design module are information about targeted learners, various course constraints, and the knowledge model -- input of these is assisted both by a powerful interface and by context-sensitive help. AGD then assists the user in building a curriculum or course structure and produces both teacher and learner guides, various didactic instruments, the knowledge model, course and lesson objectives and strategies, and the associated learning strategies and scenarios. This output comes in the form of selected reports and a global report.

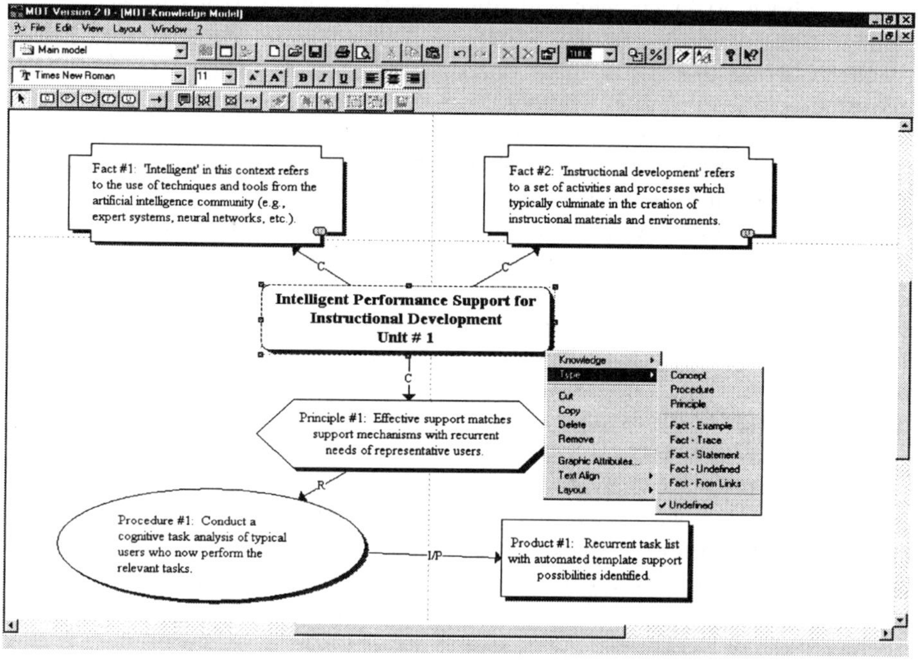

Figure 1. Sample AGD/MOT knowledge modeling screen

While AGD allows the user to adopt a variety of learning perspectives, it most easily and directly supports a version of situated learning closely resembling cognitive apprenticeship (Collins, Brown & Newman, 1989). Researchers at Télé-université have long held a bias for meaningful involvement with co-learners in realistic problem-based settings (Paquette, 1996; Paquette, Aubin & Crevier, 1994), and this bias is reflected in the language used to prompt designers to describe learning activities and scenarios. However, while presenting this quite reasonable bias, AGD does not force or constrain the user to adhere in any strong sense to a particular framework.

AGD makes its rationale explicit in the form of the interface and the advice it offers, so in this sense AGD is a glass-box support system. Glass-box support systems are appropriate when there is a desire to educate users in the process being supported, with the expectation that users will become both more knowledgeable and more expert performers after using the system for some extended period of time. What would be an interesting study would be to have expert instructional designers use MOT to create models of the practice and processes involved in instructional design, and then use those models as the basis for studies of how AGD users progressed towards ISD expertise over time while using AGD.

There are most likely recurring knowledge models, especially within a particular domain. If so, these might be identified, stored in a library of initial knowledge models, and made easily available to users so that they don't have to create new knowledge models every time. Such a library would add a case-base kind of advising to AGD. Since AGD already explicitly subscribes to the value of incorporating meaningful instructional scenarios into the design of learning environments, it would be entirely consistent to include support for scenario-building within AGD's framework.

In conclusion, AGD began as a practical aid to assist content experts in companies and university teachers in the task of preparing lessons for a distance learning setting, a relatively new and unfamiliar setting for many of them. AGD proved to be successful and was expanded to provide support for non-academic learning environments. The tool has undergone a number of modifications and has been successfully used in commercial and industrial settings as well as in distance learning university settings. AGD continues to provide a foundation for research and development at LICEF (Télé-université's Research Center, University of Québec at Montréal), and has led to the development of an important knowledge modeling tool (MOT).

2.2 GTE (Generic Tutoring Environment)

Unlike AGD, GTE grew out of an international collaboration involving academic and industrial communities in several countries, and it was focused on providing support for a very particular type of courseware, intelligent tutoring systems. Like AGD, GTE also proceeds from a particular educational philosophy perspective and explicitly adopts a particular psychological perspective involving the nature of expertise (see, for example, Ericsson & Smith, 1991; Resnick, 1989). It is this psychological perspective that is the basis for the design of GTE (Van Marcke's, 1992a, 1992b, 1998). Van Marcke (1998), the designer of GTE, has argued that teaching is primarily a knowledge-based task. Specifically, experienced teachers are able to draw on specific teaching knowledge in addition to extensive domain knowledge. As a consequence, a primary task for an intelligent tutoring system is to integrate that instructional knowledge in the system in a way that allows the system to adapt to learners just as expert teachers do.

It is noteworthy that Van Marcke confines instructional decision making in the human domain to teachers. This would seem to imply negligible or insignificant contributions by instructional designers, textbook authors, test experts, etc. Given a starting point based on human expertise, it would appear import to take into account all of the expert knowledge that comprises instructional decision making. This would be an impossible task, however. GTE should not be faulted for focusing on one type of relevant expertise. This observation is intended as a reminder that there are inherent limitations pertaining to intelligent performance support systems.

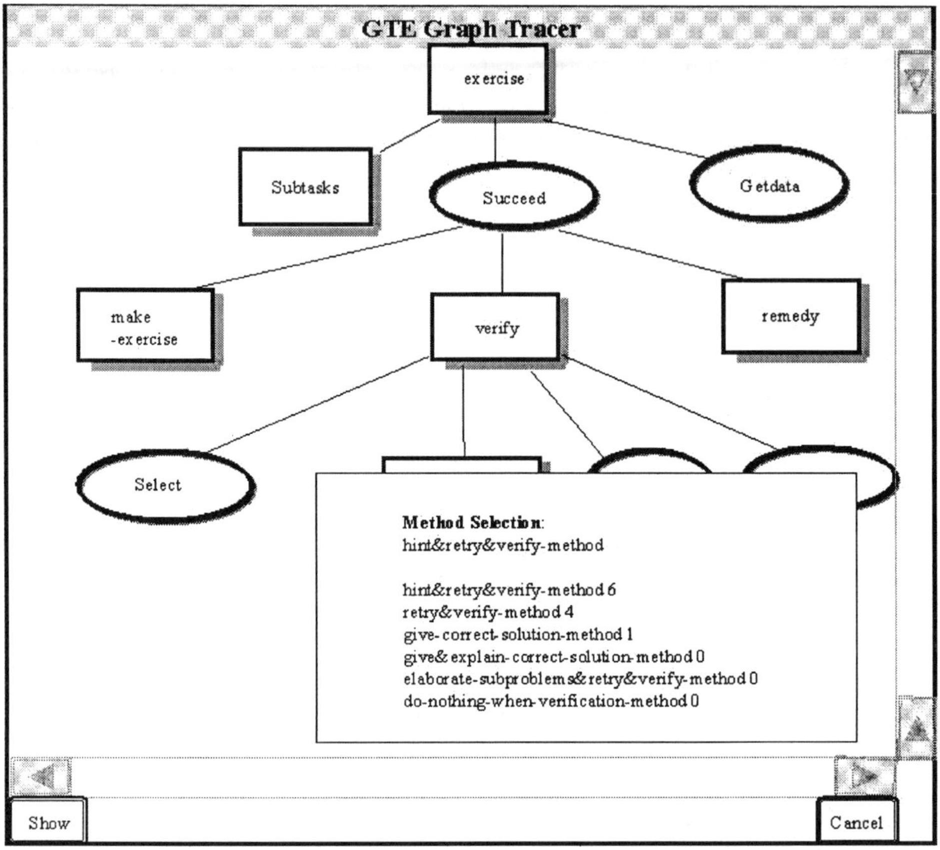

Figure 2. Sample GTE method selection screen (from Van Marcke, 1992b)

Another way to make this point is to conceive of an instructional development system, either an advising support system like AGD, or an ITS generating system like GTE, which could also determine which kinds of media to include at particular points in the instruction. Friedman (1993) has argued persuasively that this is a non-trivial task, roughly equivalent in complexity to the natural language processing problem. Perhaps it is time to admit that regardless of how powerful various artificial intelligence applications become in the domain of instructional development, there are likely to remain important tasks for human experts and designers. We should consider the AI techniques basically as powerful tools which can extend and enhance human expertise rather than replace human experts.

GTE cleverly combines a reductionist perspective with a pragmatic approach. The tutor-generation task is reduced to two tasks: (1) determining all of the relevant components and relationships (an instructional semantic network); and, (2) determining how and when to provide and combine these component to learners in ways which promote learning. One can find other elements of reductionism in GTE. For example, the domain perspective in GTE consists of a static semantic network. According to Van Marcke (1998), this network is used for purposes of sequencing material within a topic area, for indexing instructional objects, and for stating learning objectives. Like the instructional network in AGD, GTE makes use of an object-oriented network, so that components can be meaningfully combined and reused.

However, there are limits to a reductionist approach. As Van Marcke (1998) claims: (1) teaching is an inherently complex activity; (2) there are no adequate (from a reductionist point of view) theories about how people learn; and (3) generative systems should try to exploit the same basis of instructional decision making as that used by humans experts. Moreover, it is not completely clear how expert human designers work, but we do know that human experts typically use a case-based approach to complex instructional planning tasks (Perez and Neiderman, 1992; Rowland, 1992). The rationale in case-based tools is that inexperienced instructional planners lack case expertise and that this can be provided by embedding design rationale with lesson and course exemplars. Studying how less experienced instructional planners make use of such tools may eventually contribute to our understanding of instructional design expertise.

However, cases lack the granularity of the very detailed objects described by Van Marcke (1998). In fact, one could argue that it is only when a case-based approach fails or breaks down that humans revert to overtly reductionist approaches. In other words, what has been modeled in GTE is not best described as human instructional expertise. Rather, what has been modeled is knowledge about instruction that is likely to work when human expertise is not available, as might be the case in some computer-based tutoring environments, such as those which generate instruction dynamically as GTE-based tutors. This is an important distinction (Spector, 1998).

If teaching and learning are complex in the sense that they are subject to resource constraints, various kinds of uncertainty, and somewhat fuzzy goals and outcomes, then it is appropriate to be very modest with regard to claims concerning the automation of such activities. Moreover, if it is actually true that teaching is such a complex collection of activities, then we ought to have somewhat more limited expectations with regard to the extent that computer tutors are able to replace human tutors. Indeed, if one truly believed in the value of human expertise in this complex domain, then it would seem reasonable to explicitly plan for both human and computer tutoring, coaching, and facilitation as an essential aspect of a learning environment. Unfortunately, very few systems aimed at supporting the development of courseware make explicit attempts to incorporate human instructors into the environment. AGD is an exception (Paquette, Aubin and Crevier, 1994).

Perhaps the most significant contributions of GTE are in the domain of object-oriented instructional design. These contributions are a result of two aims of GTE: (1) the goal to generate courseware; and (2) the goal to provide a generic framework for doing so. Object-oriented approaches are already proving to be of significance in many other complex planning domains, most especially software design, so it is no great surprise that object orientation should prove useful to instructional design.

GTE does not directly support student modeling in the sense that this term has been used in the intelligent tutoring literature, although Van Marcke (1998) indicates that GTE's knowledge base can in principle be linked to existing student modeling techniques. It is not clear to me what would be gained by doing this. GTE contains quite a number of instructional methods with detailed elaborations and basic rules for their applicability in a dynamic generative environment. If these rules break down with novice learners, then perhaps human instructional interventions are appropriate;

if the system does not perform adequately with advanced learners, then perhaps it is time to pass more control to individual learners. In short, I believe that Van Marcke's pragmatic intuitions on this point are right. Moreover, by not directly supporting student modeling, GTE can become more generic, which is obviously one of its goals. Student modeling, of course, is highly domain specific in nature.

3. SPECULATIONS ABOUT THE FUTURE

Two very reachable possibilities pertaining to efficiency and effectiveness of intelligent performance support for courseware engineering come to mind. The first possibility concerns connecting object-oriented approaches with case-based advising, and the second concerns the creation of easily accessible, re-usable electronic databases. The key to achieving both of these possibilities revolves around the key notions of object-orientation, knowledge modeling, and instructional tagging. These key notions have been successfully demonstrated in the two systems briefly reviewed here and exist in other systems as well, including AthenaMuse.

First, lets consider connecting object-oriented design with case-based advising and guidance (for learners, instructors, and designers). Case-based advising can be made much more flexible if it is constructed within an object-oriented framework. Cases could be constructed from collections of smaller objects and could be activated according to a variety of parameters. This means that cases could be constructed as needed to suit specific and dynamic requirements rather than relying on pre-scripted cases or a case-base such as that currently found in GUIDE.

This means that there is significant promise for combining object-oriented approaches (found in AGD, GTE, and AthenaMuse) with case-based approaches (found in GUIDE). Moreover, existing courseware and simulation authoring environments could be made more intelligent if they were extended to contain annotated case illustrations or libraries of examples with appropriate explanations of when the example cases were used, why they were appropriately used in those circumstances, and what results if any have been documented.

Second, truly dramatic improvements in courseware development time and cost are now a reasonable goal. An object-oriented approach allows instructional objects to be constructed dynamically and flexibly, as ably illustrated by Van Marcke (1992, 1998). There are in existence a number of electronic databases. Some of these databases happen to be tagged (for example, with Standard Generalized Markup Language or SGML tags) which allows the database to be viewed from an object-oriented perspective. SGML is a system for organizing and tagging components of a document, especially large documents which were frequently or regularly revised and published in or using electronic media (Coombs, Renear and DeRose, 1987). SGML was first developed in 1986 for the International Organization for Standards and specified rules for formatting and tagging documents. The idea behind SGML is that a language describing particular types of document can be created so that the tags and main text can be kept independent of the system used to print or display the contents. The principle of separating content from the formatting and presenting mechanisms allows and encourages modularity and flexibility and has been successfully applied in other domains such as courseware authoring systems, for example. Figure 3 indicates what an SGML file might appear. Basically, it can be an ordinary text file with tags to indicate the type of item following the tag and links to items which might be located somewhere else in a source database.

```
<!-- Here is a conceptual sample of an SGML file: -->
<EXPLANATORY REMARK> <TITLE> SGML Note </TITLE>
<TEXT ENTRY> <PARA> This is a sample of what an SGML tagged text file
        might look like for something declared with a DOCTYPE called
        EXPLANATORY REMARK. This sample assumes that the tags such as
        TEXT ENTRY have already been properly declared. </PARA>
<PARA> You might have <BOLD> noticed </BOLD> that the tags are generally
        paired, so that the boundaries for the application for a particular tag are
        always clearly defined. </PARA>
</ TEXT ENTRY>
</EXPLANATORY REMARK>
```

Figure 3. Sample SGML file

A courseware generation system, such as GTE, could be linked to an existing electronic database, and with the assistance of a parsing engine, transform the electronic database (viewed as an existing domain model) into courseware. This particular approach has been shown to be feasible with XAIDA, with the dramatic result that courseware could be generated in a matter of minutes (Spector, Arnold & Wilson, 1996).

The most dramatic results might be obtained by a pre-cursor effort to develop an instructional tagging language. Current SGML (Standard Generalized Markup Language) and HyperText Markup Language (HTM) tags are primarily presentation oriented, with tags indicating fonts, heading levels, where to find files, etc. Figure 4 illustrates a simple HTML file. HTML has pre-defined tags which need not be separately defined in the context of a document type as is the case with SGML. Whereas SGML is a language for defining specific mark-up languages, HTML is a specific mark-up language with tags such as H3 (third level heading) and HREF (hyperlink reference) already defined and recognizable within the context of an HTML document.

```
<html> <head> <title> Mistress Magoo </title> </head>
<h1> Mistress Magoo </h1>
<A HREF = "http://www.ifi.uib.no/eist/index.html"> Educational
        Information Science & Technology Program </A> <p>
<h3> Main research interests: </h3>
*Intelligent performance support for instructional design; *System
        dynamics modeling and simulation of instructional project
        management; *Representations of instructional design expertise;
*Reusable instructional objects; *Measuring learning effectiveness in
        complex domains; and *Distance learning technologies.
   <hr> </html>
```

Figure 4. Sample HTML file

Knowledge Query Manipulation Language (KQML) is a language designed to support interaction among intelligent software agents (Finin, Labrou & Mayfield, 1995). HTML tags can be combined with KQML and extended to include instructional concerns such as 'key definition', 'good example of xxx', 'non example xxx', 'bad example of', 'moral of story', 'theme of article', 'main point of paragraph', etc. Identifying a sufficient set of instructional tags and then devising an easy way to support tagging of existing and new electronic databases would be a significant undertaking, but eminently worthwhile.

There is an active effort to create a standardized extensible markup language called XML (Connolly, 1997). This language is most similar to HTML but it is intended to provide a syntax for describing a specialized or customized mark-up language, returning to the original notion behind SGML with the advantages of a decade of experience. Basically, XML is a low-level syntax for creating new declarative representations for specific domains. Several such instantiations have been developed including MathML for mathematical expressions and SMIL for scheduling multimedia presentations on the internet. A quite natural research and development project which could be associated with the XML effort would be to create, implement and evaluate an instructional markup language using XML as the underlying mechanism.

Clearly, the use of object-oriented approaches makes it in principle possible to reuse previous courses, lessons, databases, and so on. Two long-range goals come to mind. One has to do with connecting object-oriented design with case-based advising and guidance (for learners, instructors, and designers). Case-based advising can be made much more flexible if it is constructed within an object-oriented framework. Cases could be constructed from collections of smaller objects and could be activated according to a variety of parameters. This means that cases could be constructed as needed to suit specific and dynamic requirements rather than relying on pre-scripted cases or a case-base.

Both object orientation and case libraries are making their way into commercial authoring products as well. For example, PowerSim™, an environment for creating system dynamics based systems, has tested the ability to provide model builders with partially complete, pre-constructed generic structures with all the relevant properties of reusable objects (Gonzalez, 1998). Such reusable, generic structures (adaptive templates) will most likely appear in other commercial systems as well. Creating reusable objects and case libraries makes more poignant the need for advising those who must select from these new riches relevant items.

In conclusion, it seems safe to say that the future of intelligent performance support for courseware engineering will be filled with interesting opportunities and challenges. The most interesting opportunities may arise in inter- and multi-disciplinary settings, which appear essential to making significant advances. The big challenge will be to remain humble with regard to how little we really know about human learning and intelligence.

As we develop newer and more sophisticated techniques for developing learning environments, we should remember that learning cultures are dynamic. When only mainframe, text-based systems were available, users and learners came to expect certain types of screens and interactions. Today's learners are more accustomed to highly interactive, multimedia environments, often attached to the world wide web. Learning environments that do not take into account these expectations may fall short of their intended effectiveness. Moreover, humility should remind us that machines are not likely to replace all aspects of human intelligence and expertise in the area of instructional design and development.

REFERENCES

Collins, A., Brown, J.S. and Newman, S.E., 1989, Cognitive apprenticeship: Teaching the crafts of reading, writing, and mathematics, In Resnick, L.B. (Ed.), *Knowing, learning, and instruction: Essays in honor of Robert Glaser*, pp. 453-494, Lawrence Erlbaum Associates, Hillsdale, NJ.

Connolly, D. (Ed.), 1997, XML: Principles, tools, and techniques, *World Wide Web Journal*, 2(4), 1-266.

Coombs, J.H, Renear, A.H. and DeRose, S.J., 1987, Markup systems and the future of scholarly text processing, *Communications of the Association for Computing Machinery*, 30(11), 933-947.

Duchastel, P.C., 1990, Cognitive designs for instructional design, *Instructional Science*, 19(6), 437-444.

Ericsson, K.A. and Smith, J. (Eds.), 1991, *Toward a general theory of expertise: Prospects and limits*, Cambridge University Press, New York.

Finin, T., Labrou, Y. and Mayfield, J., 1995, KQML as an agent communication language In Bradshaw, J. (Ed.), *Software agents*, MIT Press, Cambridge, MA.

Friedman, A., 1993, Designing graphics to support mental models, In Spector, J.M., Polson, M.C. and Muraida, D.J. (Eds.), *Automating instructional design: Concepts and issues*, 249-292, Educational Technology Publications, Englewood Cliffs, NJ.

Gonzalez, J.J., 1998, Internet simulators (WebSims) and learning environments, Presentation at the Annual Meeting of the American Educational Research Association, San Diego, CA, April, 1998.

Marcke, K. van, 1992a, A generic task model for instruction, In Dijkstra, S. (Ed.), *Instructional models in computer-based learning environments* (NATO ASI Series, Vol. F104.), Springer, Berlin.

Marcke, K. van, 1992b, Instructional expertise, In Frasson, G., Gauthier, G. and McCalla, G.I. (Eds.), *Intelligent tutoring systems*, 234-243, Springer, Berlin.

Marcke, K. van, 1998, GTE: An epistemological approach to instructional modelling, *Instructional Science, 26* (3/4), 147-191.

Paquette, G., 1992, An architecture for knowledge-based learning environments, *Proceedings of EXPERSYS-92*, 31-36, Ottawa, Canada.

Paquette, G., 1996, La modélisation par objets typés - une méthode de représentation pour les systèmes d'apprentissage et d'aide à la tâche, *Sciences et Techniques Educatives*, 3(1), 8-42.

Paquette, G., Aubin, C. and Crevier, F., 1994, An intelligent support system for course design, *Educational Technology*, 34(9), 50-57.

Perez, R.S. and Neiderman, E.C., 1992, Modeling the expert training developer, In Seidel, R.J. and Chatelier, P. (Eds.), *Advanced training technologies applied to training design,* Plenum, New York.

Resnick, L.B. (Ed.), 1989, *Knowing, learning, and instruction,* Hillsdale, NJ, Lawrence Erlbaum.

Rowland, G., 1992, What do instructional designers actually do? An initial investigation of expert practice, *Performance Improvement Quarterly,* 5(2), 65-86.

Spector, J.M., 1998, The role of epistemology in instructional design, *Instructional Science, 26*(3/4), 193-203.

Spector, J.M., Arnold, E.M. and Wilson A.S., 1996, A Turing test for automatically generated instruction, *Journal of Structural Learning,* 12(4), 301-313.

Tennyson, R.D., 1994a, Knowledge base for automated instructional system development, In Tennyson, T.D. (Ed.), *Automating instructional design, development, and delivery,* Springer, Berlin.

Tennyson, R.D. (Ed.), 1994b, *Automating instructional design, development, and delivery,* Springer, Berlin

Twigg, C.A., 1994, The changing definition of learning, *Educom Review*, 29(4).

Information about Authors

Adams, Ann-Marie
Director of Public Relations
Cornell Cooperative Extension
USA
Ama22@cornell.edu

van den Akker, Jan
Professor
University of Twente, Faculty of Educational Science and Technology,
Department of Curriculum
The Netherlands
akker@edte.utwente.nl

van den Berg, Ellen
Assistant Professor
University of Twente, Faculty of Educational Science and Technology,
Department of Curriculum
The Netherlands
berg@edte.utwente.nl

de Boer, Wim
Research Assistant
University of Twente, Faculty of Educational Science and Technology,
Department of Educational Instrumentation
The Netherlands
boer@edte.utwente.nl

Bos, Evelijn
Educational technologist
Open University, Educational Technology Expertise Centre
The Netherlands
evelijn.bos@ou.nl

Branch, Robert Maribe
Associate Professor
The University of Georgia, Department of Instructional Technology
USA
rbranch@coe.uga.edu

Collis, Betty
Professor
University of Twente, Faculty of Educational Science and Technology, Department of Educational Instrumentation
The Netherlands
collis@edte.utwente.nl

Foshay, Wellesley
Vice President, Instructional Design and Cognitive Learning
TRO Learning, Inc.
USA
rfoshay@plato.com

Gellevij, Mark
PhD student
University of Twente, Faculty of Educational Science and Technology, Department of Instructional Technology
The Netherlands
gellevij@edte.utwente.nl

Gettman, Dennis
Research Psychologist
United States Air Force
USA
dennis.gettman@asafa.af.mil

Gustafson, Kent
Professor
The University of Georgia, Department of Instructional Technology
USA
kgustafs@coe.uga.edu

Harper, Barry
Associate professor
University of Wollongong, Faculty of Education
Australia
barry_harper@uow.edu.au

Hedberg, John
Professor
University of Wollongong, Faculty of Education
Australia
john_hedberg@uow.edu.au

de Jong, Ton
Professor
University of Twente, Faculty of Educational Science and Technology,
Department of Instructional Technology
The Netherlands
jong@edte.utwente.nl

van Joolingen, Wouter
Senior consultant
CIBIT, Knowledge Center
The Netherlands
wvjoolingen@cibit.nl

Jury, Thomas
Division Director
WPI, Inc., OnLine Learning Division
USA
tjury@wpiusa.com

Kessels, Joseph
Professor
Leiden University, Faculty of Social Sciences
The Netherlands
kessels@rulfsw.LeidenUniv.nl

Kirschner, Paul
Senior educational technologist
Open University, Educational Technology Expertise Centre
The Netherlands
paul.kirschner@ou.nl

Kuyper, Michiel
Unilever Research, Knowledge Management Department
The Netherlands
michiel.kuyper@unilever.com

Limbach, Renate
PhD student
University of Twente, Faculty of Educational Science and Technology,
Department of Instructional Technology
The Netherlands
limbach@edte.utwente.nl

Maslowski, Ralf
PhD Student
University of Twente, Faculty of Educational Science and Technology,
Department of Educational Organization and Management
The Netherlands
maslowski@edte.utwente.nl

McKenney, Susan
PhD student
University of Twente, Faculty of Educational Science and Technology,
Department of Curriculum
The Netherlands
mckenney@edte.utwente.nl

McNelly, Terri
Consultant
Development Dimensions International
USA

Merrill, David
Professor
Utah State University, ID2 Research Group
USA
merrill@cc.usu.edu

Moonen, Jef
Professor
University of Twente, Faculty of Educational Science and Technology,
Department of Educational Instrumentation
The Netherlands
moonen@edte.utwente.nl

Muraida, Daniel
Research Psychologist
United States Air Force Research Laboratory
USA
daniel.muraida@afrlars.brooks.af.mil

Nieveen, Nienke
Assistant Professor
University of Twente, Faculty of Educational Science and Technology,
Department of Curriculum
The Netherlands
nieveen@edte.utwente.nl

Pieters, Jules
Professor
University of Twente, Faculty of Educational Science and Technology,
Department of Instructional Technology
The Netherlands
pieters@edte.utwente.nl

Plomp, Tjeerd
Professor
University of Twente, Faculty of Educational Science and Technology,
Department of Curriculum
The Netherlands
plomp@edte.utwente.nl

Preese, Frank
Vice-president of Product Development
TRO Learning, Inc.
USA
fpreese@plato.com

Reeves, Thomas
Professor
The University of Georgia, Department of Instructional Technology
USA
treeves@coe.uga.edu

Rowland, Gordon
Associate Professor
Ithaca College, Department of Organizational Communication, Learning and Design
USA
rowland@ithaca.edu

Sluijsmans, Dominique
Research assistant
Open University, Educational Technology Expertise Centre
The Netherlands
dominique.sluijsmans@ou.nl

Spector, Michael
Professor
University of Bergen,
Educational Information Science and Technology Research Program
Norway
mike@ifi.uib.no

Thompson, Bennet
PhD Student
Utah State University, ID2 Research Group
USA
bent@fs1.ed.usu.edu

Valcke, Martin
Professor
University of Gent, Department of Education
Belgium
martin.valcke@rug.ac.be

Visscher, Adrie
University of Twente, Faculty of Educational Science and Technology,
Department of Educational Organization and Management
The Netherlands
visscher@edte.utwente.nl

Visscher-Voerman, Irene
Assistant Professor
University of Twente, Faculty of Educational Science and Technology,
Department of Curriculum
The Netherlands
visscheri@edte.utwente.nl

Wright, Rob
Project director
University of Wollongong, Faculty of Education
Australia
rob_wright@uow.edu.au